NATIONAL ACADEMIES *Sciences Engineering Medicine*

NATIONAL ACADEMIES PRESS
Washington, DC

The Importance of Chemical Research to the U.S. Economy

Committee on Enhancing the U.S. Chemical Economy Through Investments in Fundamental Research in the Chemical Sciences

Board on Chemical Sciences and Technology

Division on Earth and Life Studies

Consensus Study Report

THE NATIONAL ACADEMIES PRESS   500 Fifth Street, NW   Washington, DC 20001

This activity was supported by contracts between the National Academy of Sciences and the American Chemical Society, the U.S. Department of Energy (Contract number 10004932), the National Institute of Standards and Technology (Contract number 10005235), and the National Science Foundation (Contract number 10004871). Any opinions, findings, conclusions, or recommendations expressed in this publication do not necessarily reflect the views of any organization or agency that provided support for the project.

International Standard Book Number-13: 978-0-309-68863-5
International Standard Book Number-10: 0-309-68863-9
Digital Object Identifier: https://doi.org/10.17226/26568
Library of Congress Control Number: 2022945664

This publication is available from the National Academies Press, 500 Fifth Street, NW, Keck 360, Washington, DC 20001; (800) 624-6242 or (202) 334-3313; http://www.nap.edu.

Copyright 2022 by the National Academy of Sciences. National Academies of Sciences, Engineering, and Medicine and National Academies Press and the graphical logos for each are all trademarks of the National Academy of Sciences. All rights reserved.

Printed in the United States of America.

Suggested citation: National Academies of Sciences, Engineering, and Medicine. 2022. *The Importance of Chemical Research to the U.S. Economy*. Washington, DC: The National Academies Press. https://doi.org/10.17226/26568.

The **National Academy of Sciences** was established in 1863 by an Act of Congress, signed by President Lincoln, as a private, nongovernmental institution to advise the nation on issues related to science and technology. Members are elected by their peers for outstanding contributions to research. Dr. Marcia McNutt is president.

The **National Academy of Engineering** was established in 1964 under the charter of the National Academy of Sciences to bring the practices of engineering to advising the nation. Members are elected by their peers for extraordinary contributions to engineering. Dr. John L. Anderson is president.

The **National Academy of Medicine** (formerly the Institute of Medicine) was established in 1970 under the charter of the National Academy of Sciences to advise the nation on medical and health issues. Members are elected by their peers for distinguished contributions to medicine and health. Dr. Victor J. Dzau is president.

The three Academies work together as the **National Academies of Sciences, Engineering, and Medicine** to provide independent, objective analysis and advice to the nation and conduct other activities to solve complex problems and inform public policy decisions. The National Academies also encourage education and research, recognize outstanding contributions to knowledge, and increase public understanding in matters of science, engineering, and medicine.

Learn more about the National Academies of Sciences, Engineering, and Medicine at **www.nationalacademies.org**.

Consensus Study Reports published by the National Academies of Sciences, Engineering, and Medicine document the evidence-based consensus on the study's statement of task by an authoring committee of experts. Reports typically include findings, conclusions, and recommendations based on information gathered by the committee and the committee's deliberations. Each report has been subjected to a rigorous and independent peer-review process and it represents the position of the National Academies on the statement of task.

Proceedings published by the National Academies of Sciences, Engineering, and Medicine chronicle the presentations and discussions at a workshop, symposium, or other event convened by the National Academies. The statements and opinions contained in proceedings are those of the participants and are not endorsed by other participants, the planning committee, or the National Academies.

Rapid Expert Consultations published by the National Academies of Sciences, Engineering, and Medicine are authored by subject-matter experts on narrowly focused topics that can be supported by a body of evidence. The discussions contained in rapid expert consultations are considered those of the authors and do not contain policy recommendations. Rapid expert consultations are reviewed by the institution before release.

For information about other products and activities of the National Academies, please visit www.nationalacademies.org/about/whatwedo.

COMMITTEE ON ENHANCING THE U.S. CHEMICAL ECONOMY THROUGH INVESTMENTS IN FUNDAMENTAL RESEARCH IN THE CHEMICAL SCIENCES

Members

MARK S. WRIGHTON, *Chair*, George Washington University
CATHY L. TWAY, *Vice Chair*, Johnson Matthey
ASHISH ARORA, Duke University
RAYCHELLE BURKS, American University
JOSEPH M. DeSIMONE (NAS, NAE, NAM), Stanford University
SHANTI GAMPER-RABINDRAN, University of Pittsburgh
JEANNETTE M. GARCIA, IBM
JAVIER GUZMAN, ExxonMobil
MARTHA HEAD, Amgen
RUSSELL MOY, Southeastern University Research Association (through January 2022)
KRISTALA L. J. PRATHER, Massachusetts Institute of Technology
JASON SELLO, University of California, San Francisco
BALA SUBRAMANIAM, University of Kansas
JEAN W. TOM (NAE), Bristol Myers Squibb

Staff

STEVEN M. MOSS, Study Director
LIANA VACCARI, Program Officer
JESSICA WOLFMAN, Research Associate
BENJAMIN ULRICH, Communications Associate
CHARLES FERGUSON, Senior Board Director
BRENNA ALBIN, Program Assistant
OLIVIA TORBERT, Program Assistant (through February 2021)
JEREMY MATHIS, Board Director (through September 2021)
MAGGIE WALSER, Interim Board Director (through January 2022)

Consultants

MICHAEL ZIERLER, RedOx Scientific Editing
LEE FLEMING, University of California, Berkeley
DANIEL BASCO, Vertex Evaluation and Research, LLC

Sponsors

NATIONAL SCIENCE FOUNDATION
U.S. DEPARTMENT OF ENERGY
NATIONAL INSTITUTE OF STANDARDS AND TECHNOLOGY
AMERICAN CHEMICAL SOCIETY

BOARD ON CHEMICAL SCIENCES AND TECHNOLOGY

Members

SCOTT COLLICK, *Co-chair*, DuPont
JENNIFER SINCLAIR CURTIS, *Co-chair*, University of California, Davis
GERARD BAILLELY, Procter & Gamble
RUBEN G. CARBONELL (NAE), North Carolina State University
JOHN FORTNER, Yale School of Engineering and Applied Science
KAREN I. GOLDBERG (NAS), University of Pennsylvania
JENNIFER M. HEEMSTRA, Emory University
JODIE L. LUTKENHAUS, Texas A&M University
SHELLEY D. MINTEER, University of Utah
AMY PRIETO, Colorado State University
MEGAN L. ROBERTSON, University of Houston
SALY ROMERO-TORRES, Thermo Fisher Scientific
REBECCA T. RUCK, Merck Process Research & Development
ANUP K. SINGH, Lawrence Livermore National Laboratory
VIJAY SWARUP, ExxonMobil

Staff

CHARLES FERGUSON, Senior Board Director
MEGAN E. HARRIES, Program Officer
LIANA VACCARI, Program Officer
LINDA NHON, Associate Program Officer
THANH NGUYEN, Finance Business Partner
JESSICA WOLFMAN, Research Associate
ABIGAIL ULMAN, Research Assistant
BRENNA ALBIN, Program Assistant
AYANNA LYNCH, Program Assistant
EMMA SCHULMAN, Program Assistant

Acknowledgments

This Consensus Study Report was reviewed in draft form by individuals chosen for their diverse perspectives and technical expertise. The purpose of this independent review is to provide candid and critical comments that will assist the National Academies of Sciences, Engineering, and Medicine in making each published report as sound as possible and to ensure that it meets the institutional standards for quality, objectivity, evidence, and responsiveness to the study charge. The review comments and draft manuscript remain confidential to protect the integrity of the deliberative process.

We thank the following individuals for their review of this report:

GREGG BECKHAM, National Renewable Energy Laboratory
GEOFFREY W. COATES, Cornell University
GEORGE P. COBB, Baylor University
JILL MARTIN, Dow Chemical Company
TRACY McGILL, Emory University
MELISSA PASQUINELLI, North Carolina State University
HENRY A. SODANO, University of Michigan
YING WANG, AbbVie
KATE S. WHITEFOOT, Carnegie Mellon University
LUISA WHITTAKER-BROOKS, University of Utah
JANE E. WISSINGER, University of Minnesota

Although the reviewers listed above provided many constructive comments and suggestions, they were not asked to endorse the conclusions or recommendations of this report nor did they see the final draft before its release. The review of this report was overseen by **CAROL J. HENRY**, George Washington University, and **F. FLEMING CRIM**, University of Wisconsin—Madison. They were responsible for making certain that an independent examination of this report was carried out in accordance with the standards of the National Academies and that all review comments were carefully considered. Responsibility for the final content rests entirely with the authoring committee and the National Academies.

This study would not have been successful without the assistance of many. The committee is grateful to the people who helped provide research support to the report, including analytical support for the Vertex report provided by the staff at IP Checkups, Inc., which includes Jesse Hooper, Matt Rappaport, and Mark Garner, and additional analytical support from Divya Sebastian at Duke University.

Preface

Chemistry has contributed significantly to the nation's economic prosperity, human health, national security, and overall quality of life. In fall 2020, the National Academies of Sciences, Engineering, and Medicine (the National Academies) convened a committee to consider strategies to sustain and enhance the economic activity driven by fundamental research investments in the chemical sciences. The work of the committee focused on four areas: (1) examination and definition of the role of the chemical industry in the U.S. economy; (2) assessment of how long-term investments in fundamental chemical research have contributed to national security, environmental sustainability, thriving manufacturing industries, and energy technology; (3) exploration of strategies for targeted research investments in the chemical sciences by both public and private sectors to stimulate growth and to ensure that the United States plays a leadership role in the field; and (4) consideration of options for research investments that would enhance the chemical economy and also advance environmentally sustainable practices and build a diverse workforce for the chemical economy.

The committee is diverse and has drawn on rich and extensive experiences in academia, industry, and government as researchers and as leaders of organizations that invest in the chemical sciences. The work of the committee has drawn on other leaders in the chemical sciences and those with economics expertise to provide information vital to the development of our report. We are grateful for the efforts of all those who have contributed to the information-gathering phase of our work.

The report from our committee provides a compelling rationale for why chemistry has been, and will continue to be, critical to the well-being of people everywhere. Contributions from chemists have enabled the development of life-saving pharmaceuticals; materials for structural purposes, for packaging, and for renewable energy technologies; modern microelectronics fueling our information technology infrastructure; and electrochemical devices to power transportation vehicles. Our recommendations will contribute to advancing human health and achieving environmental sustainability, while enhancing the U.S. economy and sustaining our national security.

In our work, we have identified areas where success in chemistry will contribute to addressing major global challenges. Advances in chemical instrumentation and computational power promise continuing rich returns from sustaining a long-term investment in fundamental chemical research. The daunting challenges we face affect all, and it is imperative that we apply both financial and human resources to address these problems. Chemistry is done by people. Drawing on a diverse, well-prepared workforce will be essential to make more rapid progress in the future. Proactive effort to equitably engage all of our human resources will stimulate more rapid innovation in chemistry.

While chemistry has enabled quality-of-life advances, many of these advances have had unforeseen negative consequences. Therefore, chemistry is both the source of many current global problems and will enable the potential solutions to these problems. Negative environmental impacts brought about by the combustion of fossil fuels and the proliferation of waste from plastics illustrate two such problems where advances in chemistry will be vital to the solutions for addressing the challenges. As the complementary 2022 *New Directions for Chemical Engineering* report by the National Academies also notes, we have the opportunity and responsibility to apply advances in chemistry and in chemical engineering to address the global problems we face related to energy, the environment, and sustainability.

The work of our committee has been done almost entirely via virtual meetings. Despite this mode of working, it has been a reward to come to know such a talented and dedicated group who have worked well together. Building new friendships is a result of good committee work, and it is our pleasure to acknowledge the tremendous effort from all. We are thankful, especially, for the talented team led by Steven Moss at the National Academies, which includes Liana Vaccari, Jessica Wolfman, Brenna Albin, and Benjamin Ulrich, as well as ample help from Michael Zierler, a rapporteur and consultant on the project, who all helped us complete an important report.

Mark S. Wrighton, *Chair*
Cathy Tway, *Vice Chair*
Committee on Enhancing the U.S. Chemical Economy Through
Investments in Fundamental Research in the Chemical Sciences

Contents

SUMMARY   1

1 INTRODUCTION   13
    1.1   Key Themes of the Report, 13
    1.2   Report Definitions, 16
    1.3   Imagining the Future of the Chemical Economy, 17
    1.4   Study Scope and Approach, 25
    1.5   Previous Consensus Studies Related to the Chemical Economy, 26
    1.6   Organization of the Report, 27

2 UNDERSTANDING THE ECONOMIC IMPACTS OF CHEMISTRY   29
    2.1   Brief History of the U.S. Chemical Industry, 30
    2.2   Estimating Current Size and Impact of the Chemical Economy, 32
    2.3   Research and Innovation in the Chemical Industry, 38
    2.4   Understanding U.S. Competitiveness in the Chemical Economy, 49
    2.5   Conclusions, 58

3 SUSTAINABILITY FOR THE CHEMICAL ECONOMY   61
    3.1   Basic Chemistry in Society: Contributions and Consequences, 63
    3.2   Transitioning to Sustainability and Decarbonization in the Chemical Economy, 65
    3.3   Policies to Assist in Adoption of Sustainability and Decarbonization, 74
    3.4   Fundamental Chemical Research for Sustainability, Decarbonization, and Environmental Stewardship, 77
    3.5   Conclusions, 98

4	**EMERGING AREAS IN THE CHEMICAL SCIENCES**	101
	4.1 Measurement, 102	
	4.2 Automation, 109	
	4.3 Computation, 114	
	4.4 Catalysis, 120	
	4.5 Conclusions, 131	
5	**PREPARING AND EMPOWERING THE NEXT-GENERATION CHEMICAL WORKFORCE**	133
	5.1 A Diverse and Equitable Chemical Workforce, 134	
	5.2 Mentorship and Support for Success, 137	
	5.3 Development Opportunities for Academic Institutions, 143	
	5.4 Workforce Development, 147	
	5.5 Conclusions, 152	
6	**FUNDING CHEMICAL RESEARCH**	155
	6.1 Federal Investments in Chemical Research and Education, 156	
	6.2 Corporate Funding of Chemical Research, 163	
	6.3 Philanthropic Funding of Chemical Research and Education, 167	
	6.4 Financial Responsibilities of Academic Institutions in Supporting Research, 168	
	6.5 Conclusions, 169	
7	**CONCLUSIONS AND RECOMMENDATIONS**	171
	7.1 The Importance of Chemical Research to the Chemical Economy, 171	
	7.2 Role of Chemistry in Team Science, 173	
	7.3 Chemical Research and Sustainability, 174	
	7.4 Challenging the Underlying Assumptions of Chemical Research, 176	
	7.5 Chemical Data and Analysis, 176	
	7.6 Chemical Workforce, 177	
	7.7 Funding Chemical Research, 179	

REFERENCES		181
APPENDIXES		
A	COMMITTEE MEMBER BIOGRAPHICAL SKETCHES	205
B	REQUEST FOR PROPOSAL FOR ECONOMIC ANALYSIS	211
C	LIST OF OPEN SESSION SPEAKERS	213
D	INDIVIDUAL EXPERT INTERVIEWS	215
E	CALL FOR INPUT FROM THE CHEMISTRY COMMUNITY	219

Summary[1]

For centuries, chemistry has played a central role in producing a body of scientific knowledge leading to products, processes, and technologies that have transformed health, energy, food and water production, and many other critical components of human well-being. These outcomes also had a significant economic impact, not only in the production of the chemicals and materials themselves but in other sectors of the economy that are enabled by these products. There are numerous examples that illustrate the importance of chemical knowledge, such as the production of synthetic fertilizers for agriculture, the processing of carbon to meet energy needs, the microfabrication of electronic devices, and the synthesis and production of life-saving molecules. All of these products were derived from chemical knowledge developed over decades that led to enormous impacts in the chemical economy. To continue to build upon the momentum of chemical knowledge contributing to breakthroughs and large impacts in the chemical economy, it is important to consider where new knowledge, tools, and technologies are needed in the chemical sciences, in addition to considering how chemistry can continue to innovate while having a positive impact on the environment. To accomplish this, the chemical enterprise must consider the needs for training, educating, and preparing a future workforce, and how the funding landscape can best encourage future advances (Figure S-1).

To consider the impact of fundamental chemical research on the chemical economy, and understand what strategies are needed to ensure growth in and leadership from the U.S. chemical enterprise, the National Science Foundation, the Department of Energy, the National Institute of Standards and Technology, and the American Chemical Society asked the National Academies of Sciences, Engineering, and Medicine to convene a committee of experts to consider these issues. The committee was tasked with understanding the role of the chemical industry in the chemical economy, understanding how chemical research has impacted society and the economy, and exploring strategies and options for research investments that ensure U.S. leadership while considering environmental sustainability and a diverse chemical economy workforce. Throughout the information-gathering and -synthesizing process, the committee discovered several key themes that

---

[1] Most references are not included in the Summary. Please see the associated report text for full citations.

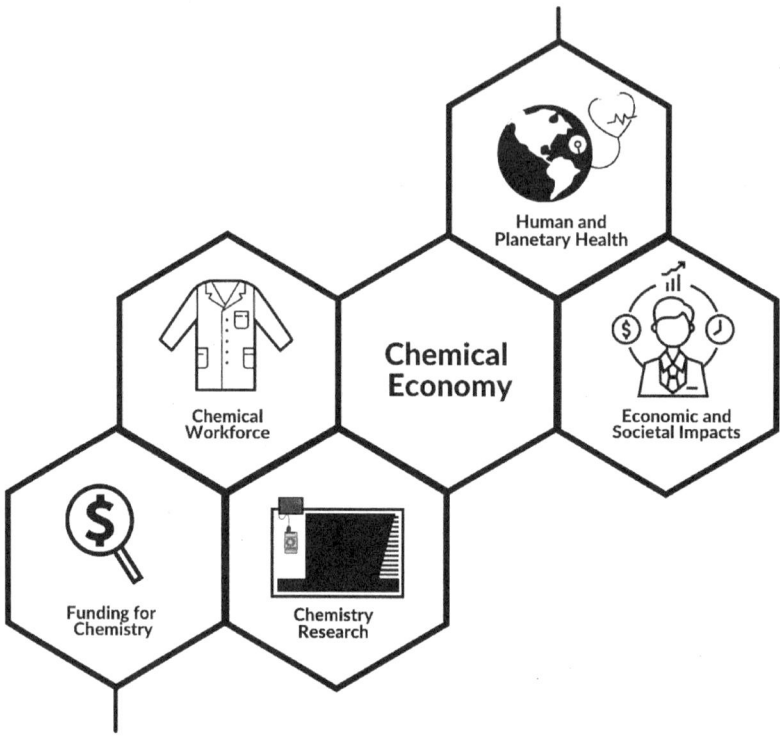

FIGURE S-1 Representation of the interconnected pieces of the chemical economy considered when developing a path forward for its success.

pervade the report: the balance of U.S. competitiveness and collaboration, a changing landscape in the chemical enterprise, emerging technologies, and a focus on sustainability (Box S-1).

ECONOMIC IMPACTS OF CHEMISTRY

The size of the chemical economy is quite expansive and is a substantial portion of the U.S. economy. All sectors reliant on the U.S. chemical economy are responsible for $5.2 trillion, or 25%, of the U.S. gross domestic product, and the entire chemical enterprise supports 4.1 million jobs in the United States. The United States is home to 10 of the top 50 chemical companies and remains very competitive in the global chemical economy. However, other countries have seen large sustained investments in their chemical and overall research enterprises, including China and several others, and their rapid advances are starting to threaten U.S. leadership in chemistry.

Fundamental chemical research has played a critical role in the size and impact of the U.S. chemical economy. This can be directly shown with examples such as the development of lithium-ion batteries, the adoption of biocatalysis in synthetic methodologies, advances related to silicon chips, and widely impactful pharmaceuticals such as oral contraceptives and those developed to fight SARS-CoV-2. Additionally, when looking at chemistry-related patents, which are used as a proxy for the economic value of fundamental chemical research, we see a spillover of chemistry knowledge and products into other areas of the economy.

BOX S-1
Key Themes of the Report

1. **Balancing U.S. competitiveness and collaboration in the global chemical economy**—Chemistry is a global enterprise and the United States is a key player. To assess how to best move forward, it is necessary to understand the historical and current positions of the United States in the global chemical economy and how innovative discoveries in fundamental chemical research are key to improving U.S. competitiveness. The role of chemistry in the economy is pervasive. Many companies beyond those in chemical manufacturing employ chemists, such as companies focused on chemical instrumentation, materials, forensics, and biotechnology. Thus, the competitiveness of the United States in many technology-based parts of the economy depends on continuing innovation in and collaboration with chemistry.

2. **A changing landscape within the chemical enterprise**—In many areas of chemistry, the landscape is changing. These changes are particularly prevalent in funding, the workforce, and training. Chemical manufacturers have long supported fundamental chemical research within their companies, but while pharmaceutical companies have maintained their research and development (R&D) programs, many chemical manufacturers have chosen to decrease the size and scale of in-house basic research programs over the past couple of decades. Recently, chemical companies have started collaborations with start-ups, university labs, and government-based research centers. The companies provide funding and industrial expertise while the other partner does the basic research that can be transferred to the company for further R&D. Increases in federal funding have been implemented in specific areas of research and have gone toward supporting training and research infrastructure. The mechanisms for training need to be continuously reevaluated because of the emergence of new technologies and the critical emphasis on equity and inclusion in the current and future chemical workforce.

3. **Emerging processes and technologies**—For the United States to enhance its competitive edge and grow its chemical economy, it must continue to innovate, and it must invest in state-of-the-art research infrastructure in the labs where people are trained and research is done. While there are many emerging areas that will drive innovation in the chemical sciences, the committee has identified four of particular importance. These are measurement, automation, computation, and catalysis. In addition, the committee strongly supports the need to expose chemistry students to how these technologies apply to chemistry and the chemical economy. For example, computational methods and data analysis have become integral to fundamental chemical research and industrial R&D.

4. **A focus on sustainability**—Both within the chemical economy and as a major global challenge that chemistry can help address, there is an urgent need to integrate sustainability into manufacturing, product usage, recycling, and product disposal. The feedstocks, processes, and products of the chemical industry and the ways in which consumers use those products have created significant environmental problems such as pollution, environmental degradation, increased scarcity of natural resources such as rare earth metals, loss of biodiversity, and climate change. The chemical industry and fundamental chemical research are uniquely positioned to innovate and develop solutions that will greatly assist in solving these problems. To do so, approaches to chemistry must also undergo a paradigm shift to incorporate greener chemical practices at every stage of R&D, including quantitative sustainability assessments to help guide chemical R&D.

Conclusion 2-1: Chemical research has an outsized economic value based on the spillover of chemical knowledge and products into other areas and the fact that chemical patents, as well as patents that rely on chemical knowledge, have a higher average value than other patents. Chemical patents accounted for 14% of all corporate patents between 2000 and 2020, but they accounted for 23% of all value in the same time period.

Conclusion 2-3: Chemistry is a foundational and central scientific discipline, and sustained investment in fundamental chemical research provides the chemical knowledge for technology development, generating unexpected discoveries that are the basis for innovation. These innovations directly influence the chemical economy, environment, and quality of life and also advance knowledge and discovery in many other scientific and technological disciplines, such as the life sciences, information technology, earth sciences, and engineering.

Conclusion 2-4: The chemical economy is critically important for our national economy and our leadership in the international chemical enterprise. This leadership relies heavily on advances in fundamental chemistry that drive the creation of new tools, technologies, processes, and products and enables environmental considerations. However, our nation's leadership in the chemical industry cannot be taken for granted, and this leadership needs continued and sustained nurturing and support.

To take action on this broad set of conclusions, the committee suggests a set of wide-ranging recommendations with an emphasis on growing and strengthening the U.S. chemical economy and U.S. competitiveness.

Recommendation 1: To foster fundamental chemical research and maintain U.S. competitiveness in the chemical economy, the U.S. chemical enterprise should support funding, workforce, and policy structures that attract international researchers and create a nurturing environment for all research talent.

Sub-Recommendation 1-1: Because it is not possible to predict where the next fundamental breakthroughs will come from, funding agencies that support the chemical sciences, such as the U.S. Department of Energy, National Science Foundation, National Institutes of Health, U.S. Department of Defense, National Institute of Standards and Technology, and U.S. Department of Agriculture, should fund the largest breadth of fundamental chemical research projects possible. This should include funding for a large range of topics in chemistry, as well as different scales of research projects, ranging from small grants for individual laboratories, to large-scale collaborations and facilities.

Sub-Recommendation 1-2: Participants in the chemical economy including chemical industry, pharmaceutical companies, and instrumentation developers should continue to invest in research and development at universities and scientific research institutions in the United States and should increase investments in broad areas of fundamental chemical research, including a focus on environmental sustainability.

Sub-Recommendation 1-3: The U.S. government should continue to produce policies that support international and open exchange of ideas in the chemical sciences and should engage policy and security experts, academic researchers, and industry professionals when considering any limitations on open engagement that are meant to mitigate economic or security risks to the U.S. chemical enterprise.

Sub-Recommendation 1-4: To help guide policy and funding decisions around chemical research, federal agencies who fund and track data related to scientific research should collaborate to collect, and make available, the tools and data needed to understand the impact of fundamental chemical research on the chemical economy. As a part of this initiative, large-scale evidence-building efforts to collect, standardize, use, and interpret these data should be funded.

CHEMICAL RESEARCH AND SUSTAINABILITY

Along with the advances in chemistry, it is important to remember that the chemical economy is also responsible for considerable negative impacts on the environment, including the production of greenhouse gases, the mismanagement of plastic waste, and cases where toxic chemicals have been released into the air or dumped in soil or water. Ironically, chemical research will be critical to solving many of these issues and help the world move toward achieving the United Nations Sustainable Development Goals.[2] To build upon previous innovation, pressures in the form of policy will need to incentivize sustainability, decarbonization, and environmental stewardship. While there are a number of mechanisms available, market- and purchasing-based policies are particularly important tools for supporting a green and circular economy and encouraging innovation in green chemistry. Some of these include national policies such as greener procurement and international policies such as regulation.

As the chemical enterprise continues to look for avenues where chemical research could make the largest impact in sustainability, there are a number of concrete areas where initial steps have already been made and further advancement is possible. The areas that are prime for chemical innovation include

- better measurements for life-cycle assessments;
- enhancement of recycling technologies and co-design of plastic products for recyclability;
- sustainable syntheses;
- sustainable feedstocks and energy sources;
- carbon capture, utilization, and storage;
- monitoring and improving air quality;
- monitoring and improving water safety; and
- monitoring and improving food safety.

To accomplish these, there are important changes to make as the chemical enterprise considers its role in sustainability.

Conclusion 3-1: To implement a circular economy, the future will require a paradigm shift in the way products are designed, manufactured, and used, and how the waste products are collected and reused. These new processes, and the use of clean energy and new feedstocks to enable these processes, will require novel chemistries, tools, and new fundamental research at every stage of design.

---

[2] See https://sdgs.un.org/goals.

Conclusion 3-2: Transitioning the chemical economy into a new paradigm around sustainable manufacturing, in which environmental sustainability is balanced with the need for products that will improve quality of life, enhance security, and increase U.S. competitiveness, will require substantial investment and innovation from industry, government, and their academic partners to create and implement new chemical processes and practices.

To accomplish this paradigm shift in the chemical sciences, many steps will need to be taken, and a concerted effort will be needed from government, industry, and academia.

Recommendation 3: The chemical industry and its partners at universities, scientific research institutions, and national labs should create opportunities to collaborate so that the objectives of fundamental research can directly assist in the design process of companies implementing new processes or practices toward environmental stewardship, sustainability, and clean energy.

While the conclusions and recommendations above take a broad view of environmental sustainability, and think through a shift of the entire chemical economy, there are specific conclusions and recommendations that apply to researchers. For chemical research to evolve with, and help advance, the moving landscape of the chemical economy toward sustainability, the committee makes two key conclusions, one of which is included below.

Conclusion 3-3: As fundamental chemical research continues to evolve, the next generation of research directions will prioritize the future of environmental sustainability and new energy technologies. Keeping sustainability principles in mind during every stage of research and development will be critical to accomplishing this goal.

To encourage academic researchers to keep environmental sustainability in mind at every stage of research, the committee noted that grant mechanisms usually do not ask researchers to consider the environmental impact of their work unless it is directly related to the grant or contract the researcher is applying for. Although not ubiquitous in grant writing, "broader impacts statements" have been an important mechanism in encouraging researchers to think through how their labs and research are interacting with the community around them. To similarly encourage all academic chemical researchers to keep environmental sustainability and stewardship at the forefront when considering all different types of research endeavors, the committee thought an optional "environmental impacts" statement would be the best way to accomplish this.

Recommendation 4: All chemistry-related research grants and proposals should have an option to explain the "environmental impacts" of the proposed research as an option under the "broader impacts" statement. The "environmental impacts" statement should include a summary of the possible environmental impacts, what is being done to mitigate those impacts, and any outcomes from the research that will directly impact environmental sustainability.

CHALLENGING THE UNDERLYING ASSUMPTIONS OF CHEMICAL RESEARCH

As a part of the paradigm shift toward sustainability, there is a transition happening in the way people collect and use energy. This change is happening rapidly and is affecting the materials, technologies, and processes that are needed to effectively implement the energy landscape. Chemical research and the chemical economy have been critical to the implementation of the current energy landscape, and there are many areas ripe for chemical discovery in a new energy landscape that prioritizes clean alternatives and decarbonization. To accomplish these scientific advances more effectively, there are several factors to take into account, such as the changing needs for metals and minerals that arise with the increase in electric vehicles and complications with acquiring metals based on shifting international politics. By understanding what is most likely for the energy landscape in the future, chemical researchers can make decisions about what the pressing needs will be to help move sustainability forward.

Conclusion 3-5: As the world moves deeper into its current energy transition—including the switch to electric vehicles, the implementation of clean energy alternatives, and the use of new feedstock sources—coupled with an increasing focus on circularity, the committee expects that decarbonization, computation, measurement, and automation will significantly alter the operations and processes of current industries, creating new opportunities and challenges that will benefit from fundamental chemistry and chemical engineering advances.

Recommendation 5: Changes in energy sources complemented by the technology and processes offered by chemical companies will lead to entire industries being created, transformed, and terminated. A group of experts from chemistry and other impacted disciplines, who represent the chemical economy and academic research, should be convened to assess the implications of these industrial shifts and understand their impacts on current chemical research paradigms. Based on the information from these discussions, funding agencies and the chemical industry should put money toward interesting opportunities for chemical research that might emerge based on these trends.

EMERGING AREAS AND NEEDS IN THE CHEMICAL SCIENCES

For chemistry to continue making advances in different areas of sustainability, emerging areas of chemical research, along with new tools and technologies, will be critical. Among the various tools and technologies that are available to scientists at present, a few are key pillars that are particularly impactful to understanding the molecular world and promoting real-world discovery:

- **Measurement**: Our ability to quantify and visualize molecules and their interactions is becoming faster and more accurate and can be accomplished on smaller instrumentation. This is driving new research with increasing accessibility to measurement capabilities and the subsequent measurement data.
- **Automation**: High-throughput techniques for measurement, synthesis, and other areas of chemistry, particularly in combination with flow chemistry, offer new avenues to researchers by enabling large numbers of chemicals or reactions to be tested, measured, and analyzed, and thus to more quickly determine new research questions to pursue.
- **Computation**: Computational chemistry is integral to fundamental research in every discipline of chemistry, and fundamental, multidisciplinary research in chemistry, physics, and engineering has played a critical role in the ongoing development of modern computing architecture.
- **Catalysis**: To establish new methods of synthesis and manufacturing that do not rely on energy-intensive processes, new advances in catalysis will be important. Promising

methods include photocatalysis, electrocatalysis, and biocatalysis, coupled with efforts to synergize theory with experimentation.

Conclusion 4-2: Measurement, automation, computation, and catalysis are the enabling tools and technologies of fundamental chemical research that will have a substantial impact on both the adoption of novel methodologies and future discoveries in the chemical economy.

ROLE OF CHEMISTRY IN TEAM SCIENCE

Major chemical discoveries that have large economic and societal impacts do not happen in a vacuum. Much scientific research in areas such as the life sciences, physics, engineering, and the social sciences contribute to every important chemical and technological advance. Additionally, the converse is true. Chemical knowledge contributes to many diverse fields of science and technology. When teams of researchers with diverse expertise gather to solve a central problem in a critical area, chemistry drives basic knowledge and practical application in order to help teams accomplish major advances. The emerging areas of measurement, automation, computation, and catalysis all rely heavily on different fields of research in order to successfully advance. The following conclusion and recommendation emphasize this.

Conclusion 4-1: Chemistry is an enabling scientific discipline that will continue to have the largest impact on society when chemists collaborate with experts from other areas such as engineering, biology, physics, computation, and data science to generate new fundamental knowledge and create translational impact at larger scales.

Recommendation 2: Research groups across all scales—small-to-medium interdisciplinary teams, large-scale collaborations, and facilities—should reflect the centrality of chemistry to science and engineering. Because of the central and enabling nature of chemistry, experts across chemistry and its subdisciplines should be considered when there are large interdisciplinary projects, highly collaborative institutions, national lab research, and other team-based scientific activities.

CHEMICAL DATA AND ANALYSIS

In assessing emerging tools and technologies, a common thread was detected: well-curated and accessible data benefit all aspects of chemistry. Although there have been some efforts to collect and share chemical data, the practice is not universal, and there is a particularly large dearth of negative data in chemistry. A large supply of both positive and negative data would be particularly helpful for developing models and for understanding different molecular properties and interactions, as well as learning how to more accurately measure chemical systems. For progress to continue, researchers must make a concerted effort to establish standards for how chemical data are collected, stored, and distributed, leading to the following conclusion and recommendation.

Conclusion 4-3: The ability to collect, document, store, share, and use chemistry-related data is needed to advance the use of new tools, such as computation and automation in fundamental chemical research, and increase the accessibility of chemical research to a larger community of practitioners. This information architecture will produce an

indispensable tool for the chemical sciences research community to increase the pace and efficiency of innovation by fully harnessing advances made with previous research investments.

Recommendation 6: The National Institute of Standards and Technology (NIST), in consultation with the International Union of Pure and Applied Chemistry, the American Chemical Society, and other global chemistry professional societies, should lead an effort to explore pathways that provide an open-source, accessible, and standardized way for chemical researchers to store, share, and use data from chemical experiments. In establishing these pathways, NIST should seek input from professional societies and stakeholders from different areas of chemical research and data science so that they can best understand the infrastructure needs of different research communities such as inorganic, organic, and analytical chemists. Once standards and data repositories are established, publishers should require researchers to submit all data related to reactions, measurements, or other chemical experiments to these established open-source repositories.

CHEMICAL WORKFORCE

When considering advances in fundamental chemical research and the chemical economy, there is an important emphasis on the individuals who are driving that work. While there are millions of employees in the chemical economy, this report focuses on the needs of those who go through chemistry and chemical engineering training programs. There are several critical components to training, preparing, and empowering the next-generation chemical workforce, including the need for a diverse workforce and equitable training practices, the need for well-developed mentorship and professional development programs, and an emphasis on educational training that is adaptable to the future needs of the chemical enterprise.

In building a *diverse workforce that is developed through equitable training practices, it is important to incorporate well-developed mentorship and professional development programs at all levels of training.* Opportunities exist at all stages of learning, including pre-college, undergraduate, graduate, postdoctoral, and beyond, to learn about chemical research and build knowledge and expertise with experienced professionals in the chemical sciences. Professional development is also a key component of these opportunities, and should be accessible and a continuous process throughout each person's career. Effective mentorship at all levels of education is an important component of professional development and a critical component of building a diverse workforce. Many programs and chemistry professional societies offer opportunities to find and network with other professionals to build a broader mentorship network that fits the needs of each individual.

In addition to professional development and networking, *training a future workforce will require chemistry curricula to be adaptable to the future needs of the chemical enterprise.* Chemistry is constantly adopting new tools, methods, and technologies, in order to better understand, measure, and build molecules and materials. It is important that chemistry curricula have the necessary flexibility to incorporate and teach new and emerging techniques. Introducing greater flexibility into chemistry and chemical engineering curricula without placing undue burden on faculty who already have a substantial number of teaching requirements will entail communication and collaboration across departmental leadership, university leadership, and accreditation bodies.

There are three conclusions that incorporate some of the ideas related to an emerging chemical workforce.

Conclusion 5-1: A skilled science and engineering workforce paired with a diverse, inclusive, and equitable science and engineering research enterprise is central to a thriving, nimble chemical economy equipped to respond to emerging challenges and maintain U.S. competitiveness.

Conclusion 5-2: The current structures and systems governing funding, promotion, retention, and professional development are in conflict and can stymie holistic career advancement for students, faculty, and research staff.

Conclusion 5-5: Creating an equitable and inclusive learning environment that exposes trainees of the future chemical workforce to new and innovative chemical tools, technologies, and instrumentation, as well as interdisciplinary knowledge and critical collaboration skills, will require a serious and sustained investment from funding agencies, universities, industry partnerships, and accreditation programs. This investment is critical because the tools and practices that enable chemical research are constantly evolving, and training programs must be able to adapt to best facilitate the learning of basic-to-advanced chemical principles that will help students succeed.

To properly address these conclusions on a practical level, the committee recommends that steps be taken to fund research in chemical education, continually reassess chemistry curricula, and continue to provide opportunities for professional development. The following recommendations lay out these ideas in more detail.

Recommendation 7-1: Funding agencies that support chemical research should put a substantial investment toward education research to continue enabling the development of innovative ways of teaching students about new and emerging concepts, tools, technologies, and instrumentation in chemistry while creating an inclusive learning environment for all students.

Recommendation 7-2: Universities, colleges, and accreditation programs should continually reassess their curriculum requirements and pedagogical practices to ensure that chemistry students in the chemical sciences are receiving state-of-the-art inclusive training and the most current chemical information and advances.

Recommendation 7-3: Universities and agencies that fund and support education in the chemical sciences should provide professional development at all levels, allowing for opportunities that are specific to the needs of each educational or career stage, such as programs that connect students with internships or resources for career exploration and providing faculty with professional development opportunities aimed at advancing their scholarship and teaching.

Recommendation 7-4: To continue progress in improving the diversity and equity of the chemical workforce, universities and chemical sciences departments should regularly assess their recruitment and retention practices related to trainees, faculty, and research staff. These assessments should be guided by relevant experts in research-informed equitable recruitment and retention practices of higher education institutions and units that also understand the nuances and details of the particular institution or entity. Institutions and units should continually take action and make meaningful investments based on their assessments. This work should be reported in a timely and transparent fashion to the institutional community.

FUNDING CHEMICAL RESEARCH

The funding landscape for chemical research in the United States is quite broad and includes a diverse set of private and public sources. One of the major advantages of this broad network of funding opportunities is that many different types and scales of chemical research are able to seek out and secure funding. Despite this wide range of funding sources, there are specific programs or changes that have a high likelihood of being impactful to the U.S. chemical research enterprise and the chemical economy. One of these is the Small Business Innovation Research and Small Business Technology Transfer (SBIR/STTR) programs in chemistry. This is one of the few government funding mechanisms that supports opportunities to convert fundamental chemical research into a product, process, or technology that will impact the broader chemical economy. The following conclusion and recommendation highlight SBIR/STTR programs.

Conclusion 6-2: Small Business Innovation Research and Small Business Technology Transfer (SBIR/STTR) programs have proven to be an important mechanism for advancing the chemical enterprise. There are many examples of fundamental chemical research being further pursued as a marketable product or process to contribute to the chemical economy through SBIR/STTR programs, and these programs also foster an emerging area of the chemical workforce where university researchers create and work in these small start-ups that are based on the grants from these programs.

Recommendation 8: Funding agencies should continue to support innovations in the chemical sciences through Small Business Innovation Research and Small Business Technology Transfer programs in order to leverage their previous investments in fundamental research and allow researchers the opportunity to bring new products or processes to market.

In the landscape of public and private funding, one area that has become more prominent over the past several years is the rise of philanthropic support. While philanthropies have contributed to fundamental chemical research, relative to federal funding for basic research, philanthropy supports far fewer schools and research projects. The future of scientific funding will include larger percentages of money from philanthropic organizations and independent donations. To ensure that science works to address big societal issues such as climate change and human health, funders will need to invest in the fundamental chemistry that informs and is critical to so many other areas of science. The following conclusion and recommendation start to address these issues.

Conclusion 6-4: In the near term, foundation and individual philanthropic support is likely to grow as a resource for innovations in chemistry. This support provides an important opportunity to use scientific evidence and exploration to address challenges that will benefit all of society such as climate change and human health.

Recommendation 9: The American Chemical Society, along with other chemistry-related professional societies, universities, and their academic leaders, should explore mechanisms to be more proactive in communicating to philanthropists and foundations about the promise of fundamental chemistry in addressing national and global problems. University

and academic leaders should emphasize the importance of funding structures between philanthropic and federal funding mechanisms that ensure balance and complementarity.

To accomplish the chemical research described in this report, there is a critical need for laboratory space, instrumental facilities and support, access to computation, and much more. Infrastructure is critical for training the next generation of the chemical workforce, as educators continually rethink and adapt curricula based on new tools and technologies that will be critical to the future of research and industry. Having infrastructure in place gives institutions the ability to train students, researchers, and other professionals on these emerging technologies. These ideas are summarized with a conclusion and a supporting recommendation.

Conclusion 6-1: Investment in the infrastructure at research universities is not well supported. This diminishes the opportunities for many talented chemical researchers to use the newest tools, technologies, and instrumentation and prevents trainees from having access to the newest technologies being used in the chemical workforce.

Recommendation 10: The federal government should invest more to support research infrastructure at research institutions to ensure that talented chemical experts and trainees with outstanding ideas can be competitive for research awards.

1

Introduction

Chemistry plays a pivotal role in the strength of the U.S. economy and in the well-being of humankind. Among many achievements, chemists have created life-saving pharmaceuticals and have been central in the development of controlled energy transformations, including nuclear, solar, and most importantly, the conversion of fossil energy to generate electricity and power the transportation system. Advances in chemistry have resulted in beneficial materials, from photovoltaic semiconductors to packaging to structural and construction materials. Contributions from chemists have been key to improving agricultural productivity since the development of the first synthetic fertilizers in the mid-18th century. Chemical processes are critical in the microfabrication of electronic devices that have revolutionized the way humans live, work, and play. In short, chemistry has contributed significantly to everyday lives of people everywhere and is critical to the nation's economic prosperity, human health, food production, energy generation, security, and sustainability.

1.1 KEY THEMES OF THE REPORT

Despite chemistry's many positive contributions to society, chemical processes and manufacturing of chemicals and materials present challenges of enormous magnitude that threaten the health of the planet. Fossil fuel combustion provides most of the energy demanded by society, but the associated production of carbon dioxide (CO_2) and other greenhouse gases is a major contributor to global warming, which is bringing about climate change with its devastating impacts. Synthetic polymers have revolutionized consumer products and packaging, including the packaging of food, thus reducing spoilage, but the proliferation of plastics is now a major environmental threat. Numerous synthetic chemicals are used in practical applications and are important contributors to the chemical economy. But a number of them have been found to be toxic or carcinogenic.

It is perhaps ironic, but further advances in fundamental chemistry are needed to address major problems arising from the use of chemistry. To transition away from fossil fuel energy, advances in the storage of electricity will be required to more fully implement the potential of solar photovoltaic electricity or of energy from wind. The needed improvement of electrochemical storage and energy conversion devices, including electrolyzers, fuel cells, and batteries, could come from fundamental advances in electrocatalysis. To eliminate single-use plastics, recycling of valuable materials will depend on advances in polymer chemistry and catalysis.

Addressing these and other global challenges will require talented people in many disciplines of science, engineering, social sciences, and the humanities. However, fundamental innovations in chemistry may well be a key to addressing important global challenges, including the overarching goal of providing the energy the population needs at an affordable cost and without adverse consequences on the environment. Success will depend on a highly trained workforce fluent in both traditional chemistry and contemporary technologies, such as data science, computing, and automation, among others. Success will also require a willingness to increase investment in basic and applied chemical research.

The outcomes of the chemical economy depend on the critical and intertwined roles played by funding of chemical research and development (R&D), the research itself, the workforce, chemistry's impact on human and planetary health, and its economic and societal impacts (Figure 1-1). As the committee discussed its charge, received input from colleagues and government agencies, and thought about these interconnected aspects of the chemical economy, the committee identified four themes on which to build this report. These themes are:

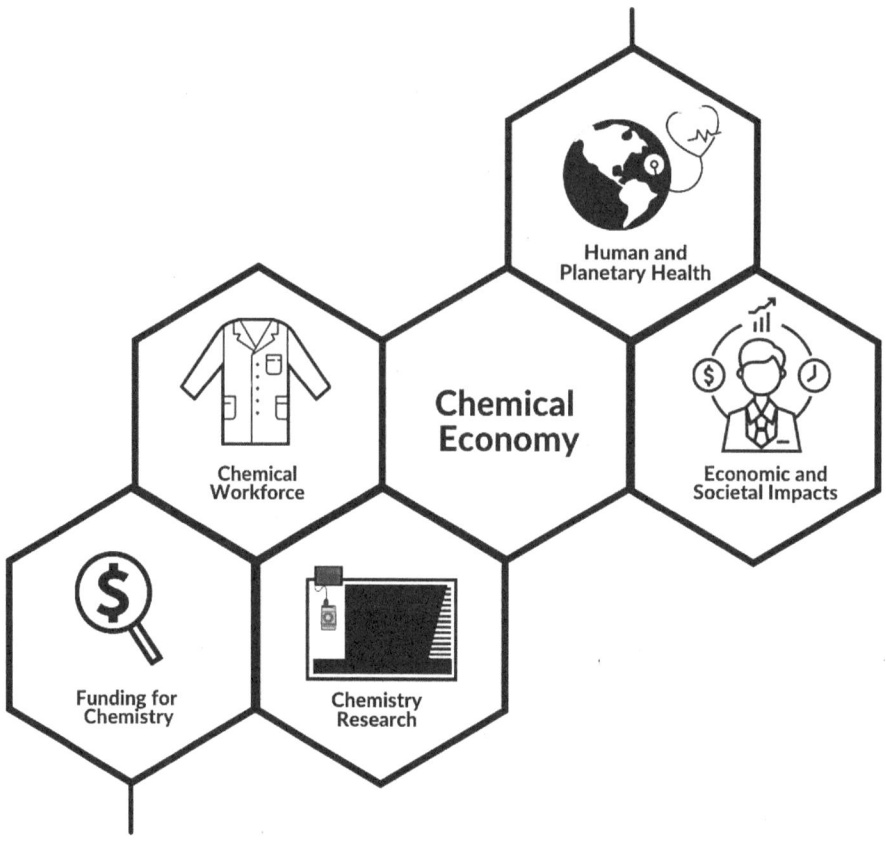

FIGURE 1-1 Representation of the interconnected pieces of the chemical economy considered when developing a path forward for its success.

1. **Balancing U.S. competitiveness and collaboration in the global chemical economy**—Chemistry is a global enterprise and the United States is a key player. To better understand this situation, the committee determined that it was necessary to understand the historical and current positions of the United States in the global chemical economy and to understand how innovative discoveries in fundamental chemical research are key to sustaining and improving U.S. competitiveness. The role of chemistry in the economy is also pervasive. Many companies beyond those in chemical manufacturing employ chemists, such as companies focused on chemical instrumentation, forensics, and biotechnology. Thus, the competitiveness of the United States in many technology-based parts of the economy depends on continuing innovation in and collaboration with chemistry.

2. **A changing landscape within the chemical enterprise**—In many areas of chemistry, the landscape is changing. These changes are particularly prevalent in funding, the workforce, and training. Chemical manufacturers long supported fundamental chemical research within their companies, but while pharmaceutical companies have maintained their R&D programs, many chemical manufacturers have chosen to decrease the size and scale of in-house basic research programs over the past couple of decades. Recently, chemical companies have started collaborations with start-ups and university labs. The companies provide funding and industrial expertise while the other partner does the basic research that can be transferred to the company for further R&D. Increases in federal funding have been implemented in specific areas of research and have gone toward supporting training and research infrastructure. The mechanisms for training need to be continuously reevaluated because of the emergence of new technologies and the critical emphasis on equity and inclusion in the current and future chemical workforce.

3. **Emerging processes and technologies**—It is an axiom in science that those with the most advanced technology are first on the forefront of science, asking and answering critical scientific questions of the day. For the United States to enhance its competitive edge and grow its chemical economy, it must continue to innovate, and it must invest in state-of-the-art research infrastructure in the labs where people are trained and research is done. While there are many emerging areas that will drive innovation in the chemical sciences, the committee has identified four of particular importance. These are *measurement*, *automation*, *computation*, and *catalysis*. In addition, the committee strongly supports the need to expose chemistry students to how these technologies apply to chemistry and the chemical economy. For example, computational methods and data analysis have become integral to fundamental chemical research and industrial R&D.

4. **A focus on sustainability**—Both within the chemical economy and as a major global challenge that chemistry can help address, there is an urgent need to integrate sustainability into manufacturing, product usage, recycling, and product disposal. The feedstocks, processes, and products of the chemical industry and the ways in which consumers use those products have created significant problems in the form of pollution, environmental degradation, increased scarcity of natural resources such as rare earth metals, loss of biodiversity, and climate change. The chemical industry and fundamental chemical research are uniquely positioned to innovate and develop solutions that will greatly assist in solving these problems. To do so, approaches to chemistry must also undergo a paradigm shift to incorporate greener chemical practices at every stage of R&D including quantitative sustainability assessments to help guide chemical R&D.

1.2 REPORT DEFINITIONS

Behind these industrial-scale chemical processes and the chemical economy lie fundamental chemical research and chemical knowledge. Our chemical knowledge stems from education, training, and research in the chemical sciences as practiced at universities, research institutes, national laboratories, and private industries. To assess the national and global impact of our chemical knowledge, the committee considered its role in the U.S. economy by defining both its contribution as a part of the U.S. chemical economy and the meaning of fundamental research.

1.2.1 Chemical Economy

For this report, the committee chose to take a broad definition of the chemical economy. To help capture the wide breadth and importance of this topic, the committee chose the following definition:

The chemical economy includes all parts of any value chain that rely on chemical knowledge and transformation processes for advancement and growth.

Industries whose processes and value chains use this knowledge to make products as a part of the chemical economy include, among others, the petroleum industry, the energy sector, materials production, pharmaceuticals, and agrochemicals. When considering the value chains of these industries, the report is referring to each step from raw material to final products that involves a chemical transformation. Thus, if a raw material must undergo a biological transformation to make a feedstock that is then chemically transformed to products, the value chain starts from the feedstock.

Much of molecular biology, biochemistry, synthetic biology, and structural biology are based on fundamental chemical principles. However, this report limits its discussion of biology to instances where chemical principles are used to study biological phenomena, and when biological entities are used or modified to perform chemical functions—for example, when microbes are used as chemical sensors or when enzymes are modified to serve as catalysts in industrial production. The committee understands that it is exceptionally difficult to parse the life sciences into chemistry, biology, and the many other disciplines that are involved, due to the interdisciplinary nature of the field, but the report makes a note of when chemistry was integral to a biological discovery or process and also seeks to acknowledge the important roles of other disciplines in any advances and their economic success.

1.2.2 Fundamental Chemical Research

A variety of definitions exist for fundamental, basic, and applied research. Some sources use "fundamental" and "basic" as synonyms whereas others distinguish between the two (See Box 1-1). The committee chose to take a broad approach to defining fundamental chemical research in order to capture a broad range of chemical discovery. For the purposes of this report:

Fundamental chemical research is basic and applied research that is made available to any interested scientific audience and which explores the structure and reactivity of atoms, molecules, and materials.

In the Stokes model of research quadrants, fundamental chemical research falls into both the Bohr (pure basic research) and Pasteur (use-inspired basic research) quadrants (Figure 1-2) (Stokes, 1997). The Bohr quadrant is important because of the unknowable influence of tools and basic discoveries and their possible widespread and unquantifiable impact on a number of areas of chemical research. However, most fundamental chemical research in the United States falls into the Pasteur

INTRODUCTION

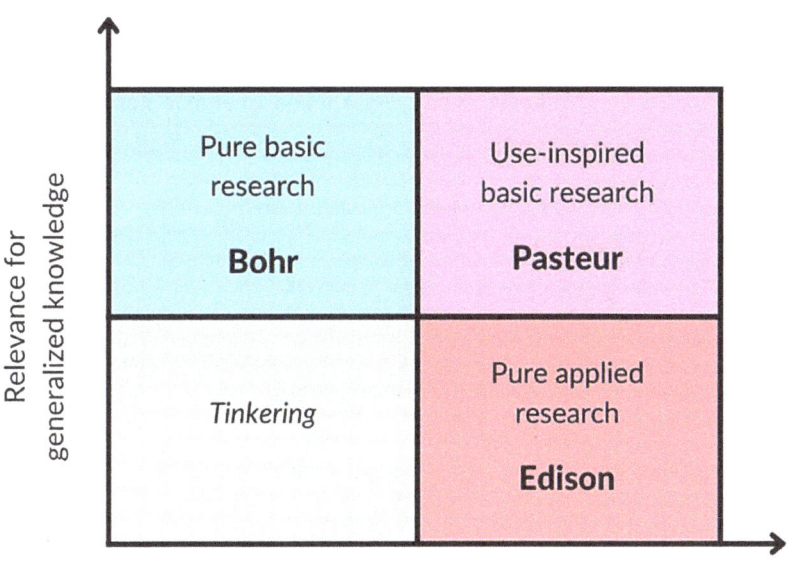

FIGURE 1-2 Stokes model of research quadrants. SOURCE: Aleahmad, 2009.

quadrant, in part because many scientific funding agencies have specific missions in areas such as health, energy, or agriculture. The variety of research in the Pasteur quadrant highlights a critical opportunity to use chemical knowledge in applications toward other areas of science, expanding the well-developed field of chemistry into an interdisciplinary research field that can be used to characterize and alter the physical properties of many more facets of the world around us.

1.3 IMAGINING THE FUTURE OF THE CHEMICAL ECONOMY

Many people view chemistry as a mature science; therefore, it is often overlooked when considering fields that are using and producing rapidly developing technologies. As the committee members continued to form their thoughts on the future of chemistry and the chemical economy, they were driven by input from experts and their own understanding of what new chemistries might be needed in different areas of research and industry. There are many scenarios to think through, but all provide a good idea of how technologies and advancements will be used and in what areas they are needed. This section highlights three scenarios that illustrate the thinking that helped the committee produce the final report.

1.3.1 Case 1: Future of Ammonia Beyond Haber-Bosch

> To catch the elusive and flirtatious floating nitrogen is one of the most cherished
> objects of present-day explorers among Nature's secrets. (Maxim, 1903)

Many traditional chemistries have been highly optimized, potentially making fundamental investigations in those areas less attractive than research in developing technological areas where more rapid advances are possible. *However, with the increasing focus on mitigating climate change, this viewpoint is primed for change because highly optimized, mature technology spaces will need to be re-created.* An example of where this has begun is in ammonia production.

> **BOX 1-1**
> **Defining Fundamental, Basic, and Applied Chemical Research**
>
> Below are a sample of quoted definitions from federal agencies and other sources:
>
> **The American Chemistry Council's *2021 Guide to the Business of Chemistry*:** "There are two broad categories of research: basic and applied. Basic (fundamental) research can be defined as any planned search for unknown facts and principles of general validity, without regard to commercial objectives. It consists of original investigation for the advancement of scientific knowledge. Applied research, on the other hand, can be defined as any investigation planned with the intent of using known phenomena or substances to accomplish a particular objective. In general, the basic research of today is the foundation for tomorrow's applied research. Once the research has been conducted, the findings (or other general scientific knowledge) must be translated into a form that is designed to meet the needs of customers. This is the 'development' part of R&D" (ACC, 2021).
>
> **Organisation for Economic Co-operation and Development:** "Basic research is experimental or theoretical work undertaken primarily to acquire new knowledge of the underlying foundation of phenomena and observable facts, without any particular application or use in view" (OECD, 2002). "Applied research is original investigation undertaken in order to acquire new knowledge. It is, however, directed primarily towards a specific, practical aim or objective" (OECD, 2015).
>
> **National Science Foundation:** "Basic research is defined as systematic study directed toward fuller knowledge or understanding of the fundamental aspects of phenomena and of observable facts without specific applications towards processes or products in mind" (NSF, n.d.).
>
> **National Security Decision Directives:** "'Fundamental research' means basic and applied research in science and engineering, the results of which ordinarily are published and shared broadly within the scientific community, as distinguished from proprietary research and from industrial development, design, production, and product utilization, the results of which ordinarily are restricted for proprietary or national security reasons" (White House, 1985).

Ammonia is currently produced via the Haber-Bosch process, one of the most important 20th-century inventions. Its inventors received two separate Nobel Prizes: Fritz Haber won the 1918 Nobel Prize in Chemistry and Carl Bosch was awarded the 1931 Nobel Prize in Chemistry, which was shared with Friedrich Bergius in recognition for his contributions to high-pressure chemistry in the unrelated technology of coal liquefaction (Nobel Prize Outreach, 2022a). Ammonia produced through the Haber-Bosch process is credited for enabling food production that feeds half the planet's population (IEA, 2021a); therefore, it has been the focus of numerous fundamental probes to better understand nitrogen activation and fixation. Although the Haber-Bosch process resulted in a substantial improvement in energy efficiency over previous attempts to artificially fix nitrogen, scientific advances in combination with engineering improvements have further reduced its energy requirements by nearly 74% over the initial Haber-Bosch synthesis energetics (Figure 1-3) (Smith et al., 2020).

Recent advances are largely incremental and are in processes outside of the ammonia synthesis step. That step is so efficient that some people argue that further work in improving the catalyst itself may be difficult to defend (Schlögl, 2003). However, despite the success in reducing the process energetics, ammonia production still accounts for nearly 2% of global energy consumption and is one of the most carbon-intensive chemical processes, accounting for nearly 1.3% of global CO_2 emissions, despite there being no carbon incorporated into the final molecule. Ammonia's CO_2 footprint is largely a result of the generation of hydrogen from fossil-derived feedstocks. With the need

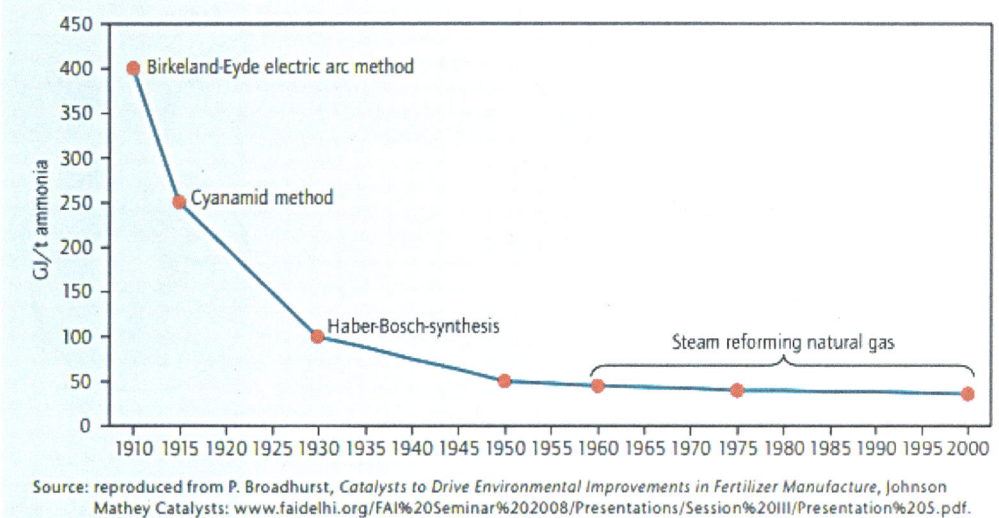

FIGURE 1-3 Timeline of changes and breakthroughs in the energy efficiency of ammonia synthesis. SOURCE: IEA, 2013.

to move toward net-zero carbon emissions, current production routes need to be improved (IEA, 2021a). While green hydrogen sourced through water electrolysis is a potential option to replace fossil-derived hydrogen, significant discoveries are needed to successfully marry green hydrogen production with the Haber-Bosch process, because these technologies are currently operated at very different scales. Advancements in catalysis, electrocatalysis, and adsorbents will be needed for this approach so that future ammonia production can be successful (Smith et al., 2020).

However, a move away from fossil-based technologies may also enable a shift away from Haber-Bosch to other ammonia synthesis options. For example, a recent paper highlighted the advantage of directly synthesizing ammonia from air and water without going through the independent synthesis of hydrogen (Schiffer and Manthiram, 2017). A wide range of approaches is being explored, including electrochemical, photochemical, biochemical, and hybrid ones (Boerner, 2019). These efforts are in early stages, with numerous needs for fundamental research advances.

Recently there has been a revitalization of ammonia investigations because the molecule has some unique properties. Ammonia can be liquefied at relatively mild conditions, readily stored, is 17.65% hydrogen by weight, and has a 45% higher volumetric hydrogen density than that of liquid hydrogen itself (Thomas and Parks, 2006). Thus, ammonia may increasingly be seen as a solution to large-scale storage and distribution of hydrogen, playing a key role in future clean energy, in addition to its traditional role of being a source for nitrogen. The potential for an ammonia-based economy, in which ammonia can be cracked to release hydrogen, burned directly in internal combustion engines, or utilized directly in fuel cells is already being envisioned (Figure 1-4) (MacFarlane et al., 2020). Although early commercial investments are being announced (Brown, 2020), full development of the envisioned ammonia-based economy will only be realized when there is a strong foundation of chemical knowledge arising from significant investment in fundamental chemical research.

1.3.2 Case 2: Needs for the Future Energy Landscape

Inorganic materials and elemental discoveries were prime chemistry research areas in the 1700s and early 1800s. During this era, electrochemistry was born, starting with Alessandro Volta's

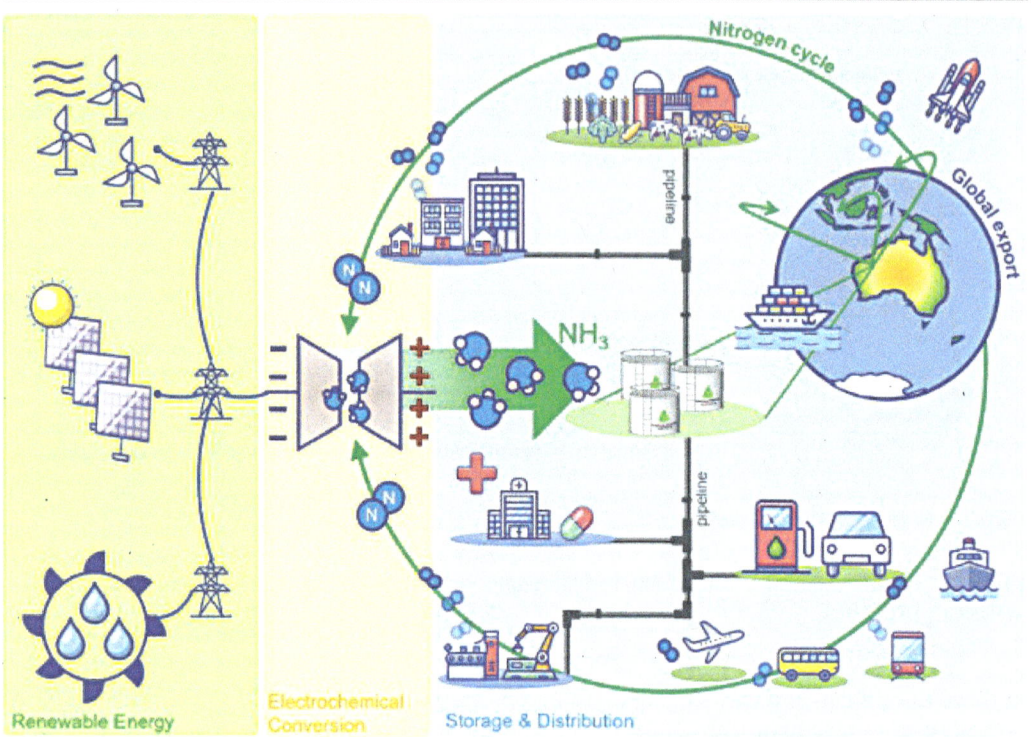

FIGURE 1-4 Representative image of an "ammonia economy," in which energy sources are all based on ammonia. SOURCE: MacFarlane et al., 2020.

invention of the first battery, the voltaic pile, followed by William Nicholson's use of the pile to begin experiments in water electrolysis (Fabbrizzi, 2019). Electrochemistry enabled Sir Humphrey Davy's discovery of potassium and sodium as well as the first isolation of lithium by Davy and W.T. Brande.[1] Although exploratory work continued in these areas, the focus of chemical research shifted to organic chemistry and ultimately the birth and development of today's petrochemical and polymer industries (American Chemical Society National Historic Chemical Landmarks, 1993; Ramberg, 2000; Stone, 2021). However, a new chemical research era is now beginning (Figure 1-5). With the increased need to shift toward renewable energy, chemistry will, in part, come full circle due to a renewed focus on inorganic molecules and electrochemistry, in addition to a shift toward activation of new carbon sources such as biomass and CO_2.

In this new era, inorganics will play an increasing role, especially as electrification expands. A recent International Energy Agency study estimated that demand for critical minerals related to clean energy technologies will quadruple, with some minerals such as lithium predicted to see a 13- to 51-fold increase in demand by 2040. Much of this increase will be needed to build out the electricity grid as well as for battery materials for electric vehicles and energy storage. Many of the new technologies will require increased mineral inputs compared with what is used in fossil fuel–based technologies. For example, an electric vehicle will require about six times the mineral inputs compared to a current internal combustion engine vehicle, and an onshore wind power plant will require nine times the minerals as a similar-size gas-fired power plant (IEA, 2021b). Such

---

[1] For more information, see WebElements, https://webelements.com/.

INTRODUCTION

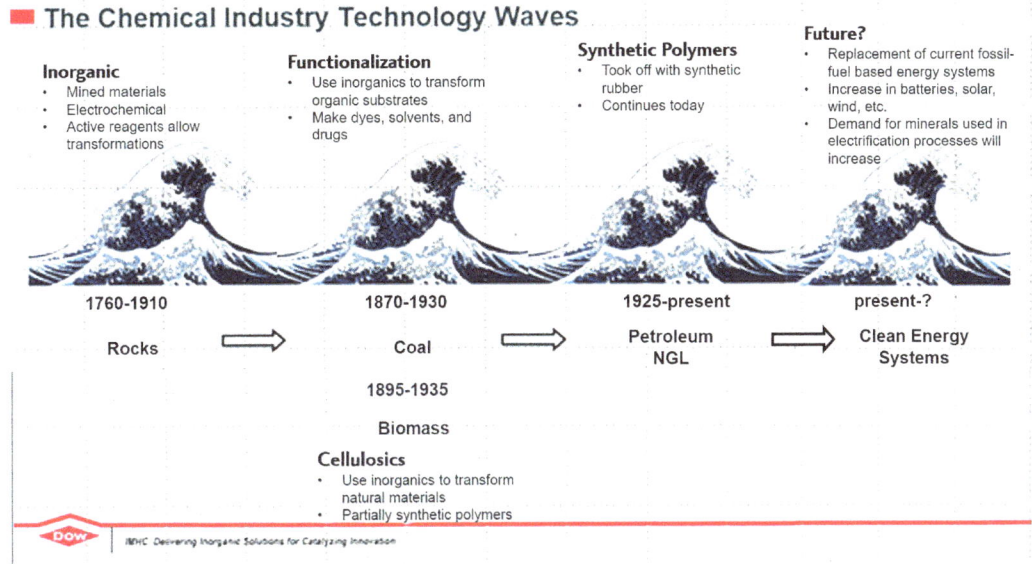

FIGURE 1-5 Representation of past technology waves in the chemical sciences. SOURCE: Modified from figure by Mark Jones, Dow Chemical Company.

comparisons led Forbes to conclude that "the energy transition will be fueled by metals" (Figure 1-6) (Mackenzie, 2020).

Mining and metals processing activities will have to rapidly increase to ensure that the needed metal is available for the energy transition, and this may exacerbate certain environmental concerns. Mining and processing of minerals are very energy, land, and water intensive and also incur a human cost. An estimate suggested that mining activities in 2018 accounted for nearly 10% of the total global energy-related CO_2 emissions (Azadi et al., 2020). Energy utilization is of particular concern in light of the decreasing ore quality of many of the needed minerals (Azadi et al., 2020; IEA, 2021b). In 2019 alone, more than 3.2 billion tonnes of ore were processed to supply metal demand (Bhutada, 2021), so decreasing ore quality is a concern from land-use and waste perspectives as well. Clearly there is a need to develop improved extraction methodologies to minimize the environmental impact of increased mining activities, and these innovations will strongly rely on fundamental research in new approaches for mineral activation and metals refining. Already, many new approaches are being reported, such as electrokinetic in situ leaching (Martens et al., 2021), electroextraction agromining (van der Ent et al., 2021), ligand-assisted displacement chromatography (Ding et al., 2020), and novel sulfidation techniques (Stinn and Allanore, 2022). In a similar vein, exploration of novel materials for metals removal from water streams is becoming an active area of research (Yang et al., 2019). Improving metal extraction and finding novel materials will need to take into account local communities and indigenous peoples who have frequently been victimized and marginalized in the pursuit of different mineral resources, including the cobalt and lithium used in most battery technologies (Frankel, 2016). A future chemical economy where environmental considerations are at the forefront cannot sacrifice human justice to accomplish other goals. As chemical researchers consider the environmental improvements associated with specific technologies, they can also consider what impact these improvements will have on communities. All these areas will require fundamental chemical research to allow for continued development. While metals recycling is well established for many current metals, recycling will need to play a large

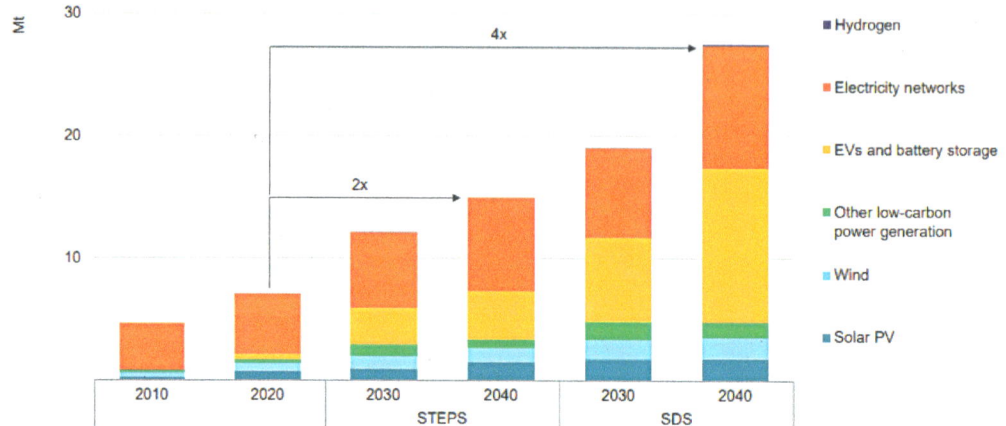

FIGURE 1-6 Total mineral demand by clean energy technology, separated by scenario. NOTE: STEPS = state policy scenario; SDS = sustainable development scenario. SOURCE: IEA, 2021b.

role in the build-out and design of battery technologies (IEA, 2021b). Although various methods are being proposed, they have not always been developed utilizing a green chemistry approach, and thus many new chemistries will be needed to ensure that the methods are sustainable (Piątek et al., 2021). Recycling metals through sustainable methods will also aid in the security of the supply chain by keeping valuable metals in the United States. Doing more to build recyclability into battery designs is an area where fundamental chemistry will play a role (Thompson et al., 2020).

In addition to battery storage of electricity, storage of electrons in chemical bonds is becoming an increasingly popular area for study either through the generation of "green hydrogen" from water electrolysis (Møller et al., 2017) or through conversion of CO_2 to more useful chemical species (Klankermayer and Leitner, 2016). These chemical transformations can be viewed as a means to ultimately pivot the chemical industry to a non-fossil feedstock base. This transition will require the development of new materials in addition to new chemical processes (Figure 1-7) (Van Geem and Weckhuysen, 2021).

1.3.3 Case 3: The Future of Distributed and Additive Manufacturing

A move away from traditional fossil-derived feedstocks will likely have far-reaching implications in how chemicals, pharmaceuticals, and materials are produced. In many chemical manufacturing facilities, feedstocks are brought to a facility to be transformed to either the building blocks or the final end products, which are later transported to consumers. This model works well because fossil-derived feedstocks have high energy densities that enable manufacturing to optimize production costs via large-scale operations. Larger-scale production is advantaged over smaller-scale operations in most situations owing to economies of scale. However, potential future feedstocks such as biomass, plastics, and municipal solid waste are bulky with low energy density, features that will make long-range transport uneconomical, challenging the centralized manufacturing model. For example, biomass can only be economically transported about 40 to 80 km (Willems, 2009). For many of the future feedstocks, a smaller-scale distributed model where manufacturing is located near the feedstock source will often be advantaged over large-scale centralized options. In turn, this manufacturing pivot will create new research opportunities. By relying on localized feedstock sources, distributed manufacturing could build sustainable communities and also mitigate supply-chain issues inherent in importing feedstocks from foreign sources for centralized facilities.

INTRODUCTION

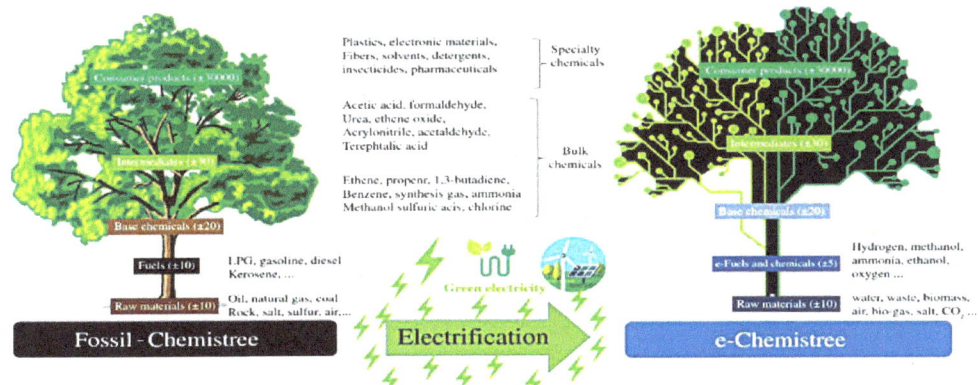

FIGURE 1-7 Needs for a transition from a fossil-based chemical industry to an electrified chemical industry.
SOURCE: Van Geem and Weckhuysen, 2021.

Entirely new engineering approaches for small-scale manufacturing plants will be needed both for improving energy efficiency and for reducing manufacturing footprints. This will be a rich research space because these methodologies will be developed while simultaneously also inventing the new chemistries for managing and activating these feedstocks (DOE Office of Science, 2017; Wang et al., 2021). These efforts will need to be highly collaborative across disciplines and are likely to lead to rapid advances in fundamental understanding of chemical systems. For example, to develop chemistries and catalysts to intensify processes where unit operations are combined, such as directly linking reaction steps with separation steps, new catalysts, evaluation methods, and measurement techniques will be needed to allow systematic studies under non-steady state conditions (DOE Office of Science, 2017). Similarly, because many of these small facilities will be more dispersed, they may not be near other industrial facilities. Development of new sensor and automation techniques will improve process operability and rapid analysis of the chemical composition of incoming feedstocks, and together with predictive technologies enabled by machine learning and artificial intelligence (AI) will provide a path toward more robust processes. Many of these approaches are being developed as part of the field of process analytical technology, and they will also open a broader landscape of new sensor R&D areas in adjacent and tangential fields as part of the so-called Internet of (Analytical) Things (Mayer and Baeumner, 2019).

While chemical processes are moving toward smaller distributed facilities, manufacturing techniques are also changing for other areas within the chemical economy where some of these same fundamental explorations will also be useful. For example, an exciting application space for process analytical technology is in the transition of pharmaceutical manufacturing to more continuous processing via flow chemistries. This shift from conventional batch processes offers many advantages including increasing pharmaceutical production efficiency and flexibility (Lee et al., 2015). Flow chemistry not only opens new manufacturing opportunities but also leads to entirely new approaches to drug discovery and synthesis by combining advanced analysis and control systems with AI, which allows chemists to probe new reaction conditions, utilize different precursors, and explore wider ranges of compositional space. This field of exploration will undoubtedly lead to improved pharmaceutical offerings and will greatly expand the knowledge of synthetic chemistry overall (Bogdan and Dombrowski, 2019).

Another advancement in manufacturing approaches lies in the future of highly customized offerings, such as those promised through additive manufacturing approaches such as 3-D printing (Figure 1-8). These techniques are already used in customized medical applications such as

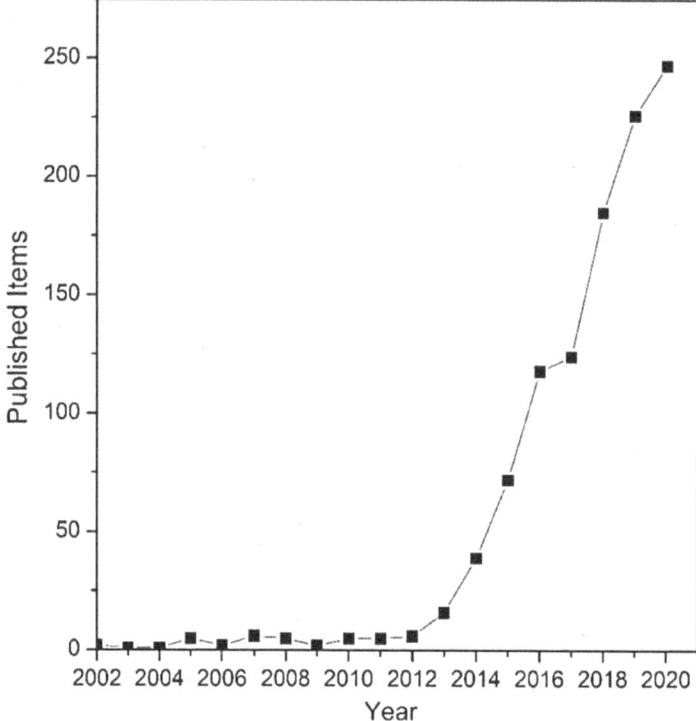

FIGURE 1-8 Number of published articles involving 3-D printing and microfluidics from 2002 to 2020. Search was done using the keywords "microfluidics" and "3-D printing." SOURCE: Alimi and Meijboom, 2021.

instrumentation, implants, and prosthetics (FDA, 2017). Other potential medical applications include 3-D printing of organs, anatomical models to assist in medical treatment, customized drug formulations where medications could be printed on demand at pharmacies, and novel drug delivery systems through controlling the release of active components via unique form factors or material approaches (Ventola, 2014). These will be promising and active areas of research for the foreseeable future.

Additive manufacturing offers the potential to create intricate articles with a high degree of manufacturing flexibility. Unlike many conventional forming techniques, additive manufacturing is not subtractive, and thus it offers the potential to reduce manufacturing wastes (Alimi and Meijboom, 2021). However, it remains more of a specialized method because additive manufacturing cannot produce articles as quickly as conventional mass production, which makes the technique inherently more expensive for large-volume applications (Dilberoglu et al., 2017). Yet in a small-scale, distributed manufacturing future, the lower output will no longer be such a limitation for additive manufacturing's broader adoption. Additive manufacturing may also help enable distributed manufacturing overall as the technique could be utilized to produce needed maintenance parts and minimize downtime of production facilities located in more remote locations (Westerweel et al., 2021). Currently the material properties of the needed parts may not be equivalent to those made with conventional techniques, and understanding the material property relationships and developing new materials for additive manufacturing are important areas of study for addressing some of these concerns (Dilberoglu et al., 2017).

Despite these concerns, distributed manufacturing methodologies and additive manufacturing are accelerating scientific research. Whether it be through direct fabrication of intricate laboratory equipment, synthesis of catalysts (Alimi and Meijboom, 2021), or design of novel chromatographic materials (Agrawaal and Thompson, 2021), recognition of the true potential of these approaches is only now being more fully recognized. And as they become more ubiquitous in research laboratories, scientific research will only continue to expand and mature.

1.4 STUDY SCOPE AND APPROACH

The central question of the study is as follows: What can be done in fundamental chemistry that will enhance our lives, strengthen and safeguard the environment, contribute to the economy, and ensure international competitiveness for the United States into the future?

Like the question above, the charge that the committee received (Box 1-2) is broad in scope. To address the charge thoroughly and in ways that will assist the sponsors in planning, decision making, and allocating resources, the committee chose to approach the Statement of Task as follows. The first two bullets were tackled by looking through and commissioning different economic analyses and doing in-depth research on specific case studies in which fundamental chemistry led to significant contributions in manufacturing, environmental protection, the energy and power sector, and national security. In approaching the other two bullets of the Statement of Task, which touch on strategies and options for research investments, the committee was careful to avoid being too specific, due to the broad nature of the charge. They chose instead to focus on tools, technologies, and methodologies that will broadly advance all of fundamental chemistry while accomplishing the specifications outlined in the Statement of Task.

Early in 2021, the committee issued a request for proposals (Appendix B) to hire an independent contractor to perform an economic analysis of the impact of the chemical economy on the U.S. economy and to analyze the impact that fundamental chemical research has had on the chemical economy. In response to this request, the committee hired Vertex Evaluation and Research, LLC to perform the analysis. The team was led by Daniel Basco who worked closely with Lee Fleming,

---

BOX 1-2
Statement of Task

An ad hoc committee of the National Academies of Sciences, Engineering, and Medicine will be convened to consider strategies to sustain and enhance the economic activity driven by fundamental research investments in the chemical sciences. The committee will:

- Examine and define the role of the chemical industry in the U.S. economy.
- Assess how investments in long-term fundamental research in the chemical sciences have contributed to such goals as national security, environmental sustainability, thriving manufacturing industries, and energy-technology development.
- Explore strategies for targeted research investments in the chemical sciences by both the public and private sectors to stimulate economic growth and to ensure the United States plays an international leadership role in the field.
- Discuss options for research investments that would enhance the chemical economy while also advancing environmentally sustainable practices and/or integrating a diverse chemical economy workforce.

a faculty member at the University of California, Berkeley, and the team closely consulted with IP Checkups, Inc. The final analysis was delivered to the committee in August, and the committee heard a final presentation from the contractors in September 2021.[2]

In addition to the economic analysis by Vertex, the committee heard from a number of people on topics germane to the study. The information-gathering process for the study took place from February 2021 to September 2021, during which the committee heard about the roles of the energy sector, materials production, pharmaceuticals, and agriculture in the U.S. chemical economy. In addition, they heard from experts in innovative chemical synthesis methods, automation, computational chemistry, and educating the next generation of chemists. There were also economists, venture capitalists, and representatives from private foundations and the U.S. Small Business Association who discussed best practices for supporting fundamental chemistry research and education. In total, 51 speakers and panelists from government, industry, academia, consulting, venture capital, and other fields spoke to the committee (for a list of speakers, see Appendix C). Several committee members also contacted individuals to ask them questions related to the study. The list of people to whom the committee spoke and the main takeaways from these conversations are listed in Appendix D.

To receive further input from the chemistry community, the committee engaged in several directed e-mail campaigns. First, a call for input (Appendix E) was sent to the listserves of the Board on Chemical Sciences and Technology, the Board on Life Sciences, and the National Materials and Manufacturing Board of the National Academies of Sciences, Engineering, and Medicine (the National Academies). Second, an e-mail was sent to the department head of every college and university chemistry department having a graduate program. In total, the call for input was sent to approximately 5,742 individuals, not accounting for individuals whose names appear on multiple listserves. Of those individuals who were contacted, the committee received 14 written responses with information to consider. These responses are available in the public access file of the report. All of these information-gathering activities greatly assisted the committee's study process and were supported by a detailed review of the relevant literature.

1.5 PREVIOUS CONSENSUS STUDIES RELATED TO THE CHEMICAL ECONOMY

During the information-gathering process, the committee also looked to previously published reports and literature on chemical research and the chemical economy. While few published reports attempt to draw a direct connection between fundamental chemical research and the chemical economy, the National Academies have published several reports that are relevant to the committee's research. Early in 2022, a complementary report, titled *New Directions for Chemical Engineering*, outlined critical future research needs for the chemical engineering community that would enhance environmental sustainability and increase the quality of life for all people (NASEM, 2022a). The chemical engineering report identified a wide breadth of focus areas where chemical engineering could make a large impact, including the decarbonization of energy systems, engineering targeted medical treatments and equitable access to medicine, and novel materials. All of the factors identified in the chemical engineering report influence the chemical economy and were therefore considered in careful detail.

There are many other relevant reports from the National Academies that helped with the exploration of ideas and concepts for this report. Some reports looked at the future of research in a particular area of chemistry, including the following:

---

[2] The full report produced by Vertex Evaluation and Research, LLC (cited as "Fleming and Basco, 2021" throughout this report) is available through the public access file for this study.

- *A Research Agenda for Transforming Separation Science*, which focused on the future of separation technologies for chemistry, as well as the chemistry that enables them (NASEM, 2019a).
- *Frontiers of Materials Research: A Decadal Survey*, which put forth a research agenda for the next 10 years of materials science (NASEM, 2019b).
- *Visualizing Chemistry: The Progress and Promise of Advanced Chemical Imaging*, a report that looked at new advancements in measurement science (NRC, 2006b).
- *Research at the Intersection of the Physical and Life Sciences*, a report detailing avenues for physics and chemistry to explore questions in biology (NRC, 2010).

Other helpful reports looked at areas such as decarbonization, environmental stewardship, human health, and business regulations, all of which were important resources because of their impact on the chemical economy. These reports include the following:

- *Innovations in Pharmaceutical Manufacturing on the Horizon: Technical Challenges, Regulatory Issues and Recommendations*, a report that recommended new pathways forward based on the ongoing changes in pharmaceutical manufacturing (NASEM, 2021e).
- *Gaseous Carbon Waste Streams Utilization: Status and Research Needs*, a report that laid out a path for carbon utilization research and current needs (NASEM, 2019c).
- *Negative Emissions Technologies and Reliable Sequestration: A Research Agenda*, a report that analyzed the usability of several different negative emissions technologies and proposed a path forward for research and usage (NASEM, 2019e).
- *Accelerating Decarbonization of the U.S. Energy Systems*, a report that laid out specific technological and socioeconomic goals for decarbonization efforts in the United States (NASEM, 2021b).
- *Preparing for Future Products of Biotechnology*, a report that looked at the regulatory frameworks around biotechnologies and identified gaps that might prevent future development (NASEM, 2017a).
- *Safeguarding the Bioeconomy,* a report that assessed the economic impact of the U.S. bioeconomy and considered national security concerns around new life sciences advances (NASEM, 2020b).

It is impossible to make an exhaustive list of all the relevant reports from the National Academies, especially since this section did not even mention reports that address chemical education, equity and inclusion in STEM, or economic analyses of the chemical industry. Many of these reports were analyzed over the course of this study, and are noted in the relevant chapters and sections.

1.6 ORGANIZATION OF THE REPORT

This report is organized into six additional chapters. Chapters 2 through 6 build the case for why it is essential for the United States to make public and private investments in fundamental research in the chemical sciences. Starting with Chapter 2, the report lays out the data for the importance of chemical research and the chemical economy while also helping to identify different metrics for U.S. global competitiveness. Chapter 3 builds a case for how chemistry is influential toward, and dependent on, society and the shifting landscape of environmental sustainability. Chapter 4 looks at chemical research, points out important enabling tools and techniques, and helps to further the case for how chemistry can use these tools in the service of fundamental chemistry for addressing environmental stewardship and sustainability. Chapter 5 assesses some key tenets that are needed for workforce development in the chemical sciences, while Chapter 6 looks at the current funding landscape in chemistry. Chapter 7 summarizes findings and recapitulates conclusions from earlier chapters, and it makes recommendations on where federal agencies, academic institutions, chemical companies, and private funders should focus resources and support of fundamental research in the chemical sciences to maximize innovation, foster a sustainable future for people and the planet, and continue to enhance the U.S. chemical economy and thus the U.S. economy overall.

Chemistry has an estimable record of accomplishments, discoveries, and innovations. Both fundamental discoveries and industrial applications have transformed societies and people's lives in innumerable ways. But the problems that need solving today are different from the problems of the past. For example, the catalytic converter is a remarkable application of fundamental chemistry that significantly reduced air pollution from automobiles and reduced the health impacts of smog, especially in cities. But now the pressing problem is burning fossil fuels, and the apparent solutions will produce cars that no longer need catalytic converters. As society's problems change, so must chemistry. New tools, new technologies, and new ways of thinking are needed, along with the adaptation of chemistry's mature tools and technologies to be used in new and innovative ways.

2

Understanding the Economic Impacts of Chemistry

Key Takeaways:

- The U.S. chemical economy is a large portion of the overall economy, but its more important feature is the influence it has on so many other sectors of the economy, including health care, manufacturing, and so much more. All of the sectors that are reliant on some aspect of the chemical economy in the United States total $5.2 trillion.
- Fundamental research is important to the U.S. chemical economy because it provides monumental changes as well as incremental steps. Incremental advances in the chemical economy have a much larger influence on the overall economy because of the spillover of chemical knowledge, products, and processes.
- It is very difficult to measure the exact impact that chemical research has on the chemical economy, but we can measure the spillover of chemical knowledge and products into other areas, and this spillover is significant.
- The U.S. chemical economy is part of a complex global ecosystem, and by some metrics, the United States is losing its competitive advantage in the chemical industry and in chemical research. Despite this decline, the United States currently remains a strong leader in the chemical economy.
- The U.S. chemical economy relies heavily on foreign-born workers and people from diverse backgrounds.

Chemical research has contributed to economic prosperity and financial security in the United States. Products from chemical companies, such as polymers, coatings, pharmaceuticals, and pesticides, play an important role in everyday life. These chemicals are all derived from products and processes that started in some way as fundamental chemical research. Of course, all products have a unique trajectory from fundamental research to final product, but this movement frequently requires many iterations, steps, and decades of complementary research discoveries. These research and development (R&D) efforts, when successful, are converted into products or processes that are used by the chemical industry to increase quality of life. The outcome can produce a novel chemical or technique, or can improve upon an entity that already exists by, for example, making a better product or decreasing the environmental impact of a process.

As the committee discussed the size and impact of chemical research on the chemical economy, it took a very broad approach to defining what is included in the chemical economy. As noted in Chapter 1, the chemical economy includes all parts of any value chain that rely on chemical knowledge and transformation processes for advancement and growth. This broad approach is important when considering fundamental research that has impacted the chemical economy, and the breadth will be reflected in any subsequent analyses and examples of success.

To evaluate the impact of chemical research on the chemical economy, as well as the U.S. economy more broadly, the committee used several approaches. Two lines of evidence used were economic output of and employment from chemical companies and the chemical sector. The committee relied on previous assessments of the chemical economy and the impacts of chemical research on the economy, as well as an economic analysis that was produced by an independent consulting firm. To try to evaluate the size of the chemical economy, as well as the impact of R&D on the chemical economy, Vertex Evaluation and Research, LLC (Vertex) assembled a team led by Lee Fleming of the University of California, Berkeley's Haas School of Business and Daniel Basco from Vertex, and included a group of collaborators from IP Checkups, an organization that specializes in patent analysis, to address these questions. In response to a Request for Proposal from the study committee (Appendix B), Vertex put together an analysis plan with three phases, which included plans to "assess the economic value of the chemical economy, assess industries where chemical research is driving employment opportunities and value creation, and assess how chemical research is contributing to sustainable economic progress" (Fleming and Basco, 2021). Most of the material related to the size of the chemical economy, and importance and impact of chemical research in the chemical economy came in their first phase of work. Most of their analysis relied on chemical patents and their valuation, which served as a proxy for chemical research since there are economic methods for valuating patents. Throughout this chapter, there are discussions of various aspects of the report from Vertex, including the interesting findings and important caveats.

The use of this report and the other collected data helped to paint a picture of the impact of the chemical industry but frequently fell short of helping the committee gain an understanding of the economic impact of fundamental research. To better understand the economic impacts of chemical research, individual case studies were selected that highlight chemical discoveries that led to a product or process with widespread implications for society.

2.1 BRIEF HISTORY OF THE U.S. CHEMICAL INDUSTRY

The chemical industry in the United States has been a prominent player on the world's stage since the early 20th century. While it is notable that the United States emerged later than many European countries, the U.S. domestic market, transportation system, and R&D infrastructure propelled it to become one of the largest chemical producers in the world. With the boost in chemical production from World War I and World War II, the United States established and maintained a strong presence in the chemical marketplace from 1920 to 1960, sometimes referred to as the "golden age of innovation in the chemical industry" (Arora and Gambardella, 2010).

While it is impossible to name all of the products and processes that were responsible for the thriving industry, it is important to note that these inventions included chemicals and materials that improved and saved lives, as well as those that wrought tremendous harm. Many of the prominent examples include chemical weapons that were researched and used during this period of invention, and include, but are not limited to, mustard gas in World War I and, later, the herbicide Agent Orange during the Vietnam War (Everts, 2015). Half a century to more than a century later, we continue to struggle with the use of chemical agents in conflicts despite widespread international support limiting their production and use (OPCW, 2020). A prominent example of a product that improved quality of life is the discovery of nylon by Wallace Carothers of DuPont. This discovery was a major breakthrough in polymer chemistry and led to the production of the first fully synthetic fibers. It was noted by Hounshell and Smith (1988) that the success of this innovation encouraged a research-based approach at DuPont. In addition to the production of nylon, basic research by industrial chemists led to the discovery and production of a large number of polymers such as polyester, various acrylics, neoprene, and many more. Underpinning the massive commercial success of these products was the research and innovation that led to them. This fundamental research also laid important groundwork in the chemical understanding of materials and their interactions.

The U.S. chemical industry was also particularly good at developing processes for the production of chemical goods. Scaling up chemical production, like in the case of sulfuric acid or fertilizers, was a fruitful collaboration between chemical researchers and engineers (Arora and Gambardella, 2010). The growing popularity of the automobile and the resulting need for refined petroleum were partially responsible for the development of chemical engineering in America (NASEM, 2022a). The ability of the United States to produce large quantities of usable feedstock, especially the refining of crude oil, and to synthesize commodity chemicals was an additional boon to the economic success of the U.S. chemical industry. These industrial capabilities were also important for laying the groundwork for developing and understanding industrial chemical processes that are critical now, and will be used and refined even as commodity chemicals need to be increasingly produced from new feedstocks.

Developing successful large-scale production processes was also important to improving human health. In 1939, the first large-scale production of the antibiotic penicillin was accomplished in Brooklyn, New York (Aldridge et al., 1999). While the synthesis of this life-saving molecule was done by biological fermentation, it was chemists and chemical engineers who developed new fermentation conditions that allowed for the extraction, purification, and stabilization of the final molecule. The design and commercial-scale production of this antibiotic led the way for the large-scale production of other specialty chemicals and pharmaceuticals.

The U.S. chemical industry was also responsible for technological advances that moved computation forward, including revolutionizing the semiconductor. In 1979, researchers at IBM developed chemically amplified photoresists, the result of R&D of new polymers capable of a photoactivated chemical "chain reaction." This new chemical amplification process gave a 30-fold improvement in light sensitivity over its predecessor (Brock, 2007). The increased sensitivity made it ideal for a growing electronics industry that was constantly searching to perform computational tasks at faster speeds and smaller scales. Chemically amplified resists are now integrated into most electronic devices and are important to billions of people around the world.

As the U.S. chemical industry continued to innovate throughout the late 20th century, chemical production grew and adversely affected the environment. This had a disproportionate impact on local and indigenous communities (Borunda, 2021; Langston, 2010). With heavy influence from these communities, a growing environmental movement sought accountability for the environmental damages that the chemical industry played a prominent role in accelerating. The interplay between environmental sustainability and the chemical sciences is covered in more detail in Chapter 3 but remains an important consideration as we look further at the history and impact of the chemical economy.

2.2 ESTIMATING CURRENT SIZE AND IMPACT OF THE CHEMICAL ECONOMY

The chemical industry has a significant direct impact on the economy through its contribution to gross domestic product (GDP), employment, and national competitiveness (Figure 2-1). An expanded analysis shows that there is also a significant indirect impact, noting that chemical knowledge is used in a wide variety of areas, and chemical products are used and consumed by almost every sector of the U.S. economy. This section highlights the direct economic impacts of the chemical economy by showing the value and employment added by the chemical sector. Several methods of indirect impact are also shown, primarily focusing on measurable impacts such as the use of chemical knowledge in other economic sectors or the reliance that patents have on chemical knowledge. Another aspect of indirect impacts is considered later in the chapter where the report notes the wide social, health, environmental, and economic impacts that specific chemical products can have on the entire national population. It can be difficult to measure the widespread impacts of every chemical product in every economic sector, but this can be displayed through specific examples as shown later in this chapter (see Section 2.3.3). Some of these examples include the widespread use and environmental potential of batteries, the economic and societal improvements afforded by oral contraception, and the immediate helpfulness and economic revival afforded by treatments for COVID-19.

2.2.1 Measuring Direct Contribution to the U.S. Economy

There are many measures of the size of the chemical economy, both domestically and internationally, and it is important to provide a short digest of some of the most prominent resources in order to get an idea of the size and impact of the chemical industry. To accomplish this, the

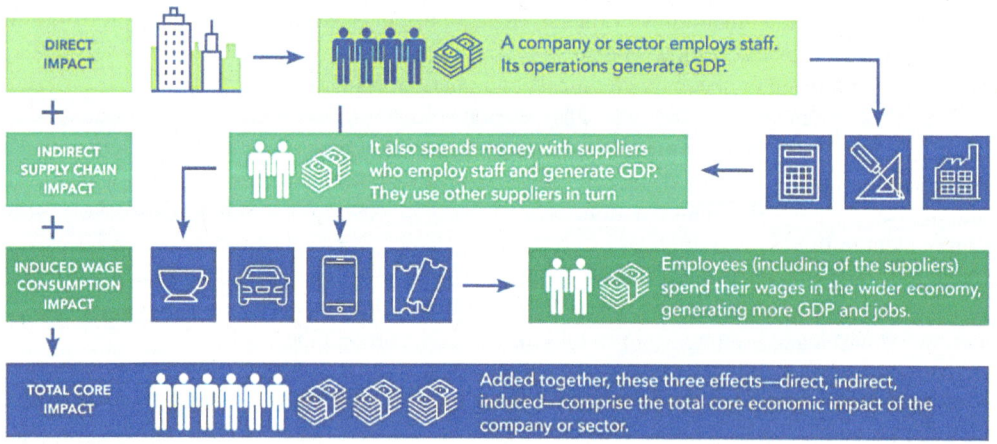

FIGURE 2-1 Representation of the different types of economic impact, including both direct and indirect impacts. SOURCE: ICCA and OE, 2019.

UNDERSTANDING THE ECONOMIC IMPACTS OF CHEMISTRY

committee looked to two prominent resources: a report from the International Council of Chemical Associations (ICCA) and Oxford Economics (OE) published in 2019, and the *2021 Guide to the Business of Chemistry*, produced by the American Chemistry Council (ACC, 2021). Here the chapter includes ICCA/OE and ACC reports of sales, imports, exports, employment, and value added from the chemical industry. Both reports measured the size of the chemical industry and the impact that it has on the overall economy. An important caveat is that these analyses exclude pharmaceuticals and medical manufacturing, and only include equipment when specified. Based on the definition of the chemical economy established by this report, pharmaceuticals would be included, and their impacts will be addressed throughout the report.

2.2.1.1 Monetary Impact of the Chemical Economy

It is first important to look at the direct value added and sales from the chemical economy, and the chemical economy's impact on the overall economy. The ICCA and OE (2019) report estimated that the chemical industry contributed $5.7 trillion to the global GDP in 2017, or approximately 7% of the world's GDP that year. The ACC reports numbers that are specific to the United States and notes that, in 2020, the "final sales" of the industry were $457 billion, of which $123 billion were intraindustry sales to other chemical producers (ACC, 2021). Of the remaining $334 billion, the leading buying sectors were rubber and plastic, pharmaceuticals, health care, other manufacturing, paper, and agriculture (Figure 2-2). Note that these were direct sales and do not include indirect uses, for example, if car manufacturers buy from tire manufacturers.

Additionally, in 2020, the value added in the U.S. economy from the chemical industry was $225 billion. The report from ICCA and OE estimated the chemical industry's contribution to North American GDP to be $866 billion in 2017, of which $235 billion is the direct impact of the chemical industry, a number comparable to what was reported by the ACC. Importantly, the ICCA and OE report did not assess individual contributions of different countries in North America.

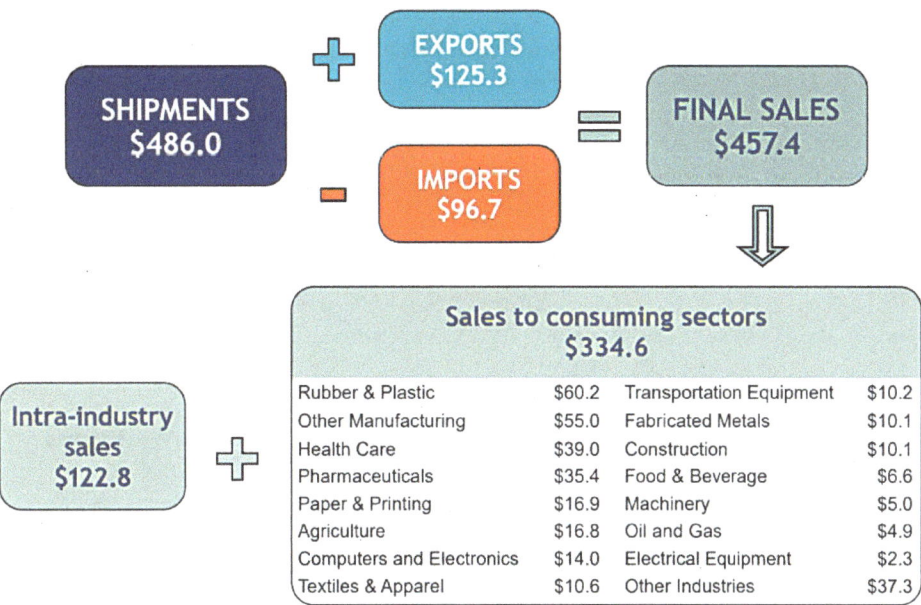

FIGURE 2-2 Economic analysis of the business of chemistry, including a breakdown of the sectors where products were sold. All dollar values are in billions. SOURCE: ACC, 2021.

The ACC report also noted that exports and imports have grown roughly threefold over the past two decades, and the United States has maintained a positive trade balance since 2016, which has been around $20 billion to $30 billion (Figure 2-3). Despite this growth, the output of the chemical industry has grown more slowly in recent years, with only a 3% increase in output between 2012 and 2019. In some ways, this slow growth indicates that the chemical industry as a whole is mature. This is especially true since innovation is mostly incremental, and changes in how products are customized and delivered to the customer are growing in importance, relative to fundamental advances.

As mentioned previously, the committee defined the chemical economy to include all products and parts of a value chain that rely on chemical knowledge, meaning that pharmaceuticals, but not biopharmaceuticals (see Section 1.2.1 for further explanation), should be included in any analysis of the chemical economy. Although most analyses do not include pharmaceutical impact, some data do describe the value of the pharmaceutical industry in the United States. When considering pharmaceutical manufacturing, which falls under the North American Industry Classification System (NAICS) code 3254, the committee was careful to exclude "Biological Product Manufacturing," which is a subsection of this sector, designated as NAICS 325414. The value of production for the pharmaceutical industry that contributed to the chemical economy, excluding biological products, was around $160 billion in 2020 (Figure 2-4). The value of production has also been steadily increasing since 2000. While it is challenging to consider these numbers within the same analysis as the rest of the chemical economy, it is important to note that the value added from the pharmaceutical industry to the chemical economy in the United States is on a scale relatively similar to the value added from the rest of the chemical industry (BLS, 2022).

2.2.1.2 Employment

Another important aspect of the chemical economy's impact is through employment numbers and employment potential. The ICCA and OE (2019) report noted that the chemical industry in 2017 supported 120 million jobs related to "all aspects of the global chemical industry." The report also notes that the North American chemical industry supported 6.1 million jobs in 2017, of which 600,000 jobs were supported directly in the chemistry industry itself. The ACC annual report documents similar employment numbers that are specific for the United States, showing that in 2020, in addition to directly employing 529,000 workers, the entire chemical enterprise supported 4.1 million individuals who were interacting with the chemical economy (ACC, 2021). Employment in the industry within the United States has remained relatively steady with direct employment numbers staying between 525,000 and 550,000 since 2016.

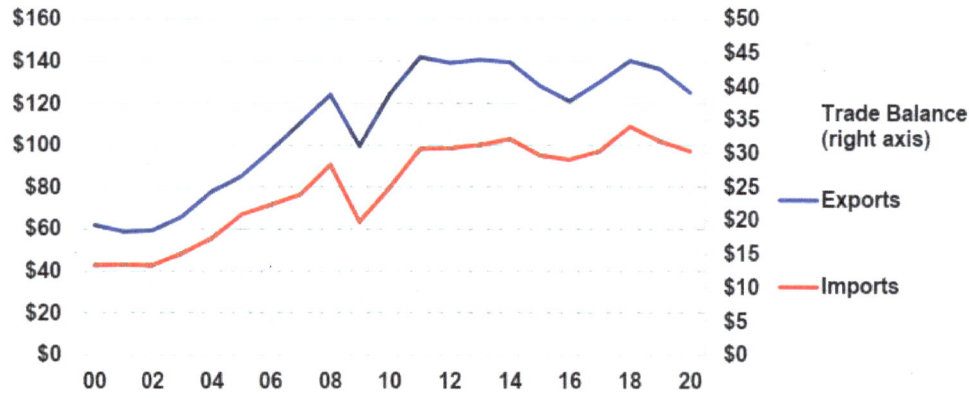

FIGURE 2-3 U.S. imports, exports, and trade balance in the chemical industry. SOURCE: ACC, 2021.

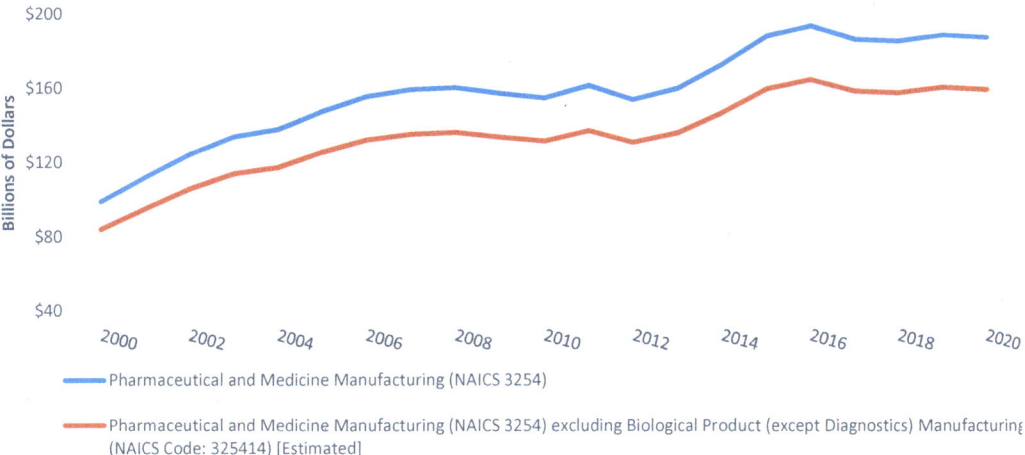

FIGURE 2-4 Value of production over time for pharmaceutical and medicine manufacturing (NAICS 3254). The blue line is the total value of production while the red line excludes an estimate of the added value from biological product manufacturing. The value for the red line was calculated using a data point from the Bureau of Economic Analysis that showed Biological Product Manufacturing (NAICS 325414) as 15.5% of the total value added from pharmaceutical and medicine manufacturing in 2012. The 2012 percentage data were used for all years, and it is therefore highly likely that this percentage is an underestimate, especially with the increased use of biologics and vaccines from pharmaceutical companies. SOURCE: Data from BLS, 2022, and the U.S. Bureau of Economic Analysis.

The North American chemical industry holds the top global position in terms of GDP/employees, when compared to other global regions (Figure 2-5). This is one of the reasons that the U.S. chemical industry pays its employees so well. The ACC (2021) report showed that employees are paid 23% more than those in other U.S. manufacturing jobs. Additionally, the job outlook for chemists and materials scientists is fairly positive, and employment is expected to grow approximately 6% over the next decade, which is similar to the average growth of all occupations (BLS, 2021).

2.2.2 Indirect Contributions of Chemistry to the U.S. Economy

To better understand the impact of chemistry on the U.S. economy, we must also consider its indirect impacts, such as the chemical knowledge that is used in a much wider range of industries. We touched on the indirect impacts in relation to economic output and employment that are supported by, but not directly within, the chemical economy. Some of the industries that are dependent on the spillover of knowledge, products, and processes from the chemical economy include strategically important industries such as semiconductors, computers, aerospace, medical equipment, and electrical equipment, in addition to other economically significant industries such as construction, food agriculture, and vehicles (ACC, 2021). Overall, the ACC report estimates that, based on dependent industries, the "business of chemistry" was responsible for adding approximately $5.2 trillion to the U.S. economy in 2020, which was about 25% of the U.S. GDP.

A report from the consulting group Vertex, which was prepared for this study (see Section 1.4 for more details), similarly noted that chemical knowledge and chemical inventions spill over into other industries (Fleming and Basco, 2021). The report first notes that chemistry patents accounted for approximately 18% of all patents filed by the U.S. Patent and Trademark Office from 2000 to 2020, with a much higher share of patents in years before 2006 (Figure 2-6). One publication

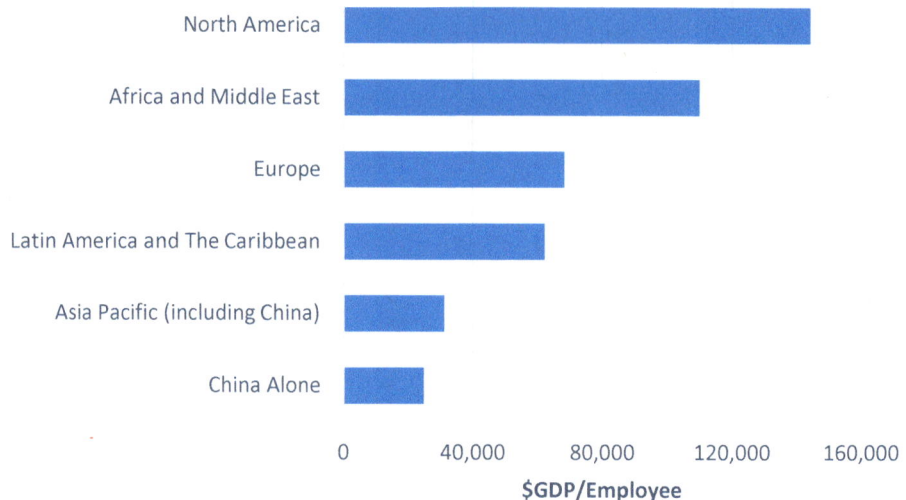

FIGURE 2-5 Chemical industry's total global economic output by region in 2017. SOURCE: Data from ICCA and OE, 2019.

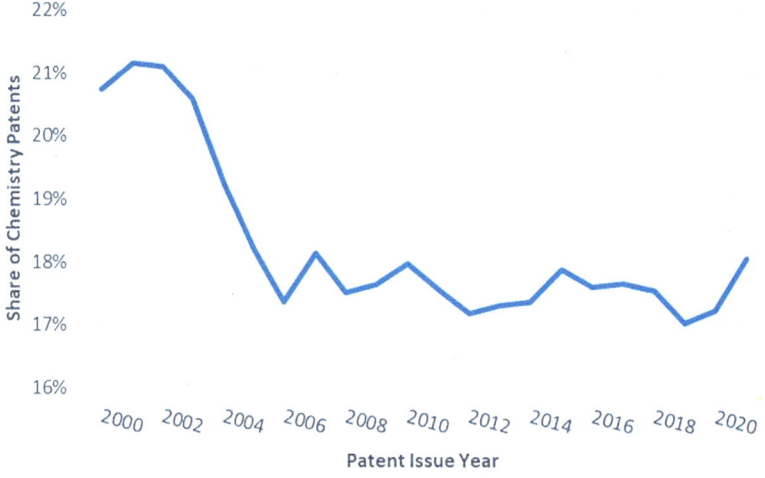

FIGURE 2-6 Share of chemistry patents in relation to all issued patents. SOURCE: Fleming and Basco, 2021.

explained this decrease by showing that, following a trend of increasing numbers of chemical patents between 1975 and 1991, the number of chemistry patents began to stagnate between 1991 and 2007 (Figure 2-7) (Autor et al., 2019). During this entire time period (1975–2007), computer patents had a boom, partially explaining the stagnation of chemical patenting as the new tech industry started to have major breakthroughs. This also explains why we see a decreasing trend in the percentage of chemistry patents in the Vertex analysis, which runs from 2000 to 2020.

In looking at chemistry-related patents, Vertex noted that a spillover of knowledge from chemical patents occurs when these patents are cited by patents from other areas such as human necessities, transport, materials, and new technology. Vertex data further suggest that 8.5% of patents rely on chemical research, as measured by citations, found within patents, to scientific publications

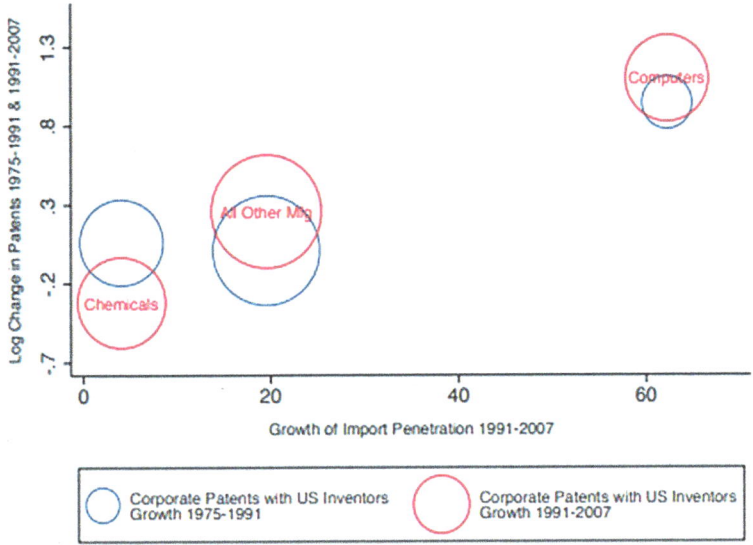

FIGURE 2-7 Patent growth in chemicals and computers in the time periods of 1975–1991 and 1991–2007.
SOURCE: Autor et al., 2019.

in chemistry-related journals (Table 2-1). In addition, of the patents that cite chemical research issued between 2000 and 2020, 73% are chemistry patents, and the rest (approximately 27%) are nonchemistry patents.

The Vertex report also presented the proportion of patents that cite chemical research in relation to patents that cite any scientific research. Figure 2-8 shows the time trend of number of patents by publicly traded U.S. corporations that cite chemistry research versus patents by the same set of corporations that cite nonchemistry scientific research. All of the cited research is designated as nonpatent literature (NPL). Among patents that cite any research, 20% of them specifically cited chemistry research over the period from 2000 to 2020, but this number has declined from 25% in 2000 to 14.4% in 2020.

To better understand the indirect impacts related to the spillover of chemical knowledge into other areas of R&D, Vertex analyzed the value of all patents that use chemical knowledge. To

TABLE 2-1 Number of Chemistry and Nonchemistry Patents That Cite Chemistry Research from Scientific Publications in Chemistry-Related Journals

Patent Cites Chemical Research	Chemistry Patents						Totals
	Yes			No			
Yes	298,911	*34%*	*73%*	109,731	*3%*	*27%*	408,642 (*8.5%*)
No	574,454	*66%*	*13%*	3,851,322	*97%*	*87%*	4,425,776 (*91.5%*)
	873,365 (*18.1%*)			3,961,053 (*81.9%*)			**4,834,418**

SOURCE: Data from Fleming and Basco, 2021.

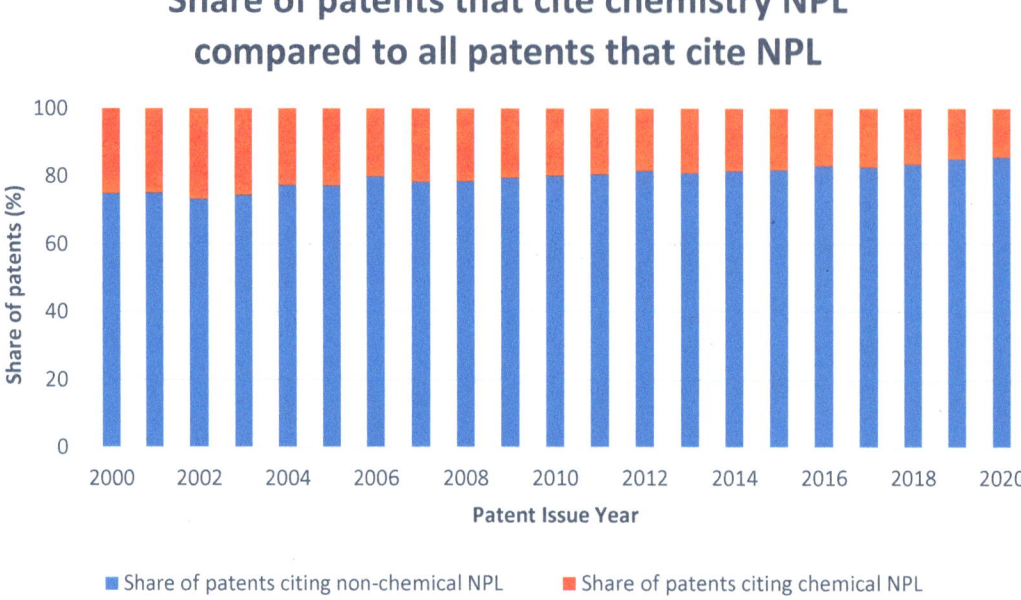

FIGURE 2-8 Share of patents that cite chemistry NPL and nonchemistry NPL. NOTE: Patents for this figure were selected conditional on having cited any NPL, such that the shares of patents citing chemical and non-chemical patents add up to 100%. These patents are also limited to those that have been assigned to public firms in the United States for purposes of evaluating the value of the patent using the methods described in Kogan et al. (2017). Within the sample period of 2000–2020, 592,831 patents cited at least one NPL and were issued by public firms. SOURCE: Data from Fleming and Basco, 2021.

perform this valuation, they matched each patent to estimates of patent values for corporate patents from Kogan et al. (2017) to calculate the value (known as the KPSS value). Importantly, the Kogan et al. estimates rely on the change in the stock market value of a company 3 days following the filing of a patent. Although this is an imperfect measurement, it is considered state of the art for global patent valuation. When considering the value of each patent, the average KPSS value of patents citing chemistry research versus patents citing other NPL can be graphed (Figure 2-9). Patents citing chemical research are more valuable on average than other patents citing nonchemical research.

These data show that the use of chemical knowledge produces products and processes with a higher valuation than the average use of other types of scientific knowledge. The ICCA and OE (2019) report points out the importance of chemical patents, and notes that "chemical products also fuel innovations and patents in other industries, such as photovoltaic cells for electricity production, lightweight vehicle parts, germ-resistant coating for medical instruments, etc." If we consider this, together with the information that chemistry supports approximately 25% of U.S. GDP, it shows that chemical knowledge and innovation have a high impact on a large portion of the U.S. economy.

2.3 RESEARCH AND INNOVATION IN THE CHEMICAL INDUSTRY

The chemical industry has historically been a hub of innovation and discovery. ICCA and OE (2019) reported that in 2017, the global chemical industry spent $51 billion on R&D, with the United States being the second-largest contributor ($12 billion) after China ($14 billion). That

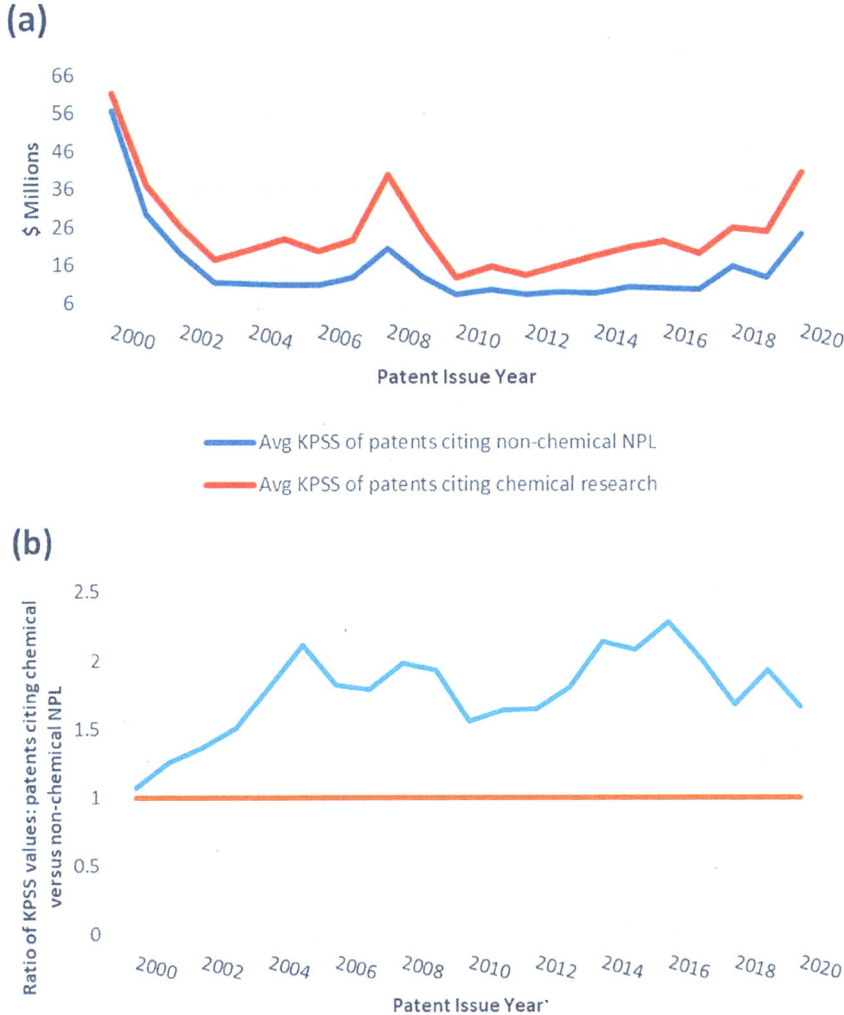

FIGURE 2-9 Comparison of average KPSS valuation of patents that cite chemistry and nonchemistry NPL. (a) Average KPSS value of patents that cite chemistry NPL in each year, and the average KPSS value for patents that cite nonchemistry NPL. (b) Ratio of the average KPSS value of patents that cite chemistry NPL over the average KPSS value of patents that cite nonchemistry NPL. SOURCE: Data from Fleming and Basco, 2021.

report also notes that the patent intensity[1] for the United States is 50% higher for chemistry than for the economy as a whole (Hu and Png, 2013). Greater patent intensity means that individual patents have a higher impact on a particular sector of the economy. These values reveal the impact of innovation in the chemical industry.

Universities spent about $4 billion in 2019 on chemistry-related basic and applied R&D (Figure 2-9) (Fleming and Basco, 2021). Funding sources of this $4 billion include about $2.1 billion from federal sources and about $250 million from "business sources" (Figure 2-10). Including the money they spent on university R&D, industry spent about $10 billion on R&D in 2019: 46% on

---

[1] According to the ICCA and OE (2019) report, "patent intensity is measured as the number of U.S. patents awarded to an industry relative to total industry sales in the United States."

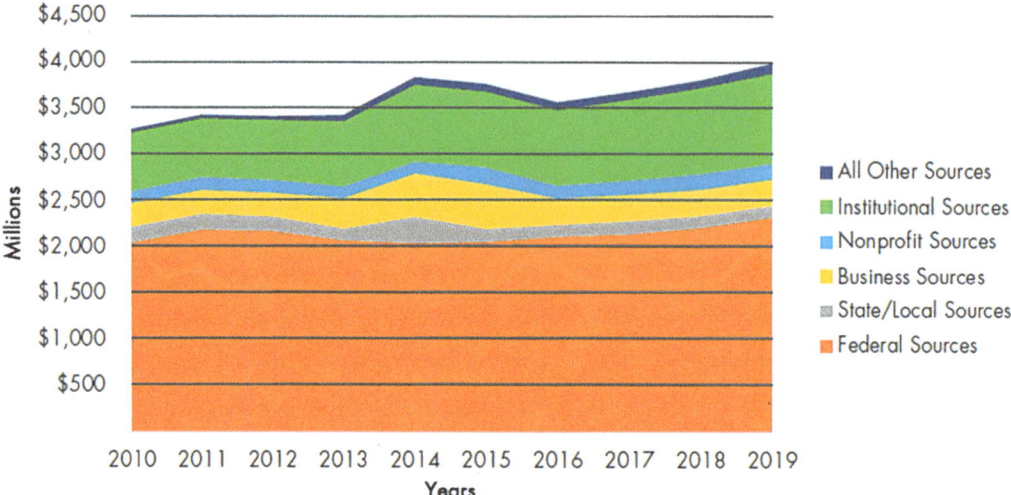

FIGURE 2-10 Chemistry-related R&D expenditures at universities by year, separated by funding sources. SOURCE: Fleming and Basco, 2021.

basic and applied research (defined as "fundamental research" in this report), and the remainder on development (ACC, 2020). This money does not include funding from the pharmaceutical sector, which spent about $72.8 billion in the United States on R&D in 2020 (ACC, 2020; Mikulic, 2021).[2] Further details on funding and R&D spending is covered in Chapter 6.

2.3.1 Measuring the Value of Chemical Research

The large expenditure on R&D clearly establishes the importance of chemical research and innovation to the chemical industry. To better address the Statement of Task, the committee sought to understand the connection between fundamental chemical research and the chemical economy. It is important to note that there are substantial challenges associated with this task. Understanding the impact of chemical research is nontrivial because chemical discoveries can take years or even decades before their economic impacts are measurable. In addition, most chemical discoveries that are considered "breakthroughs" are based on a large body of knowledge that took decades, or even centuries, to build. Continuing to build this expansive knowledge base that researchers are able to pull from is one of the most important arguments for funding chemical research. But, it makes assessing the economic impact of chemical research very challenging, because it is difficult to make a global assessment of how research dollars that build a pool of chemical knowledge translate into economic output.

Using the best information and methods available, Vertex looked at the valuation of all chemistry-related patents, using patents as a proxy for chemical research in their analysis. This is a valuation similar to what was described in Section 2.2.2, except here, Vertex considered all chemistry-related patents, not just those that rely on chemical research. Vertex estimated the value of chemical patents in 2021 dollars to be between $132 billion and $555 billion per year from 2000 to 2020 (Figure 2-11). Acknowledging the caveats of these measurements and assuming that these may be

---

[2] Research at pharmaceutical companies ranges from small-molecule drug discovery and synthesis all the way to clinical trials, and therefore does not neatly fall under the umbrella of chemical research. Additionally, pharmaceutical companies spent money in 2019 and 2020 to research vaccines, prophylactics, tests, antibody therapies, and small-molecule drugs to combat SARS-CoV-2.

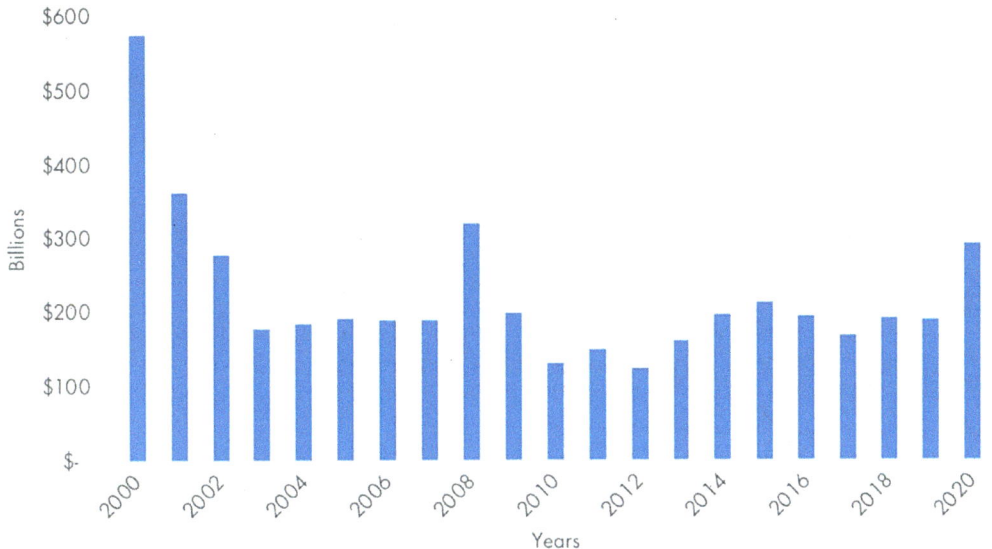

FIGURE 2-11 Value of chemistry-related patents by year using KPSS data in 2021 dollars. SOURCE: Fleming and Basco, 2021.

overestimates, Vertex deflated the values by 33%[3] and arrived at a conservative estimate of $221 billion/year to $370 billion/year, in 2021 dollars. Vertex also showed that chemical patents tend to be more valuable than the average patent (Figure 2-8). This indicates that chemical research has a higher private payoff than the average product or process that is patented after research. To quantify this, Vertex showed that while chemical patents accounted for 14% of all corporate patents between 2000 and 2020, they accounted for 23% of all value in the same time period. There are some other important caveats to this patent valuation, including

- The calculation only values corporate patents and is mute on the value of patents assigned to start-ups, universities, and other organizations not listed on the stock market.
- Based on the way the values were calculated, patents issued to a firm on the same day take the same value, even if one patent is more valuable to the firm than the others.
- These values are sensitive to stock market volatility.
- When chemical knowledge is used in the chemical industry, it is sometimes not patented and instead remains as a trade secret, so that the chemistries can only be used by the company who discovered or developed the process or product.

During their analysis, Vertex had many challenges drawing connections between R&D and economic impact. Importantly, they noted that "data limitations inhibit a comprehensive analysis." Specifically, they noted data limitations related to "patent value estimation, wide-spread availability of licensing terms data, and wide-spread availability of government grant data." To lessen the gaps, and assist future economic evaluations, the Vertex report noted several possible changes that could make an analysis more comprehensive. The first is the continued support of social science and business research by government and philanthropic organizations, especially in the pursuit of new and improved patent valuation methods. Vertex noted that the state of the art for global patent

---

[3] Based on an alternative measure that estimates value of chemical patents using a Tobin's q regression (Lindenberg and Ross, 1981).

valuation is likely a significant overestimate and improvements are needed. Second, they suggest supporting "efforts to collect data on how government research funding benefits the economy." One example they point to is Michigan's Institute for Research on Innovation and Science,[4] which is aggregating data on grants and personnel. Last, they note that if the U.S. Patent and Trade Office were to "strengthen data collection and publication," this would help any analyses on intellectual property and its connection to the economy.

2.3.2 Future Analyses to Help Understand the Impact of Chemical Research on the Economy

To continue building a full picture of how chemical research impacts the chemical economy, there are a number of other areas and analyses that could help the community gain a quantitative assessment of this relationship. Although the use of global patent valuation gives some insights, it falls short of a full understanding. Some ideas that might produce valuable results include the following:

- **An in-depth patent analysis that focuses on one or more top companies in the United States.** The committee heard from a Dutch patent attorney who put together tools for in-depth patent analyses at the companies where she was employed as a patent attorney (van Tol-Koutstaal, 2021). These analyses require a large amount of concerted time and effort for each company, but help to minimize the complications around global patent valuation metrics and the unknown sources of some patents. A valuation such as this could give a precise overall picture of how the products of research are converted into economic output but with the caveat of being specific to the sectors occupied by the company.
- **An analysis of licensing revenues generated from patents.** Although an analysis like this would require a special effort because the data are not easily available, it would provide useful data around what types of research produce patents that interest external groups. It would be most useful if this was done for a particular company or a specific patent category, and the data would likely show interesting trends on what technologies are most desired.
- **Drawing correlations between the number of patents and economic growth.** While the current study looked at direct evidence of economic performance, a concerted effort to correlate the number of patents with economic growth for a particular area or company would provide further useful information. This is a very challenging analysis, and some efforts have shown that the number of patents filed by a particular country do not necessarily relate to the impact or economic growth of that country, due to the large number of confounding factors (Elsevier Analytical Services, 2021). To figure out if a correlation exists, researchers would have to develop a way to link data from patents to the firms where they are generated, and then link that information to economic growth. These data would be helpful for understanding how research that generates patented technologies helps to directly impact economic growth.

All of these ideas provide avenues for future analysis and research as we continue to understand the impact of chemical discovery on the U.S. economy. Many, though, come with the caveat that they are analyzing a small sector or a single company. This report took a global and comprehensive view of chemical research and the chemical enterprise. Both types of information are needed for a complete picture.

---

[4] See https://iris.isr.umich.edu/.

2.3.3 Examples of Chemistry Research That Have Benefited the Economy

While we have noted the challenges in assessing the scale of the chemical economy, as well as the impact of R&D on the economy, one of the best ways to understand both concepts is through specific examples. There are many cases where chemical research has led to widespread economic and societal improvements. Many of these examples make it clear that fundamental chemical research has brought about pivotal discoveries that have led to new products or processes. Some of these examples also show the sprawling impact that such inventions have had. Six examples spanning from the 1960s to today are highlighted: rechargeable batteries, therapeutics, silicon chips, oral contraceptives, catalytic converters, and treatments for COVID-19.

2.3.3.1 Secondary (Rechargeable) Batteries

Modern secondary (rechargeable) batteries have contributed significantly to important economic, energy, and sustainability objectives. Lithium-ion and nickel-metal hydride batteries began widescale commercial production in the 1990s and were essential to the introduction of electric and hybrid electric vehicles that are competitive with their internal combustion counterparts in terms of performance and cost (Sepulveda et al., 2021). They have made possible ubiquitous mobile technologies, including telephones and portable computers, and are particularly beneficial in establishing reliable telecommunication and financial infrastructures in developing countries (NASEM, 2021b). The cost and performance characteristics of nickel-metal hydride batteries allow them to directly replace cylindrical primary (nonrechargeable) alkaline and Leclanché (carbon-zinc) batteries in virtually all applications (Revankar, 2019). The significance of lithium-ion battery research was recognized with the award of the 2019 Nobel Prize in Chemistry (*Smithsonian Magazine*, 2019).

Complementary breakthroughs in industry and academia led to the development of lithium-ion batteries. Stanley Whittingham, who worked for ExxonMobil, started working on a fast-charging battery, but the lithium and titanium combination he used was unstable. John B. Goodenough, an engineer at the University of Texas at Austin, made a cathode of lithium cobalt oxide and discovered that the battery power and capacity doubled while making it safer to use. Akira Yoshino at Meijo University in Nagoya improved the battery's capacity and safety (*Smithsonian Magazine*, 2019). Philips Research Laboratories demonstrated the first prototype nickel-metal hydride battery in 1984, building on fundamental research on metal hydrides that began in the 1800s.

The development and improvement of battery technology has important implications for environmental sustainability and plays a large role in the U.S. economy. Batteries are an important method of energy storage that enable the introduction of environmentally sustainable technologies such as electric vehicles and renewable energy sources that include solar and wind power. Additionally, a report produced for Battery Council International estimated the economic impact of the U.S. lead battery industry, and noted that the entire enterprise annually supports $10.9 billion in U.S. GDP and 92,200 jobs (EDR Group, 2019).

2.3.3.2 Biocatalysis in Synthetic Chemistry

The basic research from Frances Arnold and her lab has had a significant impact on the development of simpler and more sustainable synthetic processes to make new molecules (Turner, 2003). She uses "directed evolution" to develop new enzymes with new functionality. Directed evolution of enzymes for biocatalysis is built on a foundation of chemical knowledge from biophysical chemists and analytical chemists working together to understand the structural interactions and reaction mechanisms of different enzymes. The first protein structures of myoglobin and hemoglobin were published in 1958 and 1960, respectively, with a subsequent Nobel Prize in Chemistry awarded in

1962 (Dauter and Wlodawer, 2016). The ability to visualize protein structures allowed chemists to model and understand enzymatic reactions, leading eventually to work on directed evolution that has been widely adopted by the chemical industry. This technology is the basis of the company Codexis, which develops highly specific and efficient enzymes used in DNA and RNA synthesis and in biopharma manufacturing. Begun in 2002, Codexis had 165 employees and revenues of $64 million by 2020.

An enzyme engineered using this technology was recognized in the 2010 Presidential Green Chemistry Challenge. Merck and Codexis were given the Greener Reaction Conditions Award for their development of a second-generation green synthesis of sitagliptin, the active ingredient in Januvia, a treatment for type 2 diabetes (Codexis, 2021). Merck notes that "this collaboration has led to an enzymatic process that reduces waste, improves yield and safety and eliminates the need for a metal catalyst" (EPA, 2010). Before the use of this greener synthesis, the initial manufacturing process filed in 2005 and approved in 2006 had some inherent liabilities including inadequate stereoselectivity, which required a crystallization step; high-pressure hydrogenation (at 250 psi), which required expensive, specialized manufacturing equipment; and the need for a rhodium catalyst (EPA, 2010). Using directed evolution, a transaminase catalyst was developed by Codexis which enabled a new manufacturing process to supplant many of these concerns. The evolved transaminase improved the catalytic activity of the original enzyme by more than 25,000-fold. The challenge award website from the U.S. Environmental Protection Agency notes that

> [t]he streamlined, enzymatic process eliminates the high-pressure hydrogenation, all metals (rhodium and iron), and the wasteful chiral purification step. The benefits of the new process include a 56 percent improvement in productivity with the existing equipment, a 10–13 percent overall increase in yield, and a 19 percent reduction in overall waste generation. (EPA, 2010)

This new process has been used in manufacturing since 2012 for this critical drug (EPA, 2010).

2.3.3.3 Chemistry of Silicon Computer Chips

Modern computation is a product of many different fields of research, but chemical research has played a big part in increasing the computing speed and miniaturization of the computational technology that we enjoy today. The social, economic, and health impacts of portable computers and mobile phones are enormous.

In the mid-20th century, fundamental research at the intersection of physics, engineering, and chemistry produced innovations that led directly to modern computer architectures and the present computing ecosystem. The first patents for semiconductor devices appeared in the early 1900s (Ward, 2014), and the 1956 Nobel Prize in Physics was awarded to William Shockley, John Bardeen, and Walter Brattain for their research creating the first semiconductor-based transistor (Nobel Prize Outreach, 2022d). One of the next major advances was research in photolithography in the 1950s, which enabled printing of miniaturized integrated circuits (Computer History Museum, 2022). In particular, fundamental research in photoresists and catalysts at different length scales facilitated technology development and drove Moore's Law for computer chip manufacturing at companies such as IBM (Figure 2-12). As an example, research on photoacid generators (small molecules that react with specific wavelengths of light to form a superacid in the solid state) led to the development of material combinations needed for sub-10-nm technology. At the time of this report, a remarkable 2-nm feature size had been reached, made possible with technology based on these early discoveries derived from fundamental chemistry, polymer chemistry, and materials science. Because of these advances in chemistry, the size of computer chips has decreased steadily over the years, leading to the incorporation of more and more transistors on computer chips and concomitantly faster computation speeds (Shalf, 2020).

UNDERSTANDING THE ECONOMIC IMPACTS OF CHEMISTRY

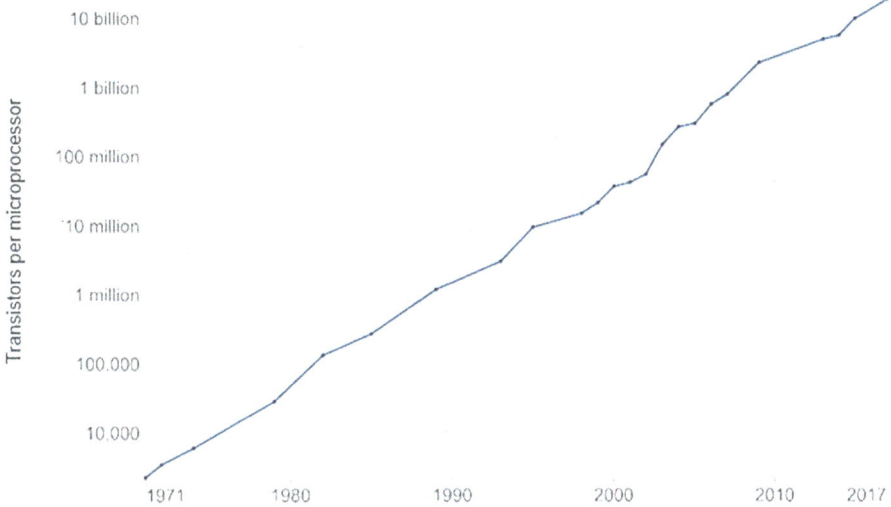

FIGURE 2-12 Display of Moore's Law showing the number of transistors per microchip in each year. SOURCE: Our World in Data, 2017.

The rapidly growing need for computational technologies has caused silicon chips to be more critical to the U.S. economy than ever before. As chemistry, engineering, and physics have allowed for increased performance of semiconductors, the semiconductor industry is now directly responsible for a large amount of the U.S. GDP. According to a report from the Semiconductor Industry Association (SIA) and OE, "the U.S. semiconductor industry is substantial, directly contributing $246.4 billion to U.S. GDP and directly employing over 277,000 workers in 2020" (SIA and OE, 2021). Similar to the chemical industry as a whole, the semiconductor industry has a wide range of indirect economic impacts due to the use of semiconductors in such a wide variety of technologies. The report notes that "300 downstream economic sectors accounting for over 26 million U.S. workers are consumers of and are therefore enabled by semiconductors" (SIA and OE, 2021).

2.3.3.4 Oral Contraceptives

Oral contraception for women has had an undeniably large effect on the economy and on the role of women in society. Norethindrone (also known as norethisterone), the original compound that was used in birth control pills in the 1950s, was still the 143rd most prescribed drug in the United States in 2019, while a combination drug of norethisterone and ethinyl estradiol was the 42nd most prescribed drug (ClinCal, 2019). Oral contraceptives provided women a level of control over their own reproductive health and the ability to make decisions about family planning that they never had before.

The adoption of the "Pill" for use in contraception took a long time. When it was introduced in the 1960s, it was only used for "cycle control" in married women. But, in the 1980s, the Pill became more widely accepted as a method of family planning, especially with the increasing number of women who were in medicine and other professional careers (Liao and Dollin, 2012). Oral contraception has also empowered women around the world, providing them with reproductive autonomy and helping to balance the power dynamic in reproductive decision making. According to the United States Agency for International Development, family planning provides a large number of benefits, including protecting women and children's health, reducing HIV and AIDS, decreasing abortions, improving educational and employment opportunities for women, and reducing poverty (USAID, 2021).

The original synthesis of norethindrone, a steroid derivative, was completed on October 15, 1951, by Luis Miramontes, a young Mexican chemist working under the supervision of Carl Djerassi at the company Syntex (Figure 2-13) (Djerassi, 2006). Syntex, located in Mexico City, was a unique company at the time, because it invested heavily in basic research around steroid hormone synthesis and was responsible for approximately 30% of the industrial publications on the topic prior to 1960 (Olofson and Gortler, 1999). Much of the reason for this heavy investment in steroid chemistry was related to the discovery that a native yam plant, *Dioscorea mexicana*, contained a substantial amount of the natural product diosgenin. In the 1930s, a chemist named Russell Marker, from Pennsylvania State University, discovered a simple synthetic route, termed the "marker degradation," to convert diosgenin to the steroid hormone progesterone (Olofson and Gortler, 1999). Progesterone was the main focus of the steroid hormone industry at the time because "of its value in treating various menstrual disorders and preventing certain types of miscarriages" (Olofson and Gortler, 1999).

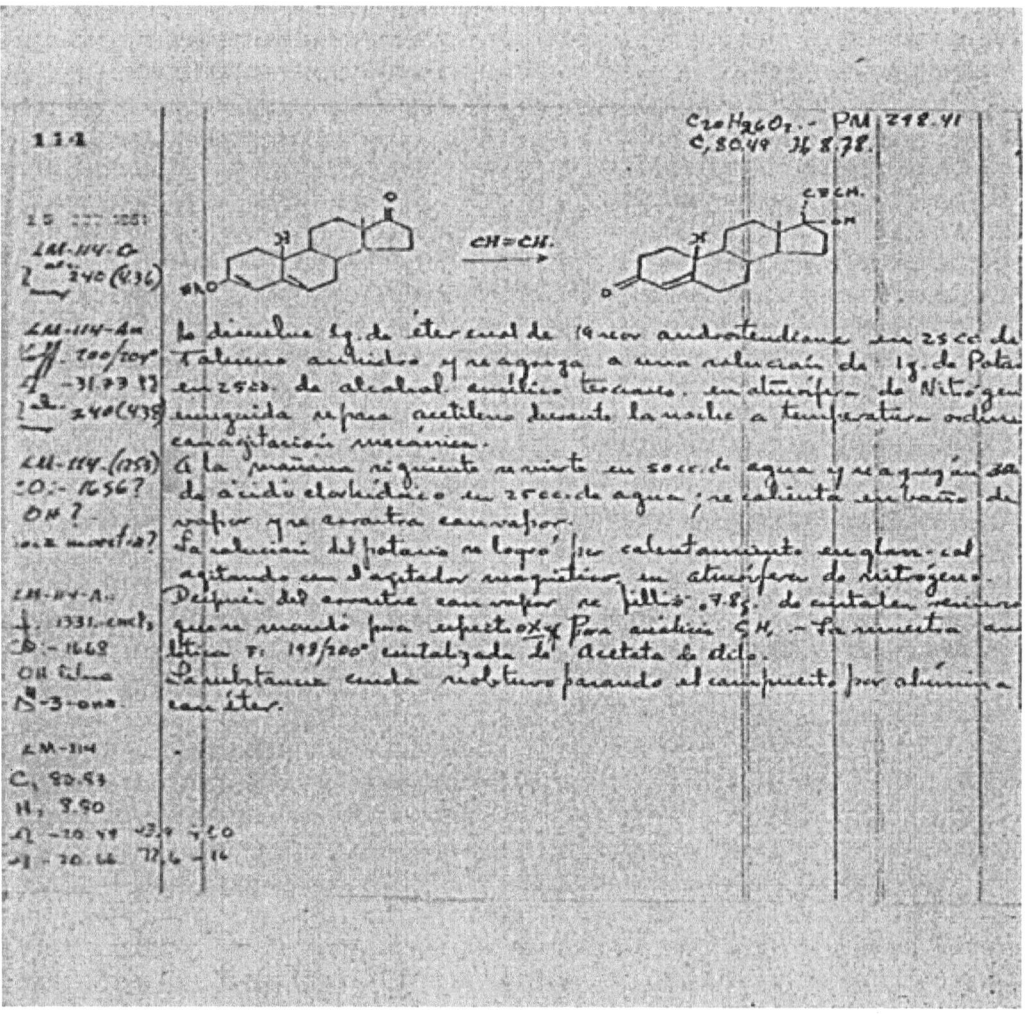

FIGURE 2-13 Lab notebook of Luis Miramontes from October 15, 1951, showing the final step in the synthesis of norethindrone. SOURCE: Djerassi, 2006.

By the time Djerassi came to Syntex in 1949, his research program was less interested in the synthesis of progesterone and instead focused on cortisone, recently discovered as a treatment for rheumatoid arthritis. While working on the cortisone synthesis from the diosgenin precursor, Djerassi was also interested in another possible line of research around progesterone derivatives. At the time, modifying progesterone was widely considered to be useless because any modifications seemed to cause the molecule to lose biological potency. However, a researcher at the University of Pennsylvania performed a really messy and impure synthesis to create a derivative of progesterone that removed a methyl group. This "19-norprogesterone" seemed to display hints of activity that were higher than progesterone and opened the door for progesterone modifications that might lead to higher potencies. After researching and trying out a large number of different progesterone derivatives, a final formulation was synthesized and then sent out for biological evaluation.

The final formulation of the "Pill" produced by Djerassi and Miramontes provided a simple synthesis derived from an abundant natural product. Norethindrone also had higher potency than progesterone, a characteristic that was critical for oral administration. Progesterone, while still effective, cannot be taken as an oral medication. The flood of research in natural products synthesis and steroid hormone development was a critical scientific endeavor that led to the final formulation of oral contraceptives and led to a simpler synthesis of many other important compounds, including progesterone, cortisone, and estrogen.

2.3.3.5 Catalytic Converters

As noted later in this report, in Section 3.4.4.1, the invention of the catalytic converter led to a significant decrease in air pollution from internal combustion engines. The introduction of the catalytic converter in the 1970s produced huge economic and environmental benefits by reducing airborne particulates, acid rain, smog, and these pollutants' concomitant health effects.

While there are several chemical strategies used inside a catalytic converter, all of them deploy some combination of platinum group metals (PGMs) platinum, palladium, and rhodium with other components to provide functions such as oxygen and hydrocarbon storage (Farrauto et al., 2019). Two separate National Medals of Technology and Innovation were awarded in 2002 related to the invention and development of catalytic converters. One went to John J. Mooney and Carl D. Keith at Engelhard (now BASF) for the invention and development of the three-way catalyst system that allowed the simultaneous control of Carbon Monoxide, hydrocarbons, and NO_x. A second award went to Haren S. Gandhi of Ford Motor Company both for his contributions toward developing these systems as well as for driving the focus on recycling of spent converters, which improved the sustainability of PGMs in their deployment (USPTO, 2002). Despite the current maturity of these catalytic systems, their performance continues to improve thanks to the application of advances from fundamental studies (Paolucci et al., 2017).

Chemical and materials sciences were critical in the development of other catalytic converter components beyond the catalysts. The bulk structure is a ceramic honeycomb typically made of cordierite, which has been extruded into a form factor that allows high gas flow while minimizing backpressure (Farrauto et al., 2019). The science needed to produce these structures was based on years of fundamental studies in rheology, surfactants, and oxide chemistries (Govender and Friedrich, 2017). Rodney D. Bagley, Irwin W. Lachman, and Ronald M. Lewis of Owens Corning (now Corning) were awarded a National Medal of Technology and Innovation in 2003 in recognition of the impact that their development of ceramic substrates for catalytic converters has had (USPTO, 2003). Chemical exploration of improved honeycomb designs remains an opportunity for fundamental research. One exciting development is the use of 3-D printing combined with advanced computational methods to design and manufacture the honeycombs (Kovacev et al., 2021). There are also great opportunities to expand the use of these materials and related substrates into other environmental research areas (Hosseini et al., 2020).

2.3.3.6 Chemical Research Toward the COVID-19 Response

A timely example of the large-scale benefits of fundamental chemical research is its impact on the response to COVID-19, the pandemic that cost nearly 1,000,000 American lives as of February 2022 (NCHS, 2019). The Congressional Budget Office estimates $7.6 trillion in lost output for a decade from COVID-19 (Cutler and Summers, 2020). Estimates that account for mortality, morbidity, mental health conditions, and direct economic losses place the economic cost at $19 trillion, assuming the pandemic had been largely controlled by fall 2021 (CBO, 2020). Response to a pandemic of this magnitude requires a multipronged approach including public health measures, vaccines that can reduce the incidence and severity of disease, and therapeutics that target multiple parts of the viral life cycle. This multipronged approach requires a multidisciplinary effort, with inputs from scientific researchers, health care workers, public health professionals, and so many more. Chemical research continues to play a significant role in this effort and has contributed to a number of areas, including the development of nonnatural nucleotides (Nance and Meier, 2021) and lipid nanoparticles (Eygeris et al., 2022) that are crucial for the stability and delivery of mRNA vaccines. Additionally, fundamental chemical research from the late 1980s onward continues to be pivotal for the discovery of oral medications that have shown efficacy in clinical trials for COVID-19, including remdesivir (Gottlieb et al., 2022), molnupiravir (Bernal et al., 2022), and nirmatrelvir (Paxlovid) (Hammond et al., 2022).

Remdesivir and molnupiravir are two in a large class of drugs that are chemically modified variations of nucleotides, the molecules that are building blocks for both RNA and DNA. Early examples of this class of therapeutics were created in the 1960s as possible anticancer drugs and in the 1980s were discovered to be promising therapies for HIV/AIDS. Modified nucleotides have subsequently been used as therapies for hepatitis, Ebola, and other viral diseases (NIAID, 2018). Fundamental research in the chemical synthesis of modified nucleotides directly enabled applied research in the use and manufacture of these molecules as drugs. As a concrete example, fundamental research in the 1980s and early 1990s discovered novel strategies to control the geometric arrangement of atoms in the modified nucleotides and therefore synthesize the specific enantiomer that is effective for safely treating disease (Wilson and Liotta, 1990).

This entire body of work enabled the discovery of molnupiravir, which was first investigated as a potential therapy for Venezuelan equine encephalitis virus (VEEV) but was later shown to have broad-antiviral activity against other diseases, including Ebola (Painter et al., 2021). Finding drugs for VEEV was a priority for the Department of Defense (DoD), and DoD, along with other government agencies, provided $35 million between 2013 and 2020 to Emory University researchers for the development of molnupiravir. This included 6 years of nonclinical testing for the drug, and the testing of its use to fight MERS-CoV (Abinader, 2021). Emory has five published U.S. patent applications directed to derivatives of N^4-hydroxycytidine, the molnupiravir parent compound, which acknowledges U.S. government funding for the development of these patents (Abinader, 2021). Emory University subsequently licensed molnupiravir to Ridgeback Biotherapeutics, which in turn collaborated with Merck for clinical trials of molnupiravir for COVID-19 and for molnupiravir manufacturing (Abinader, 2021).

In the case of molnupiravir, the U.S. Food and Drug Administration issued an emergency use authorization (EUA) in March 2022 (FDA, 2022). The drug's use is restricted to high-risk individuals who contract COVID-19, but Merck is already projecting between $5 billion and $5.5 billion worth of sales in 2022 (Dunleavy, 2022). Paxlovid was similarly issued an EUA in December 2021 (Katella, 2022), and analysts are predicting $22 billion in sales after reporting $1.5 billion in the first quarter of 2022 (Kimball, 2022). Both of these pharmaceuticals are having an immediate and noticeable economic and public health impact.

2.4 UNDERSTANDING U.S. COMPETITIVENESS IN THE CHEMICAL ECONOMY

To fully understand the economic impact of the U.S. chemical economy, it is important to evaluate its role on the global economic stage and understand changes or shifts that have occurred over time. The U.S. chemical economy is also part of a complex ecosystem that, while competitive with chemical industries in other countries, is also heavily reliant on them for collaboration, innovation, and workforce needs. These factors must be considered as a part of the same ecosystem and kept in mind in creating policies for improving the chemical economy.

2.4.1 Comparative Global Research Output in Chemistry

To analyze global competitiveness, we can compare research outputs of different countries in the form of publications, economic outputs, and several other points of analysis. It is important first to look at the research portfolio of individual countries for the purposes of understanding where the research interests are in different nations. In 2020, more than 50% of the publications from the United States were in the life sciences (37% health sciences and 14% biological and biomedical sciences) (Figure 2-14) (NSB, 2021). In contrast, only 4% of U.S. publications focused on chemistry and materials science. Inevitably, chemical knowledge informed some of the publications in the life sciences, but it is important to note that chemistry-specific publications are a low percentage of U.S. research output. China, India, and Japan each had 8% of their publication output in chemistry-specific research, with 5%, 7%, and 3% in materials science, respectively. These differing portfolio shares in chemistry are one indicator of a relative advantage of China in the area of chemistry publication (Figure 2-14).

In addition to the data showing research portfolios, it is helpful to understand the relative abundance of publications and patents in particular areas. The metrics for scientific productivity are imperfect because of the many caveats associated with using quantitation such as number of publications or number of patents in a particular area. However, these numbers, when considered as a part of a holistic picture, can help to understand trends and intensity of inventions in a scientific field during specific time periods. The National Science Foundation's (NSF's) National Science Board measures publication numbers using fractional article counts, which means that "each country receives fractional credit on the basis of the proportion of its participating authors" (NSB, 2021). The U.S. fractional share in all science and engineering (S&E) articles has declined by about half (from 31.6% to 15.5%) in the period between 1996 and 2020, and the U.S. share in chemistry-specific articles has also declined by about half over the same period (from 19.8% to 8.8%) (Figure 2-15c and d) (NSB, 2021). Interestingly, while the U.S. percentages have declined, the total fractional count of articles published by U.S. authors has increased when looking at all S&E fields (313,000 in 1996 to 456,000 in 2020) and remained stagnant when considering chemistry-specific S&E articles (between 10,000 and 20,000 over the same time period) (Figure 2-15a and b). The decrease in percentage of fractional S&E publications is therefore largely because China's total numbers of all S&E articles and chemistry-specific S&E articles increased dramatically. China's fractional publication count of all S&E articles was 34,000 in 1996 and increased 20-fold by 2020 to 670,000. Similarly, the number of chemistry articles saw an 11-fold increase, going from 4,700 in 1996 to 55,600 in 2020. This increase in total fractional counts led to China's share in overall S&E articles increasing more than sixfold (from 3% to 23%) in the period between 1996 and 2020. Additionally, China's share in chemistry has also increased more than fourfold over the same period

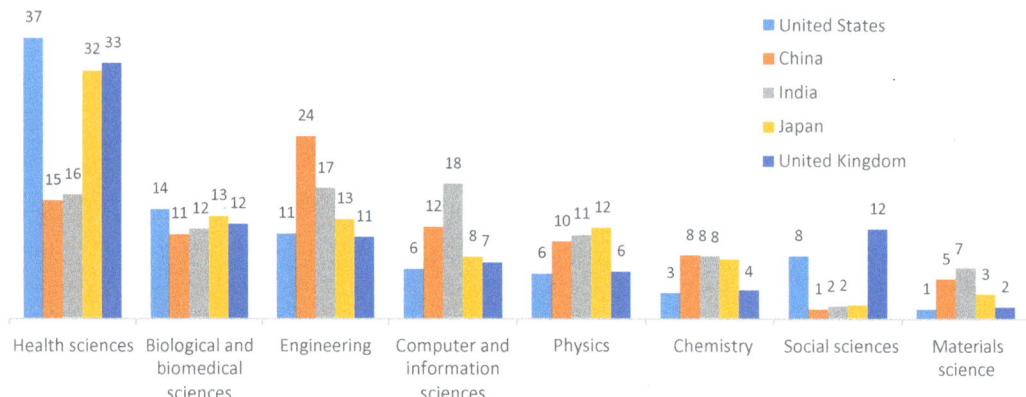

FIGURE 2-14 S&E research portfolios, by the eight largest fields of science and by selected country or economy in 2020. The numbers represent a percentage of total research, measured in publication output, for each country. SOURCE: NSB, 2021, fig. PBS-3.

(from 7% to 31%). No other country experienced such a drastic increase in publication output over this time period, although notably, India doubled its percentage of all S&E articles published. Most other countries have remained relatively static (Figure 2-15d).

Although the number of S&E articles from the United States increased from 1996 to 2020, chemistry-specific articles remained at around 10,000–20,000 articles per year (Figure 2-15a and b). There are several possible explanations for this trend. The first is that the chemical industry and its related chemical research and publication output has stagnated. Another possibility is that chemistry research in the United States has become a science that is now used to support and inform other topic areas such as the life sciences, and publications in these areas, while they might be based on chemical discovery, are not counted toward chemistry publications. Likely, it is some combination of these and other factors. As shown in Figure 2-15, China's publication rate has been steadily increasing over the past 20 years, likely due to its strong history of seeing the value in investing in the chemical sciences (Jia, 2018).

Another way to look at publications is to consider the share of S&E articles that are the most cited (e.g., top 1% cited), with citations normalized by subfield and year. Articles published in the United States in 2018 were about 60% above the world average for all S&E articles and 55% above the world average in chemistry (Figure 2-15). The only other countries to have an above-average number of highly cited papers are the United Kingdom and, in later years, Germany. This correlates well with countries that have well-established scientific research programs. When looking at China's share of the top 1% of cited S&E articles, the numbers in 2018 seem to be on par with the world average, while the share in the top 1% of cited articles in chemistry is 39% above the world average.

Looking at the trends over time for the share of highly cited articles per country, it is notable that the U.S. share in all S&E articles had been stagnant from 1996 to 2015 and then declined in more recent years, from 72% above the world average to 60% above the world average. Compared to this overall trend, the United States has had a steeper decline in chemistry, dipping from 93% above the world average to 55% above the world average (Figure 2-16).

Aside from publications, another possible metric of research productivity is to look at the share of patents published by different countries in the area of chemistry. In several different subfields of chemistry, the United States was the dominant publisher in 2017–2019 (WIPO, 2021). For example, the United States filed the most overall patent applications in biotechnology and pharmaceuticals (Table 2-2). China was more active than all other countries in basic materials chemistry, chemical

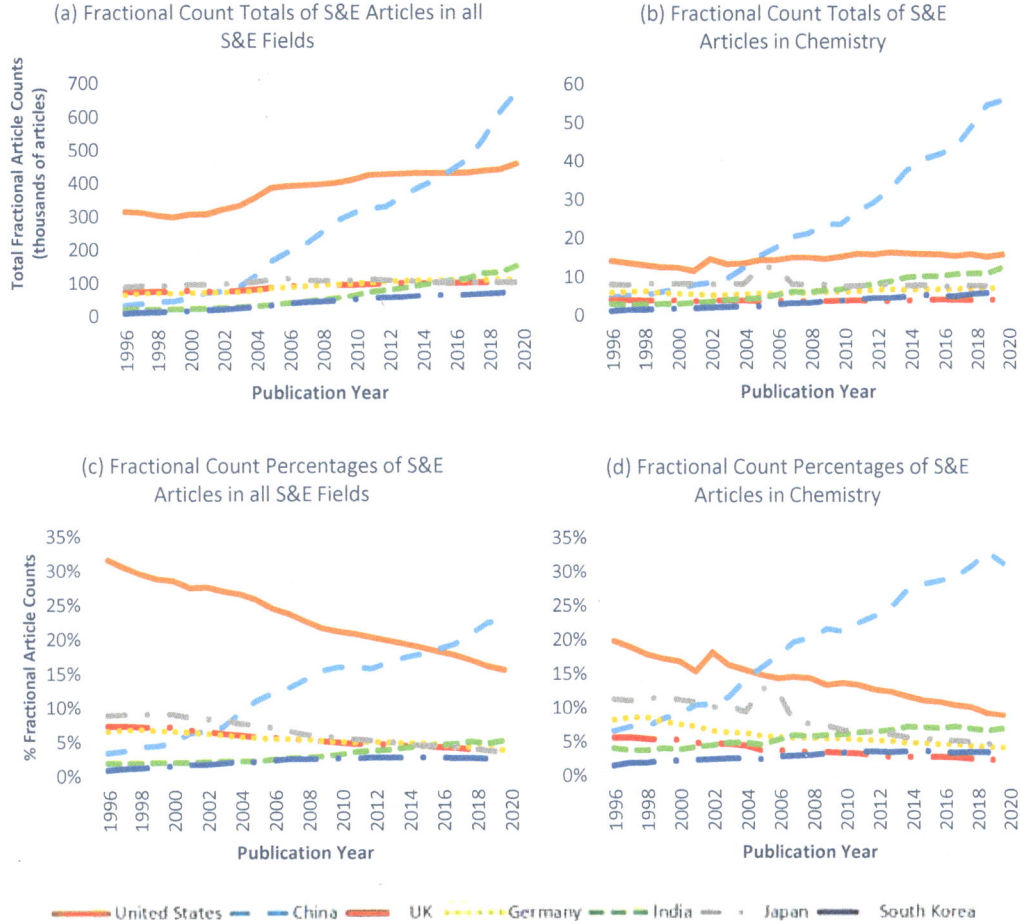

FIGURE 2-15 Fractional counts of total S&E articles and chemistry-specific S&E articles by country and publication year. The fractional count indicates that for every publication, "each country receives fractional credit on the basis of the proportion of its participating authors." (a) Total fractional article counts for all S&E articles. (b) Total fractional article counts for chemistry articles. (c) Percentage of fractional article counts as a part of all published articles for all S&E areas. (d) Percentage of fractional article counts as a part of all published chemistry articles. SOURCE: Data from NSB, 2021, table SPBS-6.

engineering, and environmental technology (Table 2-2). Several other countries, such as Russia, Germany, and Japan, were leaders in other areas of chemistry, but there are no particularly strong indicators that one country holds an advantage in patenting over another. Additionally, there are many caveats associated with patent data that were laid out in detail earlier in the chapter.

2.4.2 Comparative Global Economic Output

We can further consider the value added of chemicals and chemical products by industry in order to assess the competitiveness of the chemical economy. Between 2002 and 2010, the United States was the unparalleled leader in the share of value added by chemical industry (IHS Markit, 2022). However, throughout that period, the U.S. chemical industry saw a steady decline. In 2011, the value-added chemical and chemical products industry in China surpassed that of the United

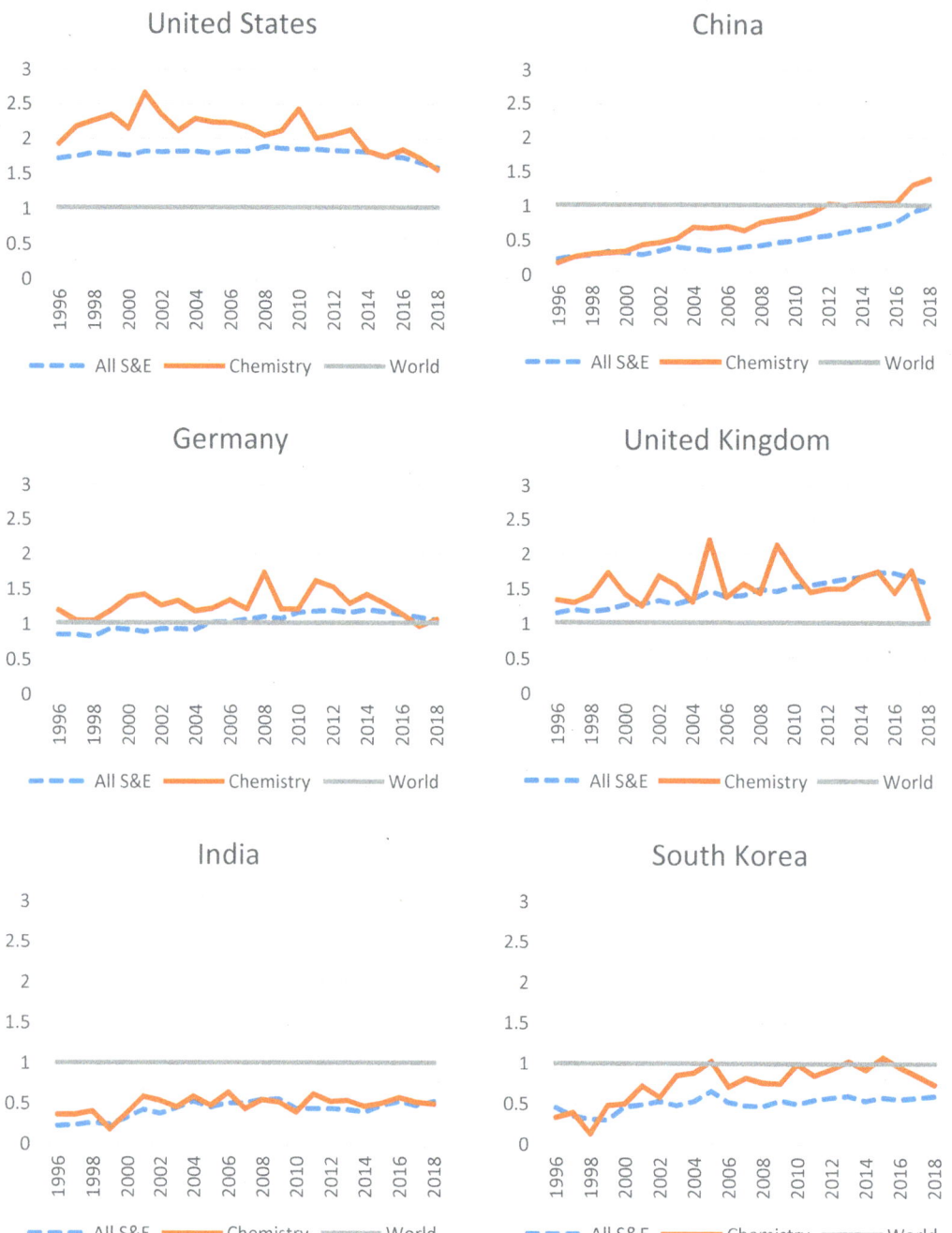

FIGURE 2-16 Fractional count of top 1% cited S&E articles compared to the world average. SOURCE: Data from NSB, 2021, tables SPBS-58 and SPBS-62.

TABLE 2-2 Share of Total Published Patent Applications by Country and Subfield, 2017–2019

Chemistry Subfield	Share of Total Published Patent Applications, 2017–2019)						Total Published Applications in 2019
	United States	Germany	Japan	China	Russian Federation	South Korea	
Organic fine chemistry	2.8	**3.0**	1.4	1.9	1.8	1.9	65,540
Biotechnology	**4.0**	1.9	1.1	1.6	1.7	1.6	70,520
Pharmaceuticals	**5.9**	2.5	1.3	2.5	4.3	2.1	96,737
Macromolecular chemistry, polymers	1.3	2.0	**2.3**	1.8	0.9	1.4	53,901
Food chemistry	1.2	0.4	0.8	3.2	**8.0**	2.1	56,343
Basic materials chemistry	2.6	3.1	2.2	**3.5**	2.8	1.8	81,429
Materials, metallurgy	1.2	1.9	2.4	3.2	**4.6**	1.8	76,570
Surface technology, coating	1.3	1.7	**2.5**	1.4	1.5	1.5	48,716
Microstructural and nanotechnology	0.2	0.2	0.1	0.2	**0.8**	0.1	5,724
Chemical engineering	2.1	2.7	1.4	**4.1**	3.9	2.3	91,855
Environmental technology	1.1	1.5	1.1	**2.9**	2.8	1.6	63,462

NOTE: Numbers in bold indicate top number of patent applications for a particular field. SOURCE: Data from WIPO, 2021.

States, following a steady increase in China's output over the entire period. Importantly, the U.S. share in the international chemical economy held steady and even saw a slight increase between 2010 and 2018, indicating that, despite China's continual increase, the United States was also experiencing growth. This was in large part due to the extraction and increased supply of natural gas (see Box 2-1). By 2018, the U.S. chemical industry value added was 0.216 trillion current U.S. dollars, second to China's value added in the chemical industry at 0.298 trillion current U.S. dollars. That year, China and the United States accounted for 50% of the world chemical output (IHS Markit, 2022).

Another metric of the economic competitiveness of chemistry in the United States and in the world is sales of the leading chemical companies globally. In 2020, as reported by *Chemical & Engineering News*, the top 50 chemical companies, headquartered in 19 countries around the world, had total combined sales of nearly $796 billion. Ten of those companies are headquartered in the United States (Table 2-3) and had combined total sales in 2020 of $154 billion (Tullo, 2021). Among the U.S. chemical companies in the top 50, only two, Dow Chemical (#3) and Lyondell-Basell (#10), are in the top 10. Nonetheless, the 10 U.S. companies claim the largest percentage of total sales (19.3%) among the 19 countries on the list.

All of these factors taken together paint a picture of several countries who are dominant in the chemical sciences enterprise, with leadership in various facets from the United States, China, Japan, Germany, and the UK. Before making decisions about international competitiveness and leadership, it is important to consider a wide swath of data and a number of different metrics. The United States continues to be an important player in chemical research and the chemical industry. To continue being a leader in this space, the United States has to consider making data-driven and aspirational investment decisions relating to the chemical sciences enterprise.

BOX 2-1
Contributions of Natural Gas to U.S. Competitiveness in the Chemical Economy

Despite continued expansion of the global chemical industry between 2000 and 2010, no new chemical plants were planned for the United States during this time period (NASEM, 2016), and North America saw a reduction in base chemical capacity growth through the early 2000s (Eramo, 2014). Using ethylene as a proxy for basic chemicals, between 2000 and 2010, nearly 90% of all new investment occurred in the Middle East and China (Eramo, 2014). A review of chemical sales showed nearly a global doubling between 2001 and 2011, with China more than tripling sales value during the same time period that North American sales values were declining (Eramo, 2019).

In addition to the eastward movement of chemical markets, the U.S. chemical industry was hampered by increasing energy and feedstock prices (Budde, 2011). However, the energy and feedstock situation improved rapidly due to the increasing supplies of low-cost natural gas and natural gas liquids as a result of the growth of hydraulic fracturing (fracking) of shale deposits (NASEM, 2016). As the shale industry rapidly increased production, some models estimated that between 2010 and 2020, nearly 18% of new ethylene capacity would be added in North America, in sharp contrast to what was seen in the previous decade where ethylene capacity was removed (Davis, 2009; Eramo, 2014). Beginning with restarts of previously shuttered capacity such as Dow's ethylene cracker in Hahnville, Louisiana (Esposito, 2013), the availability of low-cost natural gas and natural gas liquids resulting from fracking operations has resulted in $109 billion in investments in new or expanded U.S. chemical facilities since 2010 (ACC, 2022). Additional projects either under construction or in planning will bring the total investment in the U.S. chemical industry to $208 billion with nearly 70% of these projects having a foreign investor or partner (ACC, 2022). Shale gas has been a "game changer" for U.S. industries (Barteau and Kota, 2014).

The sudden emergence of unconventional gas may give the impression that it was a new technology; however, hydraulic fracturing is not new and has been utilized for more than 60 years in other applications (FECM, 2011). What is new is combining it with horizontal drilling to extract oil and natural gas from shale deposits, a practice that really emerged over the past two decades (Elliott and Santiago, 2019). While technology innovations were part of the reason for its rapid expansion in the 2000s, many other factors have also been cited as playing a role in the growth of shale gas investments, including high natural gas prices, favorable geology, U.S. mineral rights ownership, market structure and availability of financing, water availability, and the natural gas pipeline system (Wang and Krupnick, 2015). The impact of these factors coming together resulted in a significant increase in the amount of oil and gas produced in the United States (Elliott and Santiago, 2019), which has ultimately benefited the chemical industry such that North America is again quite competitive with the Middle East (CEFIC, 2022a). While the major component of shale gas extraction, methane, is used primarily as an energy source, the collateral products, consisting of light hydrocarbons, represent a copious source of alternative feedstock for the U.S. chemical industry. The challenge is to develop chemical technologies that can utilize this resource with high carbon efficiency to make everyday products.

TABLE 2-3 Top 50 Chemical Companies by Sales

2020 Rank	Company	Chemical Sales ($ millions)	Headquarters Country
1	BASF	67,491	Germany
2	Sinopec	46,656	China
3	**Dow**	**38,542**	**U.S.**
4	Ineos	31,310	UK
5	Sabic	28,792	Saudi Arabia
6	Formosa Plastics	27,711	Taiwan
7	LG Chem	25,477	South Korea
8	Mitsubishi Chemical	25,323	Japan
9	Linde	24,392	UK
10	**LyondellBasell Industries**	**23,407**	**U.S.**
11	**ExxonMobil Chemical**	**23,091**	**U.S.**
12	Air Liquide	23,089	France
13	PetroChina	21,769	China
14	**DuPont**	**20,397**	**U.S.**
15	Hengli Petrochemical	17,265	China
16	Sumitomo Chemical	15,822	Japan
17	Toray Industries	15,196	Japan
18	Shin-Etsu Chemical	14,019	Japan
19	Evonik Industries	13,919	Germany
20	Reliance Industries	13,600	India
21	Covestro	12,216	Germany
22	Shell Chemicals	11,721	Netherlands
23	Yara	11,591	Norway
24	Braskem	11,348	Brazil
25	Mitsui Chemicals	11,348	Japan
26	Syngenta	11,208	Switzerland
27	Bayer	11,204	Germany
28	Solvay	11,084	Belgium
29	Wanhua Chemical	10,636	China
30	Indorama	10,589	Thailand
31	Lotte Chemical	10,354	South Korea
32	Johnson Matthey	9,951	UK
33	Umicore	9,738	Belgium
34	Asahi Kasei	9,283	Japan
35	DSM	9,249	Netherlands
36	Arkema	8,996	France
37	**Air Products**	**8,856**	**U.S.**

continued

TABLE 2-3 Continued

2020 Rank	Company	Chemical Sales ($ millions)	Headquarters Country
38	Mosaic	8,682	U.S.
39	Hanwha Solutions	8,596	South Korea
40	Eastman Chemical	8,473	U.S.
41	**Chevron Phillips Chemical**	**8,439**	**U.S.**
42	Rongsheng Petrochemical	8,359	China
43	Borealis	7,780	Austria
44	**Westlake Chemical**	**7,504**	**U.S.**
45	Sasol	7,288	South Africa
46	Nutrien	7,156	Canada
47	Lanxess	6,965	Germany
48	Tosoh	6,864	Japan
49	DIC	6,567	Japan
50	**Corteva Agriscience**	**6,461**	**U.S.**

SOURCE: Data from Tullo, 2021.

2.4.3 Impact of International Researchers on the U.S. Chemical Economy

The U.S. chemical economy is part of a complex ecosystem that is heavily reliant on other countries for collaboration, innovation, and workforce needs. The Science and Engineering Indicators produced by NSF's National Science Board noted that in 2017, 30% of workers in S&E occupations were foreign-born. When looking specifically at the workforce in the physical sciences, approximately 34% of Ph.D.-level scientists were foreign-born (Figure 2-17). Additionally, approximately 27% and 23% of physical sciences workforce bachelors and masters recipients, respectively, were foreign-born (Figure 2-17) (NSB, 2019). Importantly, these data do not specify what sector these individuals are working in, and they do not specify chemists or chemical engineers, but the number of foreign-born workers in the United States is still large. The highest percentage of this workforce is from Asia, with China, India, and South Korea in the top three places (NSB, 2019). If you also look at those receiving degrees, 36.2% of doctoral-degree chemists and 47.3% of doctoral-degree chemical engineers in the United States between 2010 and 2020 were foreign-born citizens (NCSES, 2020b). Based on the 2020 NCSES data, approximately 73% of these graduates say they plan to stay in the United States, and these numbers held true with doctoral recipients from 2011 to 2013 staying in the United States at 71% and those recipients from 2006 to 2008 staying at 72%, when the data were collected in 2017 (NCSES, 2020b).

As an additional example, while the Nobel Prizes are an imperfect measure of impact, it is notable that the number of researchers that make up the category of "foreign-born U.S. laureate" is greater than the number of prizes won by most countries. Including the 2021 Nobel Prize, 46% of all prizes have been given to researchers in the United States, and of that, 16% of all prize recipients are foreign-born U.S. laureates (Figure 2-18). It becomes fairly clear that in a highly competitive international environment, the United States remains a desirable place to perform groundbreaking research.

UNDERSTANDING THE ECONOMIC IMPACTS OF CHEMISTRY 57

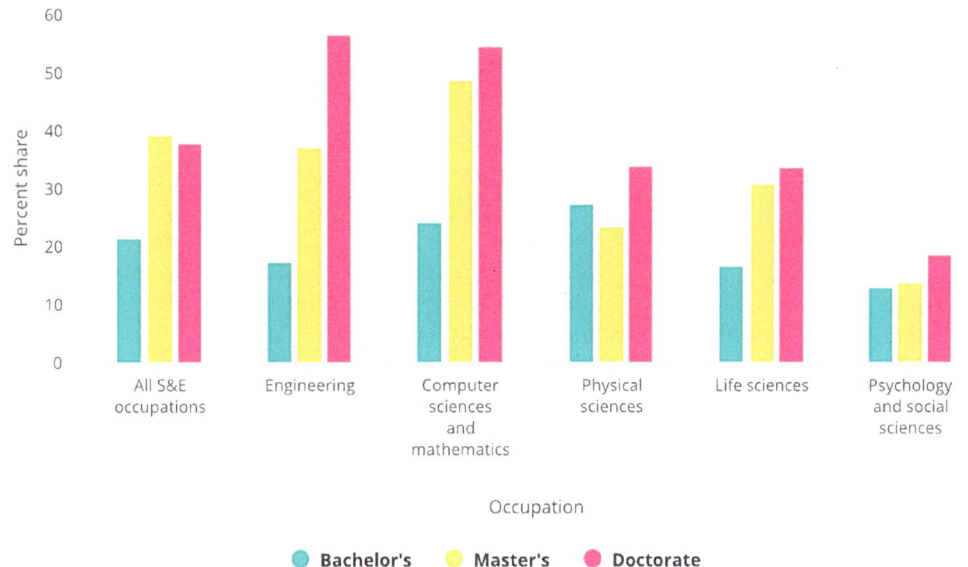

FIGURE 2-17 Foreign-born individuals in S&E occupations in the United States. SOURCE: NSB, 2019.

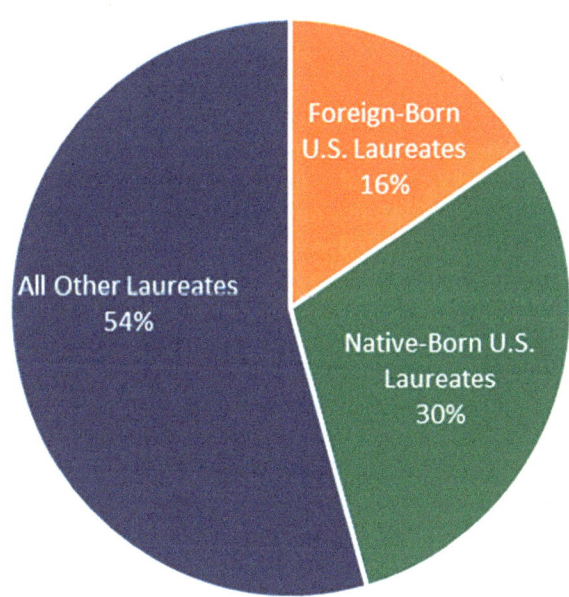

FIGURE 2-18 Analysis of foreign- and native-born U.S. Nobel Prize winners from 1901 to 2021. SOURCE: Farago and Waslin, 2021.

2.5 CONCLUSIONS

This chapter outlines the scope and impact of the U.S. chemical economy and identifies linkages between chemical research and the chemical economy using various measurable parameters and case studies. The data show that the chemical economy in the United States is still strong but could be stronger if we want to experience leadership growth. Some of the quantitative data from this chapter show that in 2020, the U.S. chemical economy, excluding pharmaceuticals, had $457 billion in sales, directly added $225 billion to the overall economy, and employed 529,000 individuals. These numbers have been relatively steady for the past 5 years. The chemical economy additionally has indirect impacts on the U.S. economy, and in 2020 the "business of chemistry" was estimated to be responsible for $5.2 trillion, or 25% of U.S. GDP, and the employment of 4.1 million individuals who in some way interact with the chemical economy.

This chapter also highlights the great difficulty in quantifying the true impact of fundamental chemical research. Despite this, the importance of chemical discovery is quite evident in specific technologies that have a huge impact on the economy and society, such as batteries, biocatalysis in synthesis, silicon chips, oral contraceptives, catalytic converters, and pharmaceuticals to fight SARS-CoV-2. In each of these cases, a large body of chemical knowledge that was built over decades or centuries contributed to a chemical discovery that enabled important advances. From the information gathered about the size and impact of fundamental chemical research and the U.S. chemical economy, the committee highlighted several conclusions.

Conclusion 2-1: Chemical research has an outsized economic value based on the spillover of chemical knowledge and products into other areas and the fact that chemical patents, as well as patents that rely on chemical knowledge, have a higher average value than other patents. Chemical patents accounted for 14% of all corporate patents between 2000 and 2020, but they accounted for 23% of all value in the same time period.

Conclusion 2-2: It is challenging to directly link chemical research to economic impact because each chemical product or process relies on a broad body of chemical knowledge and discovery that is built over decades or centuries, and chemical knowledge is also deeply integrated into other disciplines, making the specific impacts of chemistry in the broad scientific enterprise difficult to deconvolute. Additionally, analyzing the economic impact of chemical research suffers from a lack of data, including patent value estimations, widely available licensing terms data, and government grant data.

Conclusion 2-3: Chemistry is a foundational and central scientific discipline, and sustained investment in fundamental chemical research provides the chemical knowledge for technology development, generating unexpected discoveries that are the basis for innovation. These innovations directly influence the chemical economy, environment, and quality of life and also advance knowledge and discovery in many other scientific and technological disciplines, such as the life sciences, information technology, earth sciences, and engineering.

Conclusion 2-4: The chemical economy is critically important for our national economy and our leadership in the international chemical enterprise. This leadership relies heavily on advances in fundamental chemistry that drive the creation of new tools, technologies, processes, and products and enables environmental considerations. However, our nation's

leadership in the chemical industry cannot be taken for granted, and this leadership needs continued and sustained nurturing and support.

Conclusion 2-5: The success of the chemical economy relies on a large group of employees who work both within and proximal to the chemical enterprise. Continuing to attract and retain diverse talent in the chemical sciences, both internationally and domestically, is critically important to a thriving chemical economy.

3

Sustainability for the Chemical Economy

> Key Takeaways:
>
> - Basic chemistry and the chemical industry have already made significant contributions to the UN Sustainable Development Goals.
> - Companies that are leaders in sustainability can capitalize on their competitiveness in a new economic environment where sustainability and climate protection are valued.
> - Market- and purchasing-based policies and tools can be key in supporting a green and circular economy.
> - Many areas of decarbonization, sustainability, and climate improvement will benefit from advances in fundamental chemical research coupled with design and collaboration with other areas of science and engineering.

In the 21st century, society faces the urgent tasks of mitigating the global climate emergency and arresting the deterioration of global ecosystems. The Intergovernmental Panel on Climate Change's Sixth Assessment Report warned that the world is on code red because major climate-related damages are already causing significant loss of lives and livelihoods (IPCC, 2021, 2022; see also IPBES, 2019). In the United States in the past decade alone, hurricanes and wildfires exacerbated by climate change are estimated to have cost $890 billion in damages (Smith, 2021). Other climate-driven risk modeling (Hsiang et al., 2017) estimates that 1.5°C of warming compared to a 1981–2010 baseline would cost the United States –0.1% to 1.7% of its gross domestic product (GDP), and 4°C of warming would cost 1.5% to 5.6% of its GDP, with losses concentrated in the poorer counties across the country. The deterioration of global ecosystems is resulting from a multiplicity of factors. Among them are waste streams derived from products that have served human needs. Plastic pollution is estimated to reduce the benefit humans derive from global oceans by

1% to 5%, amounting to an estimated annual loss of $0.5 to $2.5 trillion (Beaumont et al., 2019). Herbicides and pesticides, while contributing to increased agriculture production, also contribute to biodiversity losses and ecosystem degradation (Beketov et al., 2013; NRC, 2000; Schütte et al., 2017). Additionally, plastics production has a large energy cost that has already produced more than 100 million metric tons of carbon dioxide (CO_2) equivalent in greenhouse gas (GHG) emissions in the United States alone, and these GHG emissions are predicted to substantially increase based on the rate of plastic consumption (Nicholson et al., 2021).

These climate and ecological crises necessitate meeting human needs while making transformative changes in a society in which sustainable development is the central paradigm—that is, the alignment of economic development with public health and environmental protection. The United States affirmed its support for sustainable development as early as the 1992 Earth Summit. The United States then reiterated its support in helping to formulate and launch the Sustainable Development Goals (SDGs) of the United Nations in 2015 (UN, 2015). These goals aim to achieve economic progress, good health, clean water, affordable renewable energy, and other human needs and aspirations for everyone, while protecting ecosystems and maintaining a stable climate, which are vital to support the former.

To reach these goals, there is a need to balance the benefits from products that meet human needs with the costs of those products to public health, the environment, and the climate. This is both a challenge and an opportunity for the chemical economy. Basic chemistry and associated industries have made tremendous contributions toward meeting the SDGs, but have also created products and processes that contribute to the climate and ecological crises. To address these crises, the chemical economy will need to make a transformative shift in which sustainability and decarbonization are central tenets.

A report from the National Research Council on *Sustainability in the Chemical Industry* commented that "sustainability is a path forward that allows humanity to meet current environmental and human health, economic, and societal needs without compromising the progress and success of future generations" (NRC, 2006a). Decarbonization, defined as the reduction in use and emissions of fossil-based carbon (NASEM, 2021b), along with sustainability, is an essential feature of the transformation needed both for industry's social license to operate and to meet pressures in the United States and abroad. Companies that lead on sustainability will be more competitive in a new economic environment that values sustainability and climate protection. Nevertheless, to transition from the business-as-usual scenario to this sustainable future, companies face a number of challenges. Market and public policies do not consistently and fully reward sustainability and climate protection, which is necessary to provide the certainty of payoffs for companies to invest in such a shift.

The growing embrace of sustainability by the public has spurred efforts in basic chemical research (as well as in commercial research and development [R&D]) to find solutions that can meet the SDGs using green chemistry and other principles, such as circular design. However, much work is still needed, and there are large gaps in achieving the world's sustainability goals that chemistry is primed to solve. All areas within chemistry have a role to play in advancing fundamental research toward decarbonization, sustainability, and environmental stewardship. In this chapter, we discuss contributions that chemistry has already made to sustainability, decarbonization, and the environment; consider industry initiatives and public policies that can provide more certainty for the payoffs that companies need to drive the transformative shift; and then lay out some important research avenues for fundamental chemistry.

3.1 BASIC CHEMISTRY IN SOCIETY: CONTRIBUTIONS AND CONSEQUENCES

3.1.1 Contributions of Chemical Research to Society

Fundamental chemical research as well as applied R&D have contributed to sustainability, including saving lives, empowering individuals, and reducing emissions. We summarize in Table 3-1 examples given throughout this report of impactful developments arising because of basic chemical research, and list SDGs supported by each example. Several of these examples illustrate how products have brought benefits, and investments to advance basic chemistry can potentially improve the products and processes by reducing the amount of energy input and the amount of waste by-products. The committee's focus, per its Statement of Task, is on the potential contributions of basic chemical research to these goals. We note that socioeconomic and political efforts, beyond technical approaches, are also necessary to meet these goals.

3.1.2 Adverse Consequences to Public Health, Ecosystems, and Climate from Chemical Products and Processes

Notwithstanding these numerous contributions to the SDGs, some of the same chemical products and processes adopted for their ability to meet specific needs have subsequently had significant public health and environmental impacts attributed to them. Debates continue on the use of these products: Should they be used in specific circumstances when benefits outweigh the costs; should they be severely restricted or banned to incentivize the search for more benign substitutes; and is a complete overhaul of the fundamental approach in that particular sphere needed or even possible

TABLE 3-1 Examples of Basic Chemical Research Contributions to the Sustainable Development Goals

Example	Description	Relevant SDGs (SDG #)
Haber-Bosch	See Section 1.3.1	Zero Hunger (2)
Distributed manufacturing	See Section 1.3.3	Decent Work and Economic Growth (8);
		Industry, Innovation, and Infrastructure (9);
		Responsible Consumption and Production (12)
Additive manufacturing	See Section 1.3.3	Industry, Innovation, and Infrastructure (9)
Rechargeable batteries (lithium-ion)	See Section 2.3.3.1	Affordable and Clean Energy (7)
Biocatalysis	See Sections 2.3.3.2 and 4.4.2	Industry, Innovation, and Infrastructure (9);
		Responsible Consumption and Production (12)
Photolithography for silicon wafers	See Section 2.3.3.3	Industry, Innovation, and Infrastructure (9)
Oral contraceptives	See Section 2.3.3.4	Gender Equality (5);
		No Poverty (1);
		Good Health and Well-Being (3)
SARS-CoV-2 antiviral pills	See Section 2.3.3.6	Good Health and Well-Being (3)
Catalytic converters	See Sections 2.3.3.5 and 3.4.4.1	Good Health and Well-Being (3)
Ozonation/chlorination	See Section 3.4.4.2	Clean Water and Sanitation (6)

(e.g., shifting from chemical-based agriculture to regenerative agriculture)? All of these are foundational questions needing further attention.

Examples include the perfluoroalkyl and polyfluoroalkyl substances (PFAS), highly fluorinated synthetic chemicals with unique properties due to their strong carbon-fluorine bonds. PFAS, which were invented in the 1930s and prized for their water- and oil-repellent properties, became widely used in firefighting products (e.g., at airports and military bases) and consumer products (e.g., nonstick pans, waterproof clothing, stain-resistant carpets). However, scientific research subsequently uncovered that PFAS[1] are biopersistent—they are environmentally mobile and do not decompose—and pose public health and environmental risks (U.S. Congress, 2019). Regulations and voluntary actions ended the manufacture of two PFAS chemicals, perfluorooctanoic acid and perfluorooctane sulfonate, in the 2000s (Hogue, 2019; NASEM, 2021d), but the United States today is still facing the challenge of redressing PFAS pollution in drinking water sources.

Another example is chlorpyrifos, one of the most widely used pesticides, applied to crops including corn, wheat, citrus, apples, and strawberries. In 2000, as more research revealed that chlorpyrifos posed neurodevelopmental risks to infants and children, environmental groups petitioned the U.S. Environmental Protection Agency (EPA) to end its use. Following the virtual removal of chlorpyrifos from homes and lawns in 2000, the chemical was banned from agricultural uses by 2007 (EPA, 2022). In 2015, the EPA under the Obama administration proposed to ban the use of the pesticide, but the decision was reversed under the subsequent administration. In 2021, in response to the Ninth Circuit Court of Appeals ruling that the EPA prove it was safe for use on food crops or to declare it illegal, the EPA announced its ban on the use of chlorpyrifos.

Although the calculation of cost versus benefit was clear in the chlorpyrifos example above, the calculation of chemistry-based products and processes even on the yardstick of public health is not always clear-cut and often is context dependent. The case of dichlorodiphenyltrichloroethane (DDT) is illustrative. The United States has banned the use of DDT for malaria control and has chosen to use alternatives for mosquito control. On the other hand, the World Health Organization (WHO) recommends the use of DDT for controlled indoor spraying as a strategy to control mosquitoes that are vectors for malaria and dengue. The WHO assessment favored DDT over alternative insecticides.

A recent study showed the extent of pharmaceutical pollution in Earth's waterways (Wilkinson et al., 2022). While the pollution was extensive, and thought to be one of the drivers of antibiotic resistance seen today, it is notable that pharmaceuticals were found in all water systems, especially in wealthier countries. The paper additionally noted that the sites of highest contamination were "associated with areas with poor wastewater and waste management infrastructure and pharmaceutical manufacturing." These discoveries were built on a body of knowledge around monitoring contamination of antibiotics and the risk of antibiotic resistance (Scott et al., 2016). On the basis of this knowledge, pharmaceutical companies are generally aware of this issue and have made strides to address pollution, but much work is still needed (EEB, 2018).

In these and other examples, the chemical innovations have addressed a pressing problem but subsequently were found to have produced negative unintended consequences. Such outcomes have been the result of incomplete foresight, and in some cases, they have been the result of incomplete transparency in public health and environmental impact evaluations of products before they enter the marketplace (McHenry, 2018). The PFAS incidents, alongside other contaminant and waste issues, underscore that industries within the chemical economy cannot simply take support from the public as a given, but will need to ensure that public good, public health, and protecting the environment are central to their mission (Mazzucato and Li, 2021).

---

[1] The PFAS umbrella comprises more than 9,000 substances with different properties, potential toxicities, uses, and potential for human exposure.

3.2 TRANSITIONING TO SUSTAINABILITY AND DECARBONIZATION IN THE CHEMICAL ECONOMY

The chemical industry, while already contributing to sustainability and decarbonization of its own and other industries, will need to adjust its processes, feedstocks, and products to make these approaches and goals central to its operations. A National Research Council report notes that "sustainable practices refer to products, processes, and systems that support this path" (NRC, 2006a). These processes will not only need to avoid harm to health and the environment but will need to be financially feasible to be widely implemented (NRC, 2006a).

The industry currently contributes to GHG emissions in two major ways. First, it uses primarily fossil fuel–based energy resources. Second, it uses feedstocks that are based on fossil fuels, whose life cycle from extraction to product to waste contributes GHGs. The chemical industry, as noted, also is responsible for some adverse impacts on public health and ecosystems from products that are toxic or in other ways hazardous.

In 2020, the chemicals value chain ranked as the third-largest industrial subsector source of GHG emissions behind cement and steel (Aden et al., 2020), and is the largest industrial consumer of energy. This is in large part because the hydrocarbons used as feedstocks are also considered fuels. Unlike when used directly for energy production, much of the carbon in those feedstocks stays locked into the products, and as a result, the short-term CO_2 emissions are lower than would be expected based on the industry's apparent energy consumption. Because many of the chemical industry's products are commodities sold on price, there has been significant prior focus on reducing production costs, including through process energy efficiency. This has resulted in a nearly 25% reduction in nonfeedstock energy use between 1977 and 2014, such that 58% of the apparent energy input is actually in the feedstocks themselves (IEA, 2018; Kätelhön et al., 2019; Levi and Cullen, 2018). Yields in these processes are quite high, with only 15% of the carbon input leaving as waste products. Thus, further reductions in carbon emissions by the chemical industry will require approaches outside of traditional optimization strategies. The World Economic Forum noted that this sector can reduce its carbon emissions by shifting to renewable feedstocks and energy, increasing energy efficiency in its processes, and discovering new innovations in products and processes (Brudermüller, 2020).

A sustainable chemical industry of the future will increasingly base its manufacturing on plant-based biomass, abundantly available CO_2, and end-of-use waste, which will play a crucial role in a circular economy (Clark et al., 2016) (Figure 3-1). It will be powered by renewable energy sources, such as solar and wind energy, and by renewable hydrogen (derived from water splitting using renewable energy sources) to eliminate the carbon footprint of fossil fuels. The widespread availability of plant- and waste-based biomass, solar, and wind energy gives us good reason to think that this grand vision can become a reality. For this to happen, an editorial by Subramaniam et al. (2021a) notes that it is important for the current industry to develop "viable technologies to make chemicals and fuels from these emerging feedstocks, finding processes that minimize resource consumption." Fundamental research in the chemical sciences is critical to meet these challenges in a prudent and holistic manner, balancing economic, social, and environmental considerations. Select research areas in support of these aims are discussed in Section 3.4.

The shift is likely to proceed when existing companies reorient their goals and new entrants that operate in the new paradigm join the chemical economy. Fundamental chemical research will play a central role in operationalizing this shift. Chemical transformations have historically been performed under harsh conditions, for example, using extreme temperatures and pressures and corrosive reagents. Negative impacts of these processes and resulting products on human and ecosystem health were often treated as unforeseeable or unavoidable by decision makers at chemical companies. Green chemistry (Anastas and Warner, 1998), popularized by Paul Anastas and John

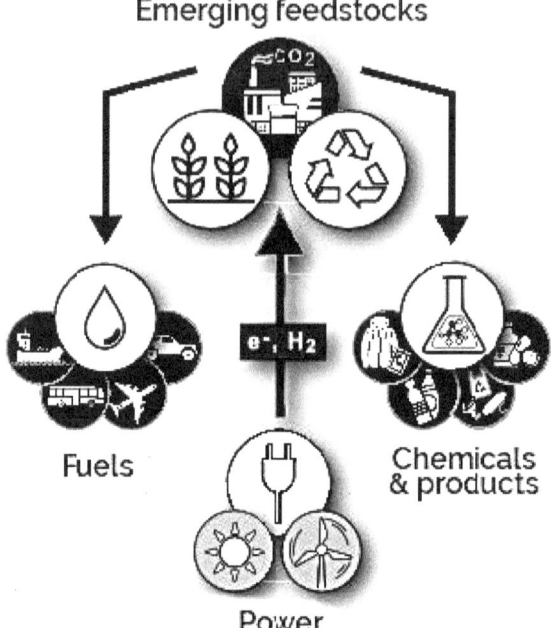

FIGURE 3-1 A sustainable chemical industry where chemicals and fuels are made from renewable and recycled carbon sources powered by renewable energy. SOURCE: Subramaniam et al., 2021a.

Warner, however, is predicated on the idea that these negative consequences are avoidable through targeted development of less toxic and less material- and energy-intensive processes. Advances in chemistry tools, such as circular chemistry modeling and prediction of chemical toxicity, have improved the ability to project the public health and environmental footprint of new innovations, thus reducing the probability of inadvertently introducing these compounds into the environment. Even so, continued monitoring of public health and the environment, using among other tools, chemistry-based instruments, and drawing on chemical knowledge to understand processes is vital to detect the unanticipated impacts of these innovations. While progress is being made in this direction, there are opportunities for basic chemical research to develop more benign substitutes and support the shift to a regenerative paradigm, for instance in agriculture (McBride, 2020).

There is also a need for thoughtful co-design for end-of-life considerations. One article notes that "increasingly bundled under the heading 'Circular Economy,' closed-loop systems keep products, components, and materials at their highest utility and value—reducing the need for extracting and processing new resources, and in the process, reducing related impacts on the environment" (Volans and United Nations Global Compact, n.d.). If these systems are well designed, they can help businesses capture untapped value by collaborating across the value chain and the wider ecosystem. A major advantage of closed-loop business models is that waste streams are more homogeneous, allowing for more effective recycling and reclamation approaches. However, products such as plastics need to be redesigned with both product specifications and effective recycling, particularly chemical recycling, in mind. This co-design process requires investments in basic research to develop new polymers and both the building blocks and processes to achieve them (Britt et al., 2019).

The paradigm shift to sustainability is by no means easy to operationalize, but on the bright side, it has been embraced by segments of the chemical sector. For instance, the Green Chemistry and Commerce Council has led in advocating for the shift to sustainable chemistry. In several cases, the shift to green chemistry has had financial payoffs as well as improved the environmental footprint for specific companies. Additionally, closed-loop business models are being pioneered by corporations who are taking full responsibility for the cradle-to-grave-to-rebirth of their products and packaging. The chemical industry's recognition of its reliance on the social license to operate has contributed to efforts to innovate such programs to reduce its public health and environmental footprint (Finger and Gamper-Rabindran, 2013; Gamper-Rabindran and Finger, 2013).

The need for decarbonization of the chemical economy has found support among a subset of companies. Presenters to the committee discussed their efforts to shift to renewable energy for their electricity and to shift away from fossil fuel–based feedstocks (Maughon, 2021; van Tol, 2021). However, according to Ernst & Young, "less than 40% of U.S.-headquartered chemical companies in the ICIS [Independent Commodity Intelligence Services] Top 100 have published net-zero climate goals or climate goals aligned with the Science Based Targets initiative" (Weick et al., 2021). Even these net-zero pledges may overstate decarbonization commitments, since some studies have shown that net-zero pledges by companies are inconsistent with their actual GHG emissions trajectories (In and Schumacher, 2021).

3.2.1 Benefits of the Move Toward Sustainability

As highlighted in Chapter 2, the chemical economy is global and highly competitive, which complicates technology investments, especially where changes will incur cost increases. We highlight some cases in which the shift to sustainability can provide companies with a competitive advantage and a profitable strategy. Basic chemical research deployed to solve challenging sustainability issues can potentially result in patentable knowledge, giving the patent holder opportunities to license to companies worldwide.

3.2.1.1 U.S. Competitive Advantage in Innovation-Intensive Products

Investing in basic chemistry R&D, plus deployment to bring products to the marketplace, provides one strategy for the United States to secure a competitive advantage in the invention and manufacturing of innovation-intensive products, which meets the growing market demand for environmental protections. For instance, the United States leads in the innovation and manufacturing of hydrofluorocarbon (HFC) substitutes. This has given U.S. chemical manufacturers a competitive advantage as countries, following the Kigali Amendment to the Montreal Protocol, are shifting away from HFCs. HFCs replaced chlorofluorocarbons (CFCs) as refrigerants. While HFCs are not ozone depleters (as CFCs were), HFCs turned out to be powerful GHGs, significantly more potent than CO_2. U.S. leadership in the innovation and manufacture of HFC substitutes was cited by congressional representatives in their support for legislation that directed the United States to sign onto the Kigali Amendment (Gamper-Rabindran, 2022).

3.2.1.2 Green Chemistry: Profits While Reducing Health and Environmental Footprint

Given the domestic and global pressures for sustainability (see Section 3.2.2) and the need for companies worldwide to implement solutions, basic research in these areas could well result in significant payoff. Innovation of green processes and products could make companies competitive and profitable, and expansion of chemical knowledge could lead to companies securing patents for eco-technologies and revenue from their licensing (Hennessey, 1996). For instance, the adoption of green chemistry precepts in the United States spurred the growth of green chemistry–related patents, though information is not readily available on the extent to which these translated into licensing opportunities (Nameroff et al., 2004).

The shift to green chemistry has both increased company profits and reduced their environmental footprint. In 1995, the EPA established the Presidential Green Chemistry Challenge Awards (now known as the Green Chemistry Challenge Awards)[2] to highlight scientific and technical advances in green chemistry. There are many illustrative examples of awardees where green chemistry has reduced inputs, environmental impacts, and in some cases production costs. We highlight a few of them:

- Merck & Co., Inc. dramatically improved on their original synthesis of a chronic cough medicine, gefapixant citrate, and were awarded the 2021 Greener Synthetic Pathways Award.[3] By innovating on at least four steps of the process, they achieved higher yield and a sixfold reduction in raw material costs and were able to replace a step that previously involved hazardous chemicals. By improving the energy efficiency of their process, they generated cost savings and reduced CO_2 and carbon monoxide (CO) emissions.
- Thermal paper (which has been widely used for receipts, tickets, and other labels) can contain bisphenol A (BPA), a potentially toxic chemical. Evidence of BPA transfer from paper to skin and absorption into the body was a motivation for Dow Inc. working with Papierfabrik August Koehler SE to develop an alternative. In 2017, they were awarded the Designing Greener Chemicals Award (EPA, 2017) for producing a thermal paper that eliminated reactive chemistries such as BPA, increased longevity of the receipt image, and was compatible with existing commercial thermal printers. This paper was designed to take advantage of the physical properties of the polymers in the paper rather than use chemical reactions to create the image.
- Life Technologies was awarded the 2013 Greener Synthetic Pathways Award (EPA, 2013) for developing a three-step one-pot synthesis process for the production of chemicals for polymerase chain reactions. This dramatically improved efficiency of conventional techniques, reducing solvent consumption by up to 95% and by 65% for other hazardous waste, and has reduced hazardous waste by 1.5 million pounds per year.
- Geoffrey Coates was awarded the Academic Award (EPA, 2012) as part of the Presidential Green Chemistry Challenge for developing catalysts that convert CO_2 and CO into polymers. This technology was the basis for Novomer Inc., a start-up company whose Converge® polyol technology was acquired by Saudi Aramco in 2016 and valued at $100 million (Aramco, 2016). Aramco is using these catalysts to manufacture coatings that require 50% less petroleum to produce. These coatings have applications in food and drink can linings and at full market penetration were projected to have the potential to sequester and avoid 180 million metric tons of annual CO_2 emissions.

3.2.2 Pressures for Decarbonization and Sustainability

The chemical industry is likely to face continuing pressure to decarbonize and to become sustainable, though policy signals in the United States have not been uniform or consistent. Keeping the average global temperature rise to below 1.5°C is necessary to avoid the worst of the adverse climate impacts. By 2030, global GHG emissions would need to be halved relative to 2005 levels if the world is to meet the aspirational goal under the Paris Climate Agreement to limit global warming to 1.5°C.

---

[2] See https://www.epa.gov/greenchemistry/green-chemistry-challenge-winners.
[3] See https://www.epa.gov/greenchemistry/green-chemistry-challenge-2021-greener-synthetic-pathways-award.

3.2.2.1 Domestic Pressure for Decarbonization

In April 2021, President Biden pledged to halve U.S. GHG emissions relative to 2005 levels by 2030 in the U.S. Nationally Determined Contributions under the Paris Climate Agreement. However, support for climate actions has been mixed among members of Congress and among state governments. Nevertheless, the ever more frequent climate-enhanced extreme weather disasters could compel more voter support for decarbonization policies (Gagliarducci et al., 2019) and prompt action at the federal and state levels as well as increase direct consumer or investor pressure on companies to reduce their carbon footprint (Breckel et al., 2021). A nationally representative survey of 1,000 respondents in September 2021 reported that "94% of liberal Democrats and 80% of moderate/conservative Democrats say global warming should be a high or very high priority for the President and Congress," while 45% of liberal/moderate Republicans and 17% of conservative Republicans say the same (Leiserowitz et al., 2021).

Investor pressure prior to the Biden administration and federal policies in the Biden administration provide some, but not uniform, signals for decarbonization. In 2020, at least 220 of Standard & Poor's 500 companies reported their climate change risks in their U.S. Securities and Exchange Commission (SEC) filings, including Chevron and Gilead Sciences (Ramonas and Rund, 2021). While the SEC has not made such reporting mandatory, investor pressure has led to this voluntary reporting. The European Union (EU) has required climate change risk reporting since 2018 (Ramonas and Rund, 2021).

SEC Chairman Gary Gensler has proposed a new rule amendment on Enhancement and Standardization of Climate-Related Disclosures (SEC, 2022). Disclosures would include

> climate related risks and their actual or likely material impacts on the registrant's business, strategy, and outlook; the registrant's governance of climate-related risks and relevant risk management processes; the registrant's greenhouse gas ("GHG") emissions, which, for accelerated and large accelerated filers and with respect to certain emissions, would be subject to assurance; certain climate-related financial statement metrics and related disclosures in a note to its audited financial statements; and information about climate-related targets and goals, and transition plan, if any. (SEC, 2022)

In October 2021, Federal Reserve Governor Lael Brainard confirmed that the Federal Reserve "is developing scenario analysis tools to model the economic risks of climate change and assess the resilience of the entire financial system" (Newburger, 2021). This could lead the financial sector to reassess its financing of the chemical sector, particularly those that are more prone to risks from direct climate impacts or from risks to existing business models from climate mitigation actions, including regulations to curb GHG emissions (Gamper-Rabindran, 2022).

3.2.2.2 Global Pressure for Decarbonization

The U.S. chemical sector is under international pressure for a lower energy footprint, including from the EU. In July 2021, the EU announced its adoption of the Carbon Border Adjustment Mechanism (CBAM), which will begin to take effect in 2023 (EC Directorate-General for Taxation and Customs Union, 2021). That tax will raise the cost of a small number of U.S. exports to the EU for affected goods that are intensive in GHG emissions (Figure 3-2). Ammonia is currently the only chemical covered by CBAM, and there are concerns that its inclusion will have unintended economic consequences for Europe, including increasing the "downstream production costs of hundreds of chemicals in the ammonia value chain" (CEFIC, 2022b). U.S. producers will still need to consider how energy efficiency for their products compares to that of other producers. Although the EU and the U.S. chemical industries focus on different markets, the EU chemical industry

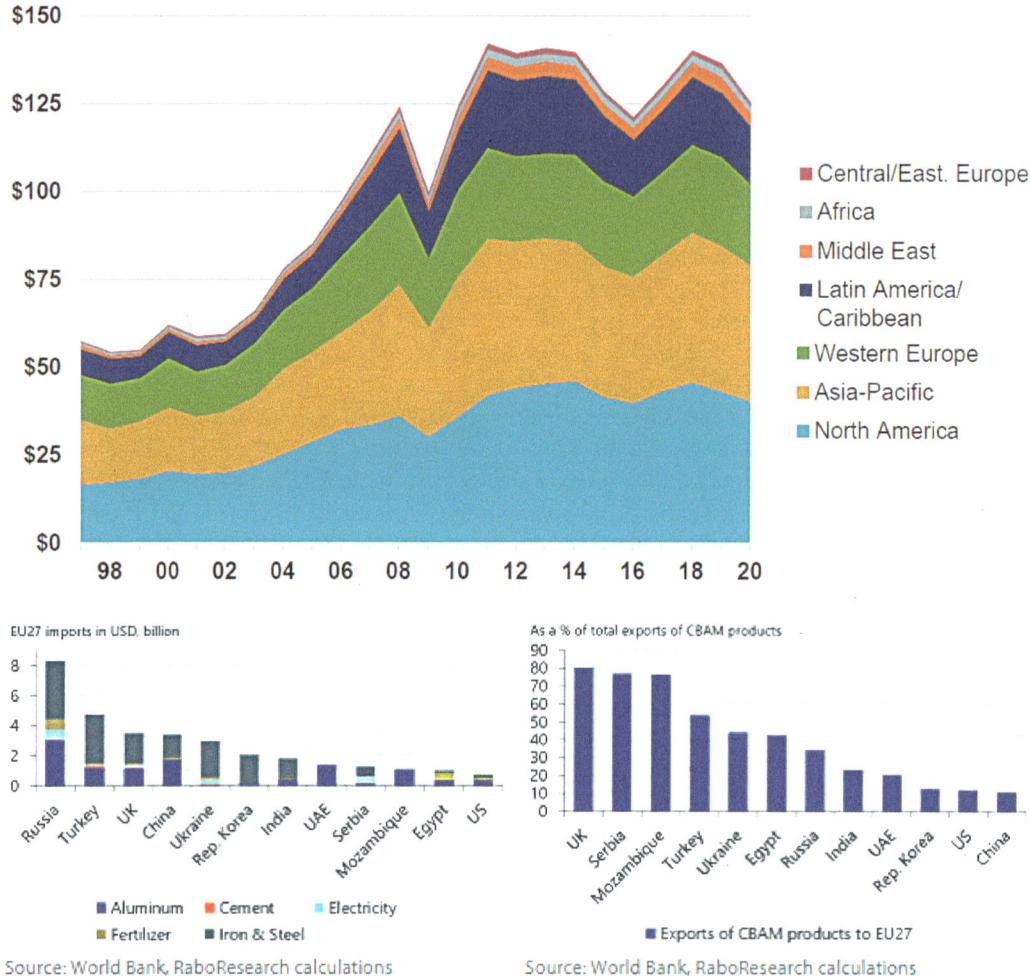

FIGURE 3-2 (Top) Value of U.S. chemical exports by region. (Bottom Left) Value of EU imports by country and industry within the Carbon Border Adjustment Mechanism (CBAM). (Bottom Right) Percentage of exports to the EU covered by CBAM. SOURCES: ACC, 2021; Dumitru et al., 2021.

reports that it made significant strides in reducing its energy intensity by 3% a year between 1991 and 2017 (CEFIC, 2020).[4]

It is difficult to predict the full impact of programs such as the recent CBAM legislation as part of Europe's Fit for 55 agenda. The European chemicals industry is concerned about how CBAM might impact its long-term competitiveness. In 2019, the EU was the largest chemical exporter globally based on value, and increased costs for their products could negatively affect their export position if other regions do not follow through with similar carbon taxes (Eurostat, 2021). There is a recognition that increased carbon taxes could have an adverse effect if not managed appropriately (Condon, 2021). Simultaneously, there is an expectation that the proposed legislation will result in increased funding for innovation that the chemical industry will need to develop different business

---

[4] These figures would capture both energy reduction from a shift out of energy-intensive products and a shift to cleaner energy sources.

models as part of its move toward a circular economy (Beacham, 2021). If similar support for fundamental research in the United States is not forthcoming, it may hinder the speed and effectiveness of the ability of the U.S. chemical industry to decarbonize and compete in an economy that values energy efficiency and reduced carbon emissions.

3.2.2.3 Domestic Pressure for Sustainability

Chemical companies have faced significant legal liability and financial penalty when their products adversely affect public health. Two examples are lawsuits against the manufacturers of glyphosate and PFAS.

Legal action related to glyphosate. Glyphosate, an herbicide developed by the Monsanto Company, was used in the first attempt at weed control and management by applying a herbicide on fields planted with genetically modified glyphosate-tolerant crops. Glyphosate has been registered as a pesticide safe for use since 1974 (EPA, 2020), and John Franz was awarded the National Medal of Technology and Innovation in 1987 for his discovery of the herbicidal properties of glyphosates, because of their significance to agricultural practices globally. However, in 2015, the International Agency for Research on Cancer (IARC) classified glyphosate as being a probable carcinogen to humans, spawning a broad review of its use in agriculture. Bayer, which purchased Monsanto in 2018, assumed ownership of and legal liability for Roundup, the brand name for a glyphosate-based weed killer (Hals, 2021). Although the IARC classified glyphosate as a probable human carcinogen in 2015, an EPA review concluded that there was no risk to human health when used according to the label (EPA, 2020). As of 2021, the product does not carry a warning label of a probable human carcinogen.

Bayer faces ongoing lawsuits from cancer patients who allege that Roundup use had caused their cancer. As of 2021, three cases had gone to trial, with the jury awarding plaintiffs millions of dollars in damages. In May 2021, Bayer announced it was reconsidering its residential sales of Roundup in the United States after a judge dismissed its proposal for responding to class action lawsuits on Roundup as unreasonable (Hals, 2021). Partly as a result of more than 125,000 lawsuits largely related to residential use of Roundup, Bayer announced on July 29, 2021, that it would discontinue offering glyphosate-based products for residential use and would reformulate those products starting in 2023 (Erickson, 2021).

The use of glyphosate and glyphosate-tolerant crops has been debated. One view, within the pesticide-is-essential-to-agriculture community, is that reduction in the use of glyphosate and glyphosate-tolerant crops will cause declines in crop production. For instance, one review of the impacts of removal of glyphosate and glyphosate-tolerant crops estimates losses of $6.76 billion in global farming incomes annually and annual reductions of the output of soybean, corn, and canola crops by 18.6, 3.1, and 1.44 million tons, respectively (Brookes et al., 2017). The review estimates a replacement with 8.2 million kg of other active herbicides, which would have a net environmental impact quotient of 12.4% relative to the use of glyphosate. A competing viewpoint, anchored in the paradigm of regenerative agriculture, points to the treadmill of glyphosate-resistant weeds (González-Torralva et al., 2012; Livingston et al., 2015), risks to pollinator populations (Motta et al., 2018), and adverse impacts on soil health (Soil Association, 2016).

Legal action related to PFAS. PFAS manufacturers have also faced several expensive verdicts and settlements. These include "an U.S.$850 million settlement with a state attorney general in 2018; a U.S.$671 million settlement to resolve 3,550 lawsuits in 2017; and a U.S.$83 million settlement in 2021 to resolve nearly 100 personal injury claims, and a cost-sharing arrangement to address up to U.S.$4 billion in PFAS legacy liabilities" (Birnbaum et al., 2021). Revelations that companies producing these compounds became aware of the health and environmental impacts of these chemicals but failed to immediately notify regulators and end their use (Richter et al., 2021;

UCS, 2019) have tarred the public perception of these companies (U.S. Congress, 2019). Such revelations also tar the social license of the chemical sector more broadly. PFAS contamination, an extreme case with billions in potential liability and high-profile contamination, has resulted in companies' stock market declines. An article in *C&EN* notes that "between January 2018 and September 2020, DuPont's shares fell 45%, 3M's dropped 31%, and Chemours's plummeted 59%. 3M and DuPont, and then the DuPont spin-off Chemours, were major producers of PFAS" (Zainzinger, 2020).

3.2.2.4 Global Pressure for Sustainability

In October 2020, the European Commission announced the EU Chemicals Strategy for Sustainability whose goals are to position "the EU industry as a globally competitive player in the production and use of safe and sustainable chemicals" and to "protect human health and the environment from harmful chemicals." The proposed requirement for the "one substance, one assessment" process aimed to strengthen the principles of "no data, no market." That proposal, which builds on the previous Registration, Evaluation, Authorisation and Restriction of Chemicals (REACH) Directive, will favor exports to the EU that meet higher standards for safety and sustainability (EC, 2020a). To support this change, a 2020 European Commission survey found that 8 in 10 EU citizens continue to worry about the impact of toxic chemicals in everyday products. Nine out of 10 people are concerned about the impact of chemicals on the environment and a slightly smaller proportion (85%) about their impact on health (Chemical Watch, 2020).

Additionally, policies discouraging the use of single-use plastics have ramifications for chemical companies, especially those that produce polyethylene and polypropylene for the consumer packaging sector. The decrease in single-use plastics consumption could also lead to overcapacity and a decrease in returns (Dickson et al., 2022). In 2019, 170 countries supported a UN resolution to reduce plastics use by 2030 (BBC News, 2019). China, Kenya, the United Kingdom, the EU, and several U.S. states (California, Connecticut, Delaware, Hawaii, Maine, New York, Oregon, and Vermont) have moved to ban some types of single-use plastics. In 2020, Canada announced its proposal to regulate "plastics as a toxic substance under the Canadian Environmental Protection Act, and to ban outright the manufacture and import of many single-use plastics by 2021" (ECCC, 2019).

Numerous countries in North America and Europe have long relied on China and a number of developing countries as the destination for their plastics waste. China implemented the National Sword Policy in 2018, which bans the import of most plastic waste material, and the 2019 Amendments to the Basel Convention list plastic waste as conditionally hazardous. Effective January 2021, signatories to the Basel Convention are restricted in what plastic waste shipments they can export to developing countries (Seay et al., 2020). The decision of China to cease accepting such waste and the Basel Convention restriction on waste shipments to developing countries means that there is no longer a cheap offshore destination for waste, and plastic waste will need to be addressed at the national or continental level. This has spurred greater action in these countries to restrict use of single-use plastics.

As a holistic approach to this problem, circular economy schemes such as extended producer responsibility (EPR) (OECD, 2006) are being proposed to shift responsibility around recycling and reuse of plastics from the public sector to the manufacturers, and to incentivize rethinking of product and packaging design. Depending on the EPR policy, this can also raise funds for more effective management of the remaining waste. In the EU, the Packaging Waste Directive (94/62/EC), implemented after limited success with voluntary take-back programs, has resulted in a variety of mandates and mechanisms in which manufacturers pay for collection and recycling of packaging materials, at least in part (Rogoff, 2014). Countries implementing EPR have seen success in material usage reduction and increased recycling (Rubio et al., 2019). More recently,

> the EU Action Plan for a circular economy, adopted in December 2015, identified recycling and reuse of plastics as a key priority. The recently adopted European Strategy for Plastic in a Circular Economy aims to increase significantly the prevention and recycling of plastics including the target to make all plastic packaging on the EU market recyclable by 2030 and to reduce the consumption of single-use plastics. (Leal Filho et al., 2019)

3.2.3 Challenges for Decarbonization and Sustainability of the Chemical Economy

Fully operationalizing the shift to sustainability and decarbonization will require reorientation of investment and production patterns for companies in the chemical sector and their entire supply chain. To make such an expensive and financially risky pivot, the private sector will need assurance that this shift can be profitable and will be applied in a harmonized fashion so they can maintain a long-term competitive position. Policies that are uniform and durable would provide these assurances and support the economywide shift to deep decarbonization and sustainability.

However, in reality, the chemical sector operates in a system in which externalities are pervasive. In most cases, prices do not capture the benefits provided by energy efficient and sustainable products and processes, nor do they reflect the costs of polluting production and processes. Additionally, the chemical sector operates in a system in which policies are not durable, which hampers long-term investments. The impact on the chemical sector of policies enacted by the U.S. government are mixed at best, and the policies or their enforcement are prone to change across administrations.

Although investors are beginning to consider environmental, social, and governance (ESG) factors in investments, narrower financial concerns still dominate (Zainzinger, 2020). ESG metrics are evolving. Thus, the failure to price externalities puts products with superior environmental attributes at a disadvantage as they are more "expensive" in a simplistic sense. There is some evidence though that ESG matters. According to an October 2020 report by the investment firm Jefferies, which surveyed more than 2,100 individual investors in the United States, the UK, Germany, and China,

> In the past decade, shares of chemical companies that ranked highly in ESG indexes from the finance company MSCI, outperformed shares of companies with low rankings by 4.8% per year…The environmental component of ESG has been the dominant driver of outperformance in the United States, while governance is the driver in Europe, which is further ahead in developing an ESG regulatory framework. (Zainzinger, 2020)

However, one piece from *C&EN* notes that "ESG goals can be more of a risk-mitigation strategy than a driver for change" (Zainzinger, 2020). The same piece quoted Ronald Köhler, head of investor relations at Covestro, as saying, "Unfortunately, it is not yet the case that companies get rewarded for being sustainable. Rather, they get punished for not being sustainable." This piece noted that there is only pressure on large chemical companies to adopt sustainability, while small and midsize ones will not implement such practices for years (Zainzinger, 2020).

There are other hurdles on the road to decarbonization and sustainability. The U.S. Government Accountability Office reviewed federal programs in support of green chemistry (GAO, 2018). That study, which interviewed experts from academia, government, and industry, identified several barriers to green chemistry, including the need to prioritize product performance; weigh sustainability tradeoffs between various technologies; risk disruptions to the supply chain when switching to a more sustainable option; address limited and expensive supplier options; consider regulatory challenges; develop a business case for sustainability investments; and address the often higher initial cost of more sustainable options. (GAO, 2018)

For companies faced with the transition, their choice is in the continuum from resisting to greenwashing to making the transition (Green et al., 2021).

3.3 POLICIES TO ASSIST IN ADOPTION OF SUSTAINABILITY AND DECARBONIZATION

Presenters to the committee from industry emphasized the need for well-designed market-based and regulatory policies that favor the more sustainable products and processes. These policies create the certainty that enables companies to take a long-term investment approach into sustainability and decarbonization. Presenters' views echo the Porter hypothesis in economics literature that "well-designed and stringent environmental regulation can stimulate innovations, which in turn increase the productivity of firms or the product value for end users" (van Leeuwen and Mohnen, 2017).

Companies that sell products directly to consumers have been able to implement business models that tie their brand firmly to sustainability and to profit from implementing that brand, for example, Adidas implementing EPR to promote recycling (Adidas, 2021). Some chemical companies have also adopted similar programs, such as Braskem's "I'm green" product line. A paper by Iles and Martin (2013) describes how the company was successful in developing standards and then using carbon measurements to demonstrate that their bioplastic polyethylene reduced environmental impact as compared to other polyethylenes.

3.3.1 Regulatory Policies

Several of the previous examples illustrate how regulatory policies that restricted or banned a process or product enabled a more benign substitute to enter and compete in the marketplace. Nevertheless, political dynamics have made it difficult to promulgate and implement regulatory policies that penalize products with harmful health and environmental consequences. Innovators favor policies that reward sustainability and decarbonization, but the established companies that produce the less sustainable and more carbon-intensive products fight to keep their market share. Incumbents have been successful at opposing policies that would restrict the use or raise the price of their products or processes.

These dynamics continue to the present day. For instance, the Union of Concerned Scientists argue that Dow Chemical, producer of chlorpyrifos, lobbied hard for the EPA under the Trump administration to reverse the EPA's 2015 decision supporting the ban on chlorpyrifos (UCS, 2017), though Corteva Agriscience, formerly Dow AgroSciences, agreed to phase out production of chlorpyrifos in 2020 after facing pressure from California and the EU (Erickson, 2020). Wagner and Steinzor (2006) document the challenges that regulatory agencies face in presenting the science to justify stricter regulations in the face of industry's strategies in the legal and scientific arenas. Wagner and Steinzor (2006) call for robust government investment in research that advances understanding of how chemical products and processes affect health and the environment. That research is fundamental to building the body of scientific evidence to enable the promulgation of regulations, which in turn can incentivize a shift to more sustainable products and processes.

3.3.2 Market-Based Policies on Carbon

Well-designed market-based policies that put a price on carbon make it less profitable for companies to use fossil fuel–intensive processes or fossil fuel feedstocks, and thus incentivize the shift to decarbonization (CBO, 2013). Carbon taxes set at an appropriate level to capture the climate costs of each ton of GHG emissions would provide such a signal. That market signal can spur companies to invest in decarbonizing their products and processes. However, the political environment at the federal level has prevented the implementation of carbon pricing policies within the United States.

3.3.3 Demand-Pull Procurement and Purchasing

The opposition to regulatory and market approaches that raise the prices or restrict the use of carbon-intensive and less sustainable products and processes means that demand-pull strategies such as procurement can offer more viable solutions. U.S. government procurement programs at the federal, state, and local levels have the potential to provide the demand-pull for energy-efficient and sustainable products (Krupnick, 2020). A number of state and local programs exist, but two federal programs are considered here: the U.S. Department of Agriculture's (USDA's) BioPreferred Program and the EPA's Safer Choice Program. The BioPreferred Program requires federal agencies and their contractors to purchase biobased products when they purchase items such as paints, lubricants, adhesives, and cleaners, among others.[5] The program saw a 93% increase between 2017 and 2019 in the number of products that the U.S. Department of Agriculture certified as BioPreferred (Golden et al., 2021). EPA's Safer Choice Program labels products that are safer for the environment, to guide consumers and businesses toward making more sustainable choices.[6] Programs like these can provide the market demand for products that are perceived as more expensive but are more energy efficient and sustainable than conventional products.

Previous studies have shown that there is a growing interest in green chemistry products—"chemical products and processes that reduce or eliminate the use or generation of hazardous substances" (Golden et al., 2021). That report, prepared for the Green Chemistry & Commerce Council, came to several conclusions about the purchasing of green chemistry products, including that "emerging government policies and investor expectations are fueling growth of the green chemistry sector." Additionally, the report showed that green chemistry products had a market growth that was more than 12 times faster than that of their conventional counterparts between 2015 and 2019 (Figure 3-3) (Golden et al., 2021). Other studies present some evidence that government procurement policies favoring products with environmental attributes correlated with innovation (Porter, 1991). Advanced market commitments have provided the demand-pull.

3.3.4 Funding R&D&D

Congress has recognized the essential role of chemical R&D—and additionally, the need for public–private sector collaboration—to support the paradigm shift toward products and processes with lower public health and environmental footprints. This is why the focus has expanded to include R&D&D, with the last "D" standing for "deployment." For example, the Save Our Seas 2.0 Act (Pub. L. No. 116-224) was passed in 2020 to address concerns about plastic pollution in coastal and marine ecosystems. This law mandates a variety of studies and reports on the impacts of microplastics and other plastic pollution, novel uses of plastic waste, and investigation of circular economy approaches and needs for material and waste management.

---

[5] See https://www.biopreferred.gov/BioPreferred/.

[6] See https://www.epa.gov/saferchoice.

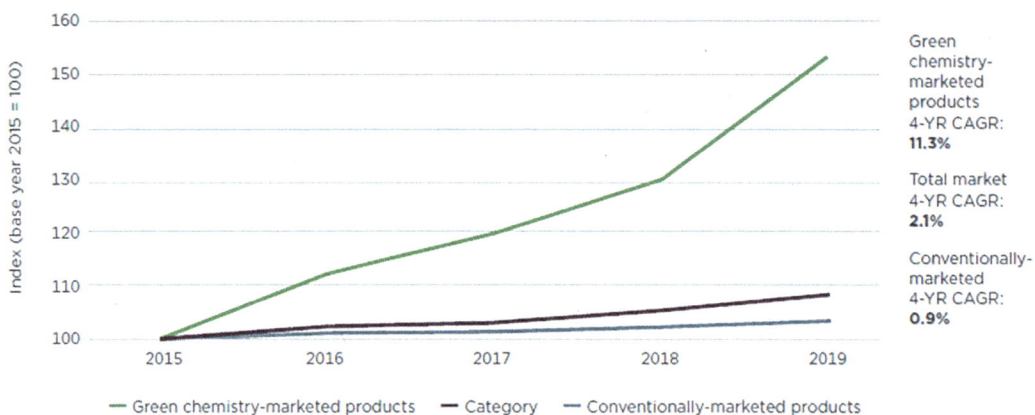

FIGURE 3-3 Growth of green chemistry market products from 2015 to 2019. SOURCE: Golden et al., 2021.

In January 2021, Congress enacted the Sustainable Chemistry Research & Development Act as part of the FY 2021 National Defense Authorization Act (Bergeson & Campbell, P.C., 2021). The Act states that "sustainable chemistry can improve the efficiency with which natural resources are used to meet human needs for chemical products while avoiding environmental harm, reduce, or eliminate the emissions of and exposures to hazardous substances, minimize the use of resources, and benefit the economy, people, and the environment." The Sustainable Chemistry R&D Act is meant to coordinate interagency efforts to accelerate U.S. innovation in this emerging area of market growth (Jalbert, 2021). The law directs the White House Office of Science and Technology Policy to "convene a multi-agency task force that will, for the first time, coordinate federal funding and promotion of sustainable (i.e., green) chemistry research" (Hogue, 2021). The interagency working group will develop a consensus definition of "sustainable chemistry" and "a framework of attributes characterizing, and metrics for assessing, sustainable chemistry" (Bergeson & Campbell, P.C., 2021).

More recently, the Infrastructure Investment and Jobs Act (Pub. L. No. 117-58, H.R. 3684, 117th Cong.) passed in 2021 directs the U.S. Department of Energy (DOE) to establish near-, mid-, and long-term targets for the hydrogen R&D&D program. In section 813, it directs the Secretary of Energy to establish at least four regional clean hydrogen hubs to aid the demonstration of production, processing, delivery, storage, and end uses of clean hydrogen. To accomplish this, $8 billion is appropriated for these activities for fiscal years 2022–2026. Regarding clean hydrogen in manufacturing, priority is given to manufacturing projects that increase efficiency and cost-effectiveness. Additionally, the act directs DOE to issue awards "for research, development, and demonstration projects to advance new clean hydrogen production, processing, delivery, storage, and use equipment manufacturing technologies and techniques" (117th Cong. H.R. 3684, 2022). For this work, $500 million is appropriated for fiscal years 2022–2026.

The act also directs the establishment of a clean hydrogen electrolysis program focused on reducing the cost of electrolytically produced hydrogen to less than $2/kg by 2026. The program will also carry out demonstrations, and seek to scale up technology including the integration of compression, drying, storage, and transportation systems. For these activities, $1 billion is appropriated for fiscal years 2022–2026.

3.4 FUNDAMENTAL CHEMICAL RESEARCH FOR SUSTAINABILITY, DECARBONIZATION, AND ENVIRONMENTAL STEWARDSHIP

The pressing need to attain the United Nations's SDGs provides numerous opportunities for chemical R&D to advance our knowledge and innovate solutions in support of decarbonization and sustainable development in the broadest sense. Table 3-2 lists future-looking chemical topics that are discussed in the remainder of this chapter and the SDGs on which these knowledge advances may have an impact.

3.4.1 Sustainability Assessments

Global environmental and safety regulations for new and existing chemicals and materials are aimed at encouraging development of technologies that are inherently safer and sustainable

TABLE 3-2 Opportunities for Fundamental Chemical Research and How They Contribute to the Sustainable Development Goals

Opportunity	Description	Relevant SDGs (SDG #)
Life-cycle assessments (LCAs)	Accurate, reliable, and transparent LCAs underpin most approaches for sustainability and decarbonization and still require harmonization and widespread deployment for effective use.	Industry, Innovation, and Infrastructure (9); Responsible Consumption and Production (12)
Chemical recycling of plastics	Innovations in chemical plastic recycling are necessary for increasing circularity and co-designing polymers and products accounting for end of life and recyclability.	Industry, Innovation, and Infrastructure (9); Responsible Consumption and Production (12)
Sustainable synthesis	Novel catalysts can support sustainability. LCA considerations and the availability and demand of critical materials should guide development.	Industry, Innovation, and Infrastructure (9); Responsible Consumption and Production (12)
Sustainable feedstocks and energy sources	Reducing emissions and dependence on fossil fuels requires alternative feedstocks and energy sources, such as biomass, CO_2, and hydrogen.	Affordable and Clean Energy (7); Sustainable Cities and Communities (11); Climate Action (13)
Carbon capture, utilization, and storage (CCUS)	CCUS must be scaled up considerably to have an impact on climate change mitigation and provides a potential feedstock for materials and fuels of the future.	Affordable and Clean Energy (7); Climate Action (13)
Monitoring and improving air quality	Catalytic converters have dramatically improved air quality. Innovations in chemistry, including measurement and automation, will be required to reduce emissions and pollutants of the future.	Good Health and Well-Being (3); Sustainable Cities and Communities (11); Climate Action (13)
Water monitoring and safety	There are opportunities for chemistry and measurement to improve distributed water safety as well as to detect and remove contaminants not previously identified.	Good Health and Well-Being (3); Clean Water and Sanitation (6); Sustainable Cities and Communities (11); Life Below Water (14)
Food safety	More detailed understanding and measurement of interactions of food with possible preservatives and packaging will support food safety and reduce waste.	Zero Hunger (2); Good Health and Well-Being (3); Responsible Consumption and Production (12); Climate Action (13); Life on Land (15)

(EC, 2020b; U.S. Congress, 2019). A U.S. Government Accountability Office report (GAO, 2018) concludes that quantitative sustainability assessment is vital for timely implementation of sustainable chemical technologies. The 2019 United Nations Environmental Programme report (UNEP, 2019) concurs with this sentiment. Life-cycle assessment (LCA) tools are increasingly available to quantify the ecological impacts of a chemical product during its complete life cycle, which includes manufacturing, use, and disposal ("cradle-to-grave") or recycling ("cradle-to-cradle") phases (Figure 3-4). The environmental burdens during the manufacturing phase stem from not only unit operations within the chemical plant but also the extraction of the raw materials that often occurs far away from the chemical plant.

LCA is essential to provide rational guidance for implementing circular chemistry, which refers to a framework that promotes sustainable practices throughout the life cycle of a chemical product. Circular chemistry is characterized by near-total carbon atom economy and minimal adverse impacts on the environment and human health (Keijer et al., 2019). To perform quantitative sustainability assessment, metrics such as atom economy, process mass intensity, and the environmental factor are available to determine if feedstock resources are being efficiently used to make the desired products (Allen and Shonnard, 2001). Several software tools that incorporate such metrics are available in the open domain for estimating environmental impacts of chemical products and processes. For example, the American Chemical Society's Green Chemistry Institute has made available several tools to guide the selection of green reagents, including the Green Chemistry Innovation Scorecard Calculator.[7] The EPA has introduced online tools to estimate the effects of chemical processes and products on the environment and human health. These include the environmental fate, bioaccumulation, and toxicity of chemicals.[8] New tools continue to evolve, incorporating predictive toxicology, product biodegradability, and even social dimensions as part of a comprehensive sustainability assessment (Zimmerman et al., 2020).

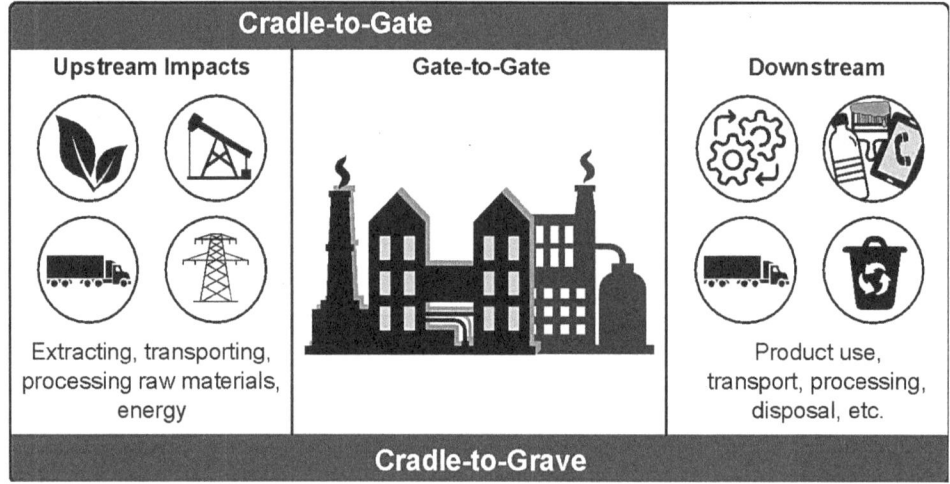

FIGURE 3-4 Scope of a life-cycle assessment study including activities upstream and downstream of the manufacturing process. SOURCE: Subramaniam et al., 2021b.

---

[7] See https://www.acs.org/content/acs/en/greenchemistry/research-innovation/tools-for-green-chemistry.html.
[8] EPI Suite™-Estimation Program Interface, https://www.epa.gov/tsca-screening-tools/epi-suitetm-estimation-program-interface.

To compare the environmental impacts of conventional processes and alternative chemistries or technologies, LCA tools such as SimaPro,[9] GaBi,[10] Materials Flows through Industry (MFI),[11] and GREET[12] are available. The Economic Input-Output Life Cycle Assessment (EIO-LCA),[13] developed at Carnegie Mellon University, estimates the material and energy consumption of a chemical product, along with its associated environmental impacts in various economic activity sectors across the supply chain. The Department of Energy's National Energy Technology Laboratory also provides a suite of reports and tools on LCA of energy technologies and pathways.[14] When LCA methods are used early in development, and throughout a process life cycle, the assessments can be useful in guiding research and public policy decisions (Bento and Klotz, 2014; Sharp and Miller, 2016). Incorporating sustainability metrics to guide research and process development is essential to achieve sustainable processes and products in a timely manner. To be useful in comparing tradeoffs between alternatives, LCA relies wholly on reliable life-cycle inventory metrics for the processes or products modeled and a rigorous process model that accurately accounts for inputs and outputs, both in terms of mass or material flows and energy balances. Advances in fundamental chemistry will continue to inform LCA processes, and vice versa, through better analytical tools and a deeper understanding of chemical matter.

Another emerging aspect of sustainability assessments is the use of systems-level thinking to take a holistic view of what impact a chemical or process might have on the environment or health. Systems thinking is generally defined as "the ability to understand and interpret complex systems" (Evagorou et al., 2009) and includes "visualizing the interconnections and relationships between the parts of the system; examining behavior that changes over time; and examining how systems-level phenomena emerge from interactions between the system's parts" (Orgill et al., 2019). In relation to sustainability, the idea is to incorporate more information into standard LCAs, sometimes termed a life-cycle and sustainability assessment. LCAs are already a systems-based analysis approach, and the incorporation of broader systems thinking encourages other considerations, including "revealing macro-level impacts, consideration of social and economic impacts, and taking into account underlying mechanisms" (Onat et al., 2017). This approach is particularly important in chemistry due to the impact that chemistry and chemical methods have on many other areas of science and manufacturing. As shown in this chapter and Chapter 2, many different aspects of society, policy, economics, and the environment impact, and are impacted by, changes in the chemical enterprise. Current supply-chain concerns, for example, touch all facets of society. Continued research on the interconnectedness of chemicals, the environment, and society remains critical, and advancing basic understanding of individual chemicals and methodologies will increase our ability to understand the whole enterprise.

3.4.2 Resource-Efficient Chemistry

Feedstocks that come from greener sources and allow for more efficient reactions will be vital to the green chemistry landscape. New fundamental research will be needed to explore the use of circular feedstocks and for altering reaction conditions to be more energy efficient.

---

[9] Pre Consultants, SimaPro Life Cycle Assessment Software, www.pre-sustainability.com/simapro.
[10] GaBi Life Cycle Assessment Software, www.gabi-software.com/america/index/.
[11] See https://mfitool.nrel.gov.
[12] Argonne GREET Model, http://greet.es.anl.gov/.
[13] Carnegie Mellon University Green Design Institute's Economic Input-Output Life Cycle Assessment, www.eiolca.net/.
[14] See https://netl.doe.gov/LCA.

3.4.2.1 Plastics and Polymer Production and Recycling

Plastics production has grown to more than 100 lbs of plastic per person every year,[15] and plastics are needed by nearly every manufacturing industry (Figure 3-5). Global production of polyolefins (polyethylene and polypropylene) is greater than any other plastic because they are derived from an inexpensive feedstock such as natural gas. Additionally, they have extraordinarily useful properties: low density, resistance to degradation, resistance to swelling from both water and common hydrocarbons (grease, oil, cleaning solvents), and easily molded or blown into packaging film. Beyond polyolefins, the other plastics with high annual production numbers include polyesters, polyamides, and polyurethanes.

Plastics were discovered in the early 1900s when the first synthetic material, Bakelite, was developed serendipitously by Leo Baekeland. Plastic quickly became known as the material of "1,000 uses" and was manufactured on a mass scale for everything from telephones to jewelry. Today, polymeric materials can be found in almost everything that we touch and see, from food packaging to single-use disposable medical equipment to excipients in pills. Plastics are an important and well-integrated part of the chemical economy. However, the very properties that make plastics so useful have caused severe stress on the environment: microplastics fill our oceans, sachets clog waterways in Africa leading to flooding and disease, and microscopic fibers from clothes can be carried by the wind and found in soil samples across the world. Estimates for the naturally occurring breakdown of plastic in the environment is 1,000 or more years. There exists a need, therefore, to collect waste plastics prior to environmental dispersion and to develop waste separation techniques and waste treatment plants to recover the valuable chemicals and embodied energy contained within the covalent bonds of the materials.

During the transition to a circular plastics economy and for dealing with recalcitrant plastics, biodegradable plastics have been proffered as a part of the solution to uncollected plastic waste.

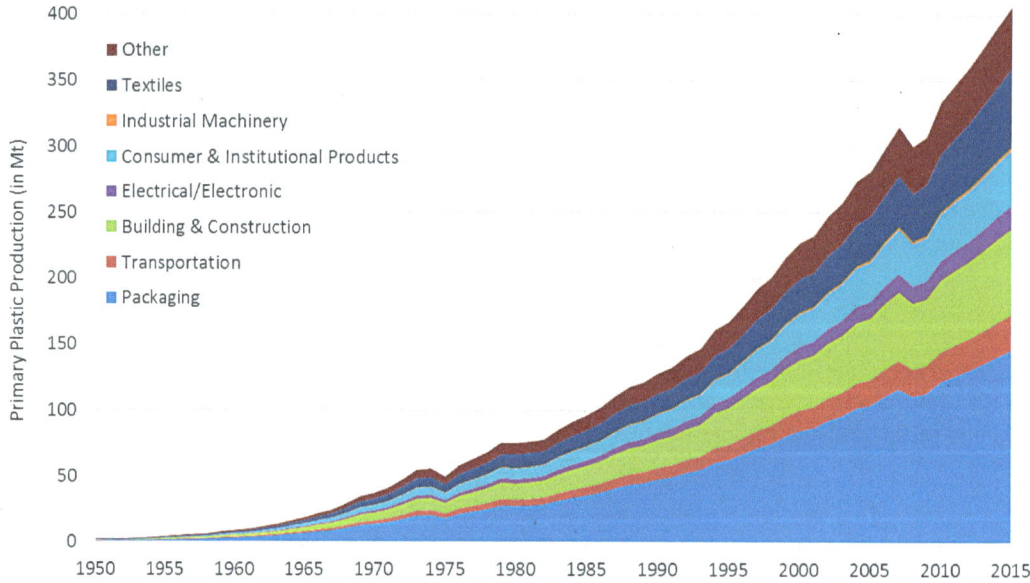

FIGURE 3-5 Global primary plastics production (million metric tons) according to industrial use sector from 1950 to 2015. SOURCE: Geyer et al., 2017.

---

[15] Calculated from 7.753 billion worldwide population in 2020 and 368 million metric tons of plastics produced.

However, currently available biodegradable plastics have limited uses due to their often-brittle characteristics (Narancic et al., 2018). They may be suitable for some packaging applications, in single-use items, and in some agricultural uses, being mindful, though, of potentially hazardous degradation by-products (Dilkes-Hoffman et al., 2019; Flury and Narayan, 2021). Importantly, if the feedstocks for biodegradable plastics are not renewable, biodegradable plastics can be a vehicle for higher carbon emissions than the alternative in the short term (Zheng and Suh, 2019). Emissions can arise from both their production and accelerated degradation. Successful biodegradation often requires industrial composting and therefore extra energy, and degradation intrinsically yields carbon emissions. There is active development of biodegradable plastics, but research is needed to support improved materials properties, optimize degradation conditions, and understand the environmental and life-cycle implications and tradeoffs (Flury and Narayan, 2021). Despite the potential value of biodegradable plastics, reducing the volume of uncollected plastic waste and thereby decreasing the loss of material and energy remains the preferable approach.

Unlike paper, discussed in Box 3-1, plastics are a structurally diverse group of polymers with different chemical properties. For example, polyolefins are long-chain molecules containing only carbon and hydrogen atoms. Therefore, they have few "chemical handles" to use to chemically break down the polymer via common chemical reactions into monomers that can be used as building blocks for new polyolefins. Polyesters, polyamides, and polyurethanes, on the other hand, because they contain functional groups, are easier to degrade chemically and convert to small molecules and starting materials. Economically viable recycling processes are being developed for polyamides (Suntinger, 2020), polyurethanes (Laird, 2021), and especially for the polyester used commonly in bottled-water containers (Smalley, 2021; Tudball, 2022; Tullo, 2021). Through advances in fundamental and applied research in catalysis, it has been demonstrated that even polyolefins can be pyrolyzed back to oils, which can be refined and processed in steam reactors to make building-block olefins for polyolefins in a circular fashion (Gebre et al., 2021). Major plastic makers, ExxonMobil, Dow, SABIC, Eastman, and Chevron Phillips Chemical, have all established efforts to bring capacity for chemical recycling of sorted plastic waste streams. The largest example of this to date is ExxonMobil's, which is scheduled to come online by the end of 2022. This recycling plant in Texas expects to process 30,000 metric tons of plastic waste per year back into chemical feedstocks for the generation of polyolefins (Tullo, 2021). This is a positive development that was only possible through R&D investments in catalysis and chemical process intensification. This is only the beginning, as this recycling represents less than 0.01% of the plastics produced globally each year. Additionally, it is important to explore and research other chemical recycling methods beyond pyrolysis and to use process and life-cycle analyses to determine which methods are providing the most benefit.

3.4.2.2 Sustainable Synthesis

For the chemical industry, sustainability means safer chemistry, sustainable products, and circular chemistry with carbon efficiency. Sustainable chemical synthesis is one key part of a sustainable chemical economy. Improving reaction rates and selectivities inherently decreases the resources used per reaction and promotes sustainability. Principles of green chemistry and green engineering provide guidelines for choosing reaction components, such as reagents, solvents, catalysts, and operating conditions that minimize adverse health and environmental effects.

Catalytic processes often involve the use of other substances such as water, organic solvents, acids, and bases. Product and catalyst separation steps are typically solvent and energy intensive. The use of certain harmful solvents, such as chloroform and benzene, is banned by international

BOX 3-1
Plastic Recycling and Lessons Learned from Cardboard Recycling

Single-use plastics make up an increasing and unsustainable percentage of the waste disposal stream. To address this problem, there have been massive new investments in research encouraging novel chemical approaches; however, chemical solutions in isolation are inadequate to fully correct this problem. The municipal waste stream collection process—in which a wide range of materials are mixed at the point of collection, many of which are not recyclable or are otherwise contaminated—is also part of the reason society has a major polymer waste stream proliferation issue.

Our nation's current approach to consumer product waste does not hold responsible the companies that make and sell products. Those companies rely on community trash collection for disposal and recycling of their products and packaging. In terms of reuse and recycling, this waste management approach has had limited success, since nationwide less than 10% of plastic waste is recycled (EPA, 2021). If all corporations that produced a certain threshold of polymeric products were responsible for reclaiming those products and recycling them, it would fundamentally change business models, practices, and material design to support closed-loop and circular economy approaches. Some companies have started to integrate circular economy principles into their business model and practices. Adidas designed a shoe from a single recyclable material made without glue and promotes a take-back program, where returned shoes are ground up and used to make a new generation of the same shoe (Adidas, 2019). Eastman Chemical Company has developed a yarn made from cellulose esters that is produced in a near-closed loop process.[a]

Absent thoughtful co-design, plastic recycling is a complex process. The large number of different types of polymers and their use as mixtures in many consumer products make it difficult to reuse them. For example, a bag for potato chips can be a composite material made of multiple different thin polymer layers, often including a metal layer. The use of plastics for food packaging also increases the level of organic contamination of single-use plastics in the waste stream, thus requiring cleaning. Thus, plastics recycling requires multiple steps, each with different energetic and logistical costs: the plastic waste is first collected, then sorted, often by hand, and processed before a pure feedstock is available to be reused.

Beyond the inefficient collection step, which relies on end consumers and municipalities, the process is imperfect. Separating composite materials that consist of many layers can be difficult, if not impossible; only certain types of plastics (e.g., PET [polyethylene terephthalate], HDPE [high-density polyethylene], LDPE [low-density polyethylene]) are good candidates for the melt-and-remold process that most recycled plastics currently endure; and large differences in the chemical composition of polymeric materials require different treatments unique to the polymer type. Thermoset plastics (e.g., polyester, silicone, and epoxy), as opposed to the more meltable thermoplastics mentioned above, are not easily recycled and pose a significant end-of-life challenge of their own. Therefore, generalized recycling treatments cannot be applied to mixed waste.

In contrast to plastic waste, paper products are collected at a much higher rate and around 60% of consumed paper is recycled. The recovery rate of cardboard boxes is particularly high because there are concentrated sources for collection, for example, at stores and warehouses, and the need for cardboard boxes is growing with increased online shopping. Compared with plastic recycling, paper recycling is easy (it is largely mixed with water and turned into pulp), economically favorable (there are markets at each step of the cycle), composed of mature and comparably uniform processes, and tolerant of contaminants. Paper recycling can be improved, though, by reducing energy demand through a better understanding of the vaporization of water and a better understanding of hydrogen bonding. Additionally, wax coatings and other additives have been used for increasing water resistance in order to replace plastic products but in doing so reduce recyclability. Research into alternative methods of making paper water-resistant would reduce this tradeoff.

The comparative successes in paper recycling can be a lesson for both the chemistry community and those responsible for collection and processing of municipal waste. Given the variation and complexity of plastic use, it is unlikely that plastics recycling will achieve the level of cardboard recycling. However, more thoughtful co-design that limits the use of composite materials and makes more use of plastics that are good candidates for melt and remold processes can allow plastics recycling to take advantage of some of the same factors that allow for higher levels of cardboard recycling.

---

[a] See https://naia.eastman.com/.

regulations such as Europe's REACH. Thus, the choice of greener solvents and greener acids and bases is essential to sustainable synthesis (Byrne et al., 2016; Henderson et al., 2015).[16]

The production of catalysts itself may impose environmental damage. For example, catalyst production associated with biofuels can be >10% of the cradle-to-grave GHG emissions for the entire process (Benavides et al., 2017). Homogeneous metal catalyst complexes are often synthesized via multistep procedures that use organic solvents and mineral acids. The production of TiO_2 (used in photocatalysis and as catalyst support) from ilmenite ($FeTiO_3$) ore is energy intensive, requiring temperatures of 1,500–2,000 K. It also requires corrosive reagents, such as chlorine, and generates large amounts of waste (Gázquez et al., 2014). In all of these cases, there is a need for research to find benign manufacturing processes with reduced environmental and health impacts.

The increased demand for elements that are widely used as catalytic metals is also raising sustainability concerns. For example, palladium (Pd) catalysis is enormously important in the synthesis of pharmaceutical precursors due to its versatility, selectivity, and robustness in a number of C–C and C–X bond-forming reactions. A widely held notion is that the continued use of such platinum group metal (PGM) catalysts is not sustainable (Hayler et al., 2019). However, this view may not take into account the shift away from internal combustion engines (ICEs). As transportation becomes electrified, there will be less need for catalytic converters, which rely on PGMs to function. In 2020, the automotive catalyst industry accounted for approximately 90%, 85%, and 32% of the worldwide Pd, rhodium (Rh), and platinum consumption, respectively (Cowley, 2021). The decline in ICE demand related to the shift toward electric vehicles will result in a commensurate drop in the demand for PGMs in automotive emission control catalysis. A steadily increasing inventory in Pd and Rh accompanied by declining prices is forecast, especially after 2030 (Hochreiter, 2021). Recognition of this impending trend might encourage renewed exploration of new areas of chemistry for PGMs, especially Pd and Rh.

There is significant work on exploring Earth-abundant transition metal catalysts (iron and nickel) as cost-effective and more sustainable replacements for PGMs (Plett and Bernales, 2016). Such alternative formulations often require much higher catalyst loadings and expensive ligands. Higher metal loadings increase the possibility of residual metals contamination that is unacceptable for the end-product use, and may require cleaning procedures that are environmentally deleterious (EMA, 2019). Furthermore, nickel is likely to be in short supply in the future as a result of the twofold increase in its demand to build secondary battery technologies (IEA, 2021b). One possible way to address this is to further consider chemical capabilities for recovering and reusing critical metals from batteries. Some work is being done to include this as a solution (Sensiba, 2021), but continued chemical research is needed to increase the efficiency of recovery and bring these technologies to scale (Petranikova et al., 2022). Clearly, evolving demand trends, where cost and availability can be dependent on emergent technologies and international policy (Blas, 2022; Rosevear, 2022), must be carefully considered to rationally guide the quest for new catalytic materials. These considerations would benefit from using comprehensive process and life-cycle analyses to help chemists and chemical engineers focus on materials and processes that will have the least adverse environmental impact.

3.4.3 Greenhouse Gas Reduction Strategies

GHG emissions—primarily CO_2—are largely the result of fossil fuel combustion. Fundamental chemical research can play critical roles in helping to establish renewable energy sources, electrification of engines, and improved methods of carbon capture, utilization, and storage (CCUS).

---

[16] See also ACS's Solvent Selection Tool, https://www.acs.org/content/acs/en/greenchemistry/research-innovation/tools-for-green-chemistry/solvent-selection-tool.html.

3.4.3.1 Renewable Feedstocks for Energy

The chemical industry currently derives its raw materials and energy primarily from fossil sources. To make the transition to sustainably manufacture its myriad products, the chemical industry has at its disposal sources such as plant-derived biomass, sequestered CO_2, and end-of-use waste as alternative feedstocks to promote a circular economy. The chemical industry is already moving to renewable power such as solar and wind energy as well as renewable hydrogen to reduce its carbon footprint. However, major challenges remain in developing new chemistries and viable technologies for converting these alternative feedstocks to petrochemical-equivalent chemicals and fuels.

Biomass processing involves biological material originating from either terrestrial or marine sources including cellulose, hemicellulose, lignin, chitin, lipids, and polysaccharides. Global energy demand is expected to grow from approximately 600 quadrillion BTU (quads) per year in 2019 to 900 quads per year by 2050 (EIA, 2019). Currently, only a small fraction of energy is derived from biomass resources (EIA, 2020). However, energy and materials derived from biomass could increase significantly in the future. Projections indicate that up to a billion tons per year of crop residues, herbaceous energy crops, and woody crops and wastes are available in just the United States (DOE, 2016). Algae-based biomass could augment this feedstock, especially if cultivated with the help of concentrated CO_2 sources, such as those emitted from ethanol plants, coal-fired power plants, and natural gas–fired power plants.

Biomass resources are both physically and chemically different from common fossil fuel–based feedstocks such as crude oil and natural gas. Crude oil consists primarily of hydrocarbons ranging from light molecules to heavier ones. In petroleum refineries, the crude oil is first cracked to yield many fractions which are separated by distillation into various gas and liquid streams for further processing. In gas-processing plants, the various components of natural gas liquids are cryogenically separated. In sharp contrast, future biorefineries will rely on different processes, many yet to be developed, to convert solid and liquid biomass forms into useful chemicals. For example, fossil sources often require the addition of oxygen atoms to produce valuable feedstocks. This is done by selective oxidation of C–H and C–C bonds. In contrast, biomass is already oxygenated and hence is suitable for making chemical precursors without the addition of oxygen. However, to make hydrocarbon fuels, biomass often requires the addition of hydrogen to selectively remove the oxygen atoms. Mapping a transition for the existing chemical economy to evolve into a sustainable one is discussed in Sections 3.1–3.3.

The subject of biomass as a feedstock to produce chemical feedstocks has received increased R&D attention since the release of DOE's report that identified the top value-added chemicals that can potentially be made from various fractions of biomass (Holladay et al., 2007). Several reviews have tracked progress in the development of new chemical pathways for converting biomass-based feedstocks into either petrochemical equivalents or new products with superior functional properties (Bender et al., 2018; Biddy et al., 2016; Chheda et al., 2007; Corma et al., 2007; Deuss et al., 2014; Fiorentino et al., 2017; Huber et al., 2005; Jing et al., 2020; Kunkes et al., 2008; Schutyser et al., 2018; Shanks and Keeling, 2017; Sun et al., 2018; Takkellapati et al., 2018).

As shown in Figure 3-6, carbon-rich biological material can be "refracted" into a variety of products through appropriate processing steps. For achieving circularity, the end-of-use products must be either recycled or reprocessed to make new products or returned to the biological cycles, such as soil nutrient recycling. Biomass feedstocks often require pretreatment to fractionate components such as cellulose, hemicellulose, lignin, chitin, and lipids. These fractions are then chemically converted into precursors for making fuels, chemicals, and materials. Implementation of practically viable biorefineries requires the development of (1) robust and reliable feedstock supply chains and (2) cascade processes that valorize the rich diversity of fractionated biomolecules into a diverse

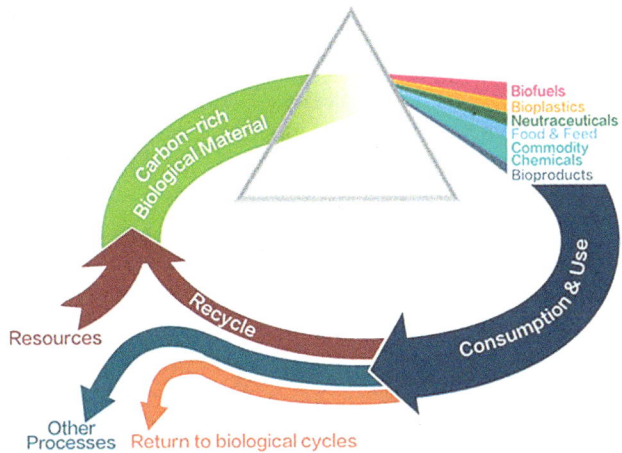

FIGURE 3-6 Model of a circular biomass-based biorefinery. SOURCE: Engelberth et al., 2021.

product portfolio, from specialty bioactive compounds and multifunctional materials to platform chemicals and energy. To implement such a "no carbon left behind" vision (Figure 3-6), integrated green chemistry and engineering approaches are needed that valorize all fractionated biomass components through use of renewable energy, regenerable green solvents, and catalysts (Clarke et al., 2018).

Moving from fossil-derived carbon to renewable carbon as a feedstock for the industry will be a challenge. Because of the predicted volume needed, it will not be possible to supply the industry solely with carbon based on biomass and recycled waste streams. The shift to biomass will need to be sensitive to competition between land for food crops and for industrial feedstocks. One scenario modeled by the Nova Institute suggests that 25% of the feedstock base for the chemical industry will be based on CO_2 (Kähler et al., 2021). The report recognizes that the use of renewable and recycled feedstock does not guarantee sustainability. Environmental and health impacts that may result during their processing must be properly accounted for and mitigated. Estimating the carbon footprint is commonly practiced today (Ubando et al., 2020), and a similar analysis is needed for feedstocks such as plastic waste and CO_2 to assess the sustainability of these processes (see Section 3.4.1 for more details).

In addition to relying on biomass, plastic waste, and CO_2 feedstocks, the future chemical industry will also depend on a different network of processes and infrastructure than the existing ones. Even partially replacing traditional fossil fuel feedstocks with renewable sources and recycled carbon will require advances in catalysts, reactors, separations, and process intensification. The transition to new feedstocks will require new approaches and new chemistries, which will come from investments in fundamental chemical research. According to a recent report from the National Research Council,

> Industrial catalysis will continue to require chemical engineers to take the work of chemists, and increasingly biologists, and run it efficiently on a larger scale. Improvements in chemical and biological catalytic selectivity and activity under varying conditions has the potential to reduce the energy intensity of the CPI [Chemical Process Industry], in addition to other aspects of achieving sustainability. (NRC, 2006a)

Other emerging technologies, such as artificial intelligence and computational models, will also be essential in accelerating the pace of discovery (see Chapter 4 for more details).

3.4.3.2 Hydrogen

Hydrogen is gaining in importance as a CO_2-free energy system and therefore is a key substance for decarbonizing the energy and chemical sectors. On a weight basis, hydrogen (net calorific value of 33.3 kWh/kg) carries almost three times as much energy as gasoline. It can be burned either directly (producing water as a product) or converted into electricity in fuel cells (Bockris, 2013). The chemical industry alone is predicted to produce (and consume) more than 60 million tons of hydrogen each year (Holladay et al., 2009). Hydrogen is currently mainly produced from coal, oil, and natural gas. All of these processes produce CO_2 as a by-product. For example, 1 ton of hydrogen produced by methane reforming generates more than 10 tons of CO_2 (Dufour et al., 2009). CO_2-free hydrogen can be produced from electrolysis ("green" hydrogen) and stored for use as needed, thereby circumventing the intermittent nature of the outputs of photovoltaic plants or wind farms that are dependent on weather and the time of day.

However, hydrogen gas is highly explosive. For storage and transportation, it must be either cooled below –250°C to store it as a liquid or kept under very high pressure as a gas because of its low density. Both options require complex and costly infrastructure. Liquid organic hydrogen carrier technology offers a safe method for storing and transporting hydrogen (Preuster et al., 2017). Ammonia, made economically from green hydrogen, is also a promising candidate as a liquid hydrogen carrier (Wan et al., 2021). Ammonia can be readily decomposed to liberate hydrogen for use in fuel cells. It also has the added advantage of globally mature transportation and storage networks.

Hydrogen as a reagent will be especially important for processing emerging feedstocks such as CO_2 and biomass. The hydrogen produced by steam reforming of methane is termed "gray" hydrogen, which produces CO_2 as emissions. If carbon capture and storage (CCS) technologies are deployed to reduce these emissions, the hydrogen thus produced is termed "blue" hydrogen. However, a recent report suggested that the life-cycle GHG emissions of blue hydrogen could be quite high when the release of fugitive methane is taken into account. More detailed evaluations demonstrate the complexities in making these assessments and support the hypothesis that blue hydrogen can indeed be produced in a low-carbon fashion (Bauer et al., 2021). Other works have also highlighted the challenges associated with carbon accounting when considering methane and demonstrate the need for ongoing research in LCA and carbon accounting methodologies (Allen et al., 2021; Roman-White et al., 2021; Rosselot et al., 2021). In addition to the environmental impact, the financial costs of moving to hydrogen-based systems are unclear. It is widely acknowledged that the cost of hydrogen is falling, but the rate at which the price is falling, especially for greener forms of hydrogen, is unclear (DiChristopher, 2021). Some estimates already put fossil-based hydrogen below $2/kg, a cost goal set by recent U.S. legislation (see Section 3.3.4), but it is unclear how long it will take for greener forms of hydrogen to get to a similar cost (DiChristopher, 2021). Given the importance of hydrogen to the energy transition and decarbonization, other sources to obtain CO_2-free hydrogen, such as methane pyrolysis and light alkane dehydrogenation, will also continue to garner attention.

3.4.3.3 Carbon Capture Utilization and Storage

As the world seeks ways to decarbonize energy systems, a key technology for decreasing CO_2 emissions in the atmosphere will be CCUS. As described in the recent National Academies of Sciences, Engineering, and Medicine's report, *New Directions for Chemical Engineering*, "Stopping the growth in concentration of atmospheric CO_2 does not require zero anthropogenic CO_2 emissions, but rather that anthropogenic CO_2 emissions are balanced by natural and anthropogenic CO_2 sinks, to achieve net-zero emissions" (NASEM, 2022a). CCUS is essential to achieve this goal because it has the potential to greatly reduce CO_2 emissions from hard-to-abate industries. The

International Energy Agency projects that CCS could mitigate up to 15% of global emissions by 2040 (IEA, 2020). In another recent National Academies' report, *Accelerating Decarbonization of the U.S. Energy System* (NASEM, 2021b), CCUS is highlighted in the report's recommendations, noting that the Departments of Transportation and Energy should plan and construct an interstate CO_2 transportation system to move captured CO_2 from sources to sites of storage and utilization. Figure 3-7 illustrates the flow of CO_2 in the CCUS schema. CCUS elements are briefly described below.

Carbon capture and storage. CCS is a set of technologies that separate or "scrub" CO_2 from smokestack emissions or capture CO_2 from industrial process emissions, compress and transport the CO_2, and then pump it into geologic formations, typically 800–1,500 meters below ground or under the ocean. CCS technologies have been field tested and refined since 1972, but then and in many subsequent cases, the captured CO_2 has been used in a tertiary technology (enhanced oil recovery [EOR]) for extracting oil and gas from underground reserves. Although the captured CO_2 *is* trapped underground, the extracted oil and gas ultimately generate CO_2 emissions through refining and use, and so this scenario is not carbon neutral. However, CO_2 storage facilities are being developed and brought online at an accelerated rate, including some that are not used for EOR. The Global CCS Institute has 135 facilities (operational and under construction) in its database as of 2021, and the U.S. National Energy Technology Laboratory notes that more than 200 CO_2 capture and storage operations are in place worldwide.[17] The International Energy Agency lists 21 CCS facilities operating globally that have the capacity to capture 40 million tons of CO_2 annually (IEA, 2020). To achieve CCS at the gigaton scale necessary to aid in keeping global temperature rise below 1.5°C, hundreds of facilities, large and small, will need to be built globally. This will require large investments and extensive collaborations among chemists, chemical engineers, geochemists, geologists, policy makers, civic leaders, and politicians from local to federal. In a recent announcement, 14 petrochemical companies operating in Houston, Texas agreed to collaborate on the development of

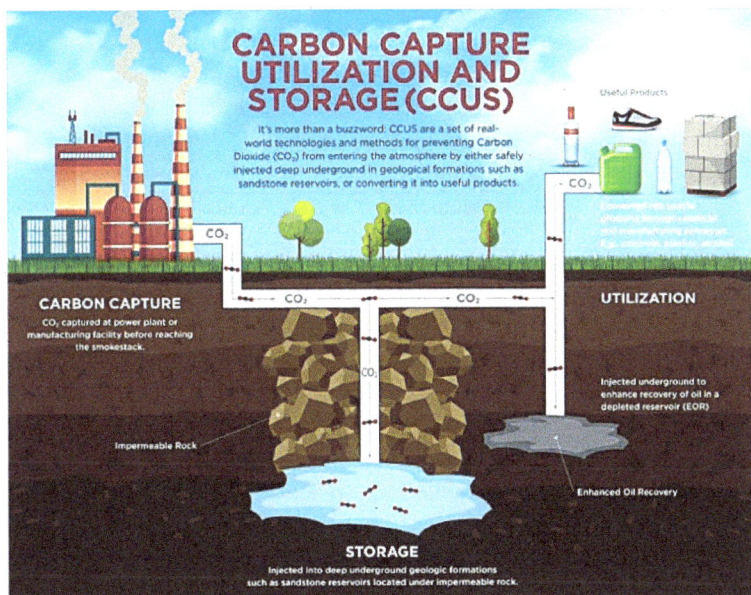

FIGURE 3-7 Flow of CO_2 through carbon capture, utilization, and storage. SOURCE: Mah, 2021.

---

[17] See https://www.netl.doe.gov/coal/carbon-storage/faqs/carbon-storage-faqs.

a very large CCS effort that, at full scale, could capture and store 100 million metric tons per year (ExxonMobil, 2022). In addition to an investment of ~$100 billion (Davis, 2021), the project will need policy and regulatory incentives and substantial scientific innovation.

Fundamental chemistry research can play an important role in CCS, especially in the area of CO_2 capture. Capturing CO_2 is most cost-effective at point sources where the concentration of CO_2 in a given waste stream is high, such as at large carbon-based energy facilities and industries with major CO_2 emissions. For example, the waste stream from ammonia manufacture is >98% CO_2 and that from cement, steel/iron, and glass production contains 20–35% CO_2 (NASEM, 2019c). The dominant CO_2 capture technology is absorption, or carbon scrubbing, with amines in solution, a process patented in 1930 to separate CO_2 from natural gas (Bottoms, 1930). Alkanolamine solutions are used most often, due to their high absorption capacity, low price, and favorable kinetics, among other reasons. However, they degrade in the presence of heat, CO_2, and O_2. In addition, as CCS expands to a wide variety of CO_2 emission sources, each with different concentrations of gas and varying amounts of contaminants, finding more efficient absorbents, including new generations of liquid and solid amines, is highly desirable. For example, metal-organic frameworks functionalized with tetraamines have been developed recently for use in conditions such as natural gas–fired power plant emissions, in which CO_2 concentration is only ~4% and is contaminated with high percentages of O_2 and H_2O (Kim et al., 2020). Other new materials being tested for CO_2 adsorption include water-lean solvents (Heldebrant et al., 2017), metal oxides (Wang et al., 2011), encapsulated liquid sorbents (Vericella et al., 2015), and porous solid adsorbents such as activated carbons, zeolites, metal-organic frameworks, and porous organic polymers (Creamer and Gao, 2016; Furukawa et al., 2013; Siegelman et al., 2021; Tian and Zhu, 2020).

Carbon utilization options. In addition to sequestering captured CO_2, there are opportunities for captured CO_2 to be used in industrial process streams, either directly as CO_2 or after chemically transforming it to other molecules. The National Academies' report *Gaseous Carbon Waste Streams Utilization: Status and Research Needs* (NASEM, 2019c) divides CO_2 utilization into three pathways: mineral carbonation, chemical conversion to produce fuels and chemicals, and biological conversion into fuels and chemicals; these are aligned with DOE's Carbon Utilization Program (Figure 3-8). The 2016 CO_2 Sciences report uses similar categories when it identifies high-value markets for captured-CO_2 utilization. These include the cement, concrete, and aggregate industries, and the formation of commodity chemicals, specialty chemicals, polyols, polycarbonates, and fuels. While these markets have begun to be commercialized, there is room for significant growth that is dependent on, among other things, fundamental research in catalysis, reaction optimization, and separation technologies.

Using captured CO_2 in building materials. Cement production accounts for 8% of global annual CO_2 emissions, releasing up to 0.95 ton of CO_2 per ton of cement produced (Bellona, 2015; Plaza et al., 2020). Approximately 65% of these emissions are process emissions (Figure 3-9), formed when limestone is thermally decomposed to calcium oxide and CO_2. The cement, concrete, and aggregate industries could noticeably reduce global CO_2 emissions, because these industries are significant emitters of CO_2 and are large global industries that are expected to continue to grow as development and population demands increase, and because the conversion of CO_2 into mineral carbonates is a thermodynamically favorable reaction. However, to achieve a net reduction in CO_2 emissions, more fundamental research is needed in LCA within these industries (Ravikumar et al., 2021). In Box 3-2, several companies are profiled that are using captured CO_2 and renewable energy to directly reduce or offset CO_2 emissions during the formation of cement, and via carbonation during the production of concrete and aggregate.

Converting CO_2 to chemicals and fuels. After fossil fuels are combusted and release their energy, they produce lower-energy, fully oxidized carbon species in the form of CO_2. When considering CO_2 utilization pathways (Figure 3-10), it is possible to select applications where the

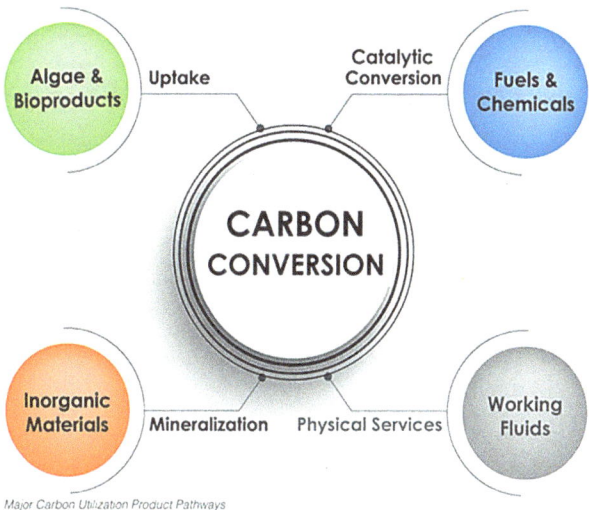

FIGURE 3-8 U.S. Department of Energy CO_2 utilization pathways. SOURCE: https://netl.doe.gov/coal/carbon-utilization.

carbon can be utilized in its fully oxidized form. These pathways are less energy intensive, and as a result, nonreductive CO_2 utilization pathways tend to be further advanced commercially than those where the CO_2 needs to be reduced. The previous examples of using captured CO_2 in the cement and concrete industries are examples of nonreductive CO_2 utilization pathways (Grim et al., 2020).

Utilizing CO_2 to form organic molecules requires an input of energy, sometimes substantial amounts. As shown in Figure 3-11, many of the target product molecules are higher in energy and their carbon(s) is less oxidized than the carbon in CO_2; thus, converting CO_2 to many chemicals and fuels requires reducing its carbon oxidation state. CO_2 reduction can be done via addition of hydrogen and/or electrons through processes such as thermocatalytic, biological, or electrochemical transformations. Ultimately, for captured CO_2 to be the feedstock for the formation of chemicals and fuels without incurring a net emissions cost, the energy used to reduce the carbon in CO_2 has to come from zero-carbon or low-carbon sources, such as wind, solar, nuclear, or biomass. Likewise, when H_2 is used for carbon reduction, it needs to be derived from a low-carbon process or through water electrolysis.

For thermocatalytic transformations of CO_2, chemistries to create CO (Daza and Kuhn, 2016), methanol (Bowker, 2019), and methane (Frontera et al., 2017) are well established. Conversions of CO_2 to either methanol or CO are quite attractive, as they are useful intermediates to allow further upgrading to a range of products through well-known and optimized processes. Methanol can be used directly to form olefins, aromatics, and oxygenated products; used in fuel cells; or put into the fuel pool (Olah, 2005), making it a very flexible molecule. Future supply chain strategies built around methanol, known as "the Methanol Economy," have long been proposed (Figure 3-12) (Olah, 2005). Likewise, conversion of CO_2 to CO makes it amenable to the entire range of syngas-based chemistries including Fischer-Trøpsch processes to make olefins, hydrocarbons, and fuels (Panzone et al., 2020). However, reducing CO_2 to methane requires shifting the carbon oxidation state from +4 to –4, which is a much more energy-intensive process than making methanol or CO from CO_2. Because the resultant methane is an energetic molecule, there are proposals to use the conversion of CO_2 to methane as a means to store excess renewable energy utilizing the current natural gas infrastructure (Schaaf et al., 2014).

> **BOX 3-2**
> **Carbon Capture and Utilization in the Cement and Concrete Industries**
>
> Other than water, concrete is the world's most used material (GCCA, 2020). In 2020, ~34 billion metric tons of concrete were produced from 4.2 billion metric tons of cement (GCCA, 2020). The scale of these two industries and the amount of CO_2 emissions they generate mean that mineral carbonation in concrete and aggregate products and captured-CO_2 replacements during cement production have the potential to reduce CO_2 emissions by 1 to 5 gigatons annually by 2030, while creating annual revenues of $150 billion to $550 billion ($CO_2$ Sciences, 2016). Below are some examples of start-up companies that are having success using CCUS in the cement, concrete, and aggregates industries.
>
> **CarbonCure** injects CO_2, captured from smokestacks and industrial waste, into the mixture as concrete is being formed. As the pressure from the tank to the injection nozzle drops, the liquid CO_2 is released into the mixture as solid and gaseous forms and reacts with calcium ions in cement to form calcium carbonate, which embeds in the concrete. The process reduces CO_2 emissions in multiple ways. The captured CO_2 that is used to make concrete is never emitted into the atmosphere and is now stored as $CaCO_3$. The CO_2-curing process strengthens the concrete and thus permits less cement to be used per batch of concrete, which means fewer emissions at the cement plant.
>
> **Solidia Technologies** produces a proprietary alternative to Portland cement in kilns at lower temperatures than the 1,450°C used conventionally, which lowers GHG emissions by ~30% compared with Portland cement production. The reformulated system in combination with CO_2 incorporation during curing makes it possible to reduce CO_2 emissions by 70% compared with the production and use of Portland cement to make concrete.
>
> **Carbon8 Systems** optimized a chemical process that upcycles industrial waste and captured CO_2, converting it into carbonated aggregate, which can be bound in concrete or used as a substrate for infrastructure, such as pipe bedding or road and airport runway beds.

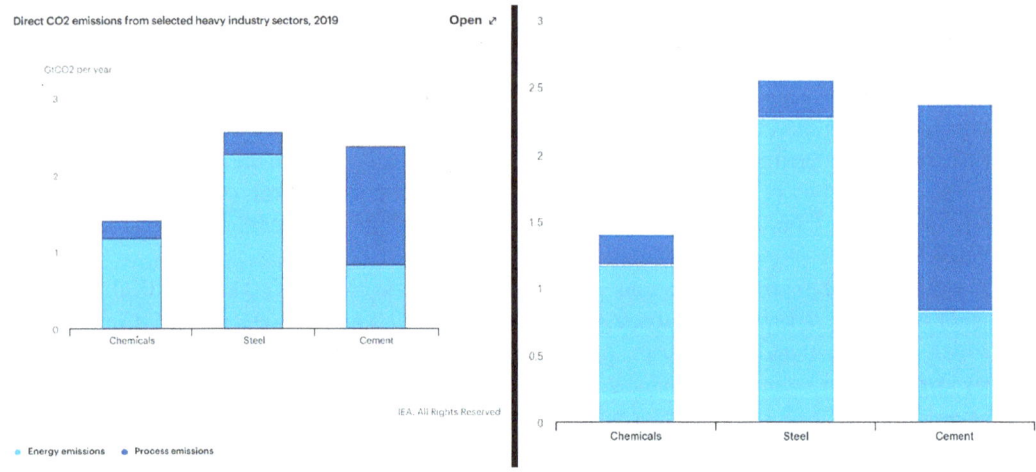

FIGURE 3-9 Direct CO_2 emissions from selected heavy-industry sectors (chemicals, steel, and cement) in 2019 from energy use and industrial process. SOURCE: IEA, 2019.

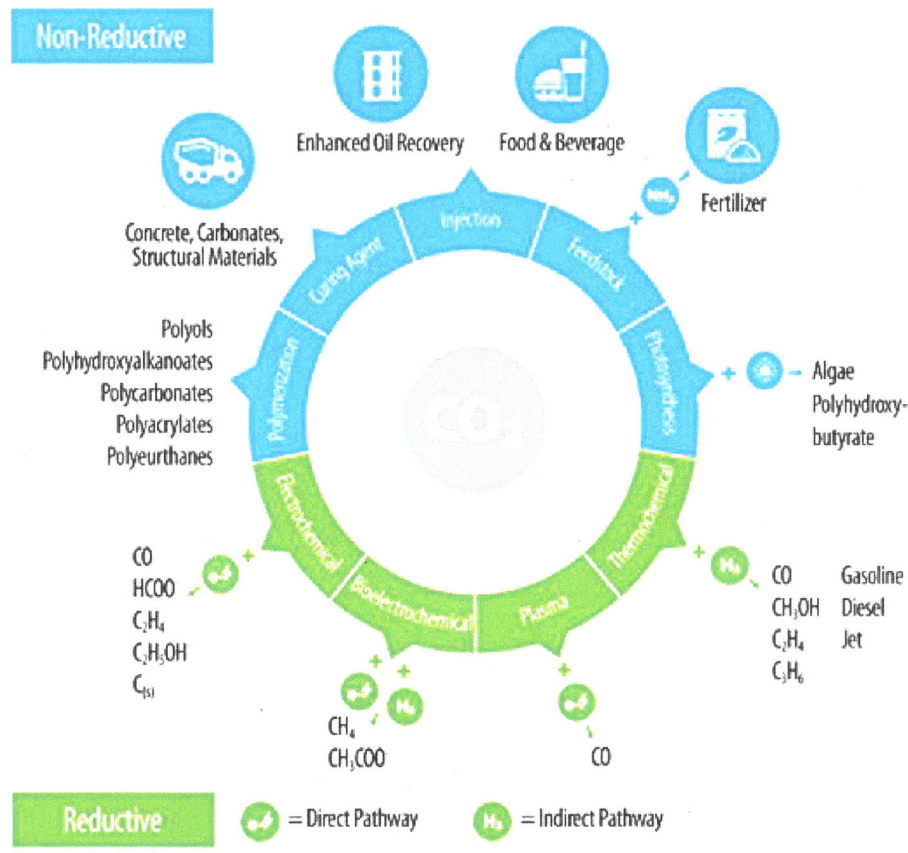

FIGURE 3-10 Reductive and non-reductive pathways for CO_2 utilization shown with major products.
SOURCE: Grim et al., 2020.

Conversion of CO_2 to methane may have less utility than conversion to other chemicals, as there are currently fewer large-scale chemical processes based on methane as a feedstock outside of its reoxidation to syngas or its use for making hydrogen cyanide. However, researchers are exploring methane as a source of carbon for structured carbon materials such as graphite, carbon nanotubes, nanofibers, and nano-onion structures through catalyzed methane decomposition reactions. In these reactions, a low-carbon H_2 is also produced, adding to the utility of the approach. Unlike steam methane reforming, the carbon is not converted to CO_2, and therefore the H_2 that is produced is low carbon, even when made from conventional methane sources (Qian et al., 2020). CO_2-derived methane could be used for these syntheses, but will be more expensive and require more energy overall. There are also new routes for producing carbon nanostructures, utilizing molten salt electrochemical systems where the CO_2 capture, carbon reduction, and carbon nanostructure synthesis steps are combined. Many experimental factors can be used to tailor the final structures, and further research is needed before moving to the development phase (Yu et al., 2020).

While many of the thermocatalytic processes are somewhat mature, further research in their development is still needed. For example, hydrogenation of CO_2 to methanol generates a lot of water, which can lead to catalyst instability. Creating more hydrothermally stable methanol conversion catalysts remains an active area of research (Guil-López et al., 2019).

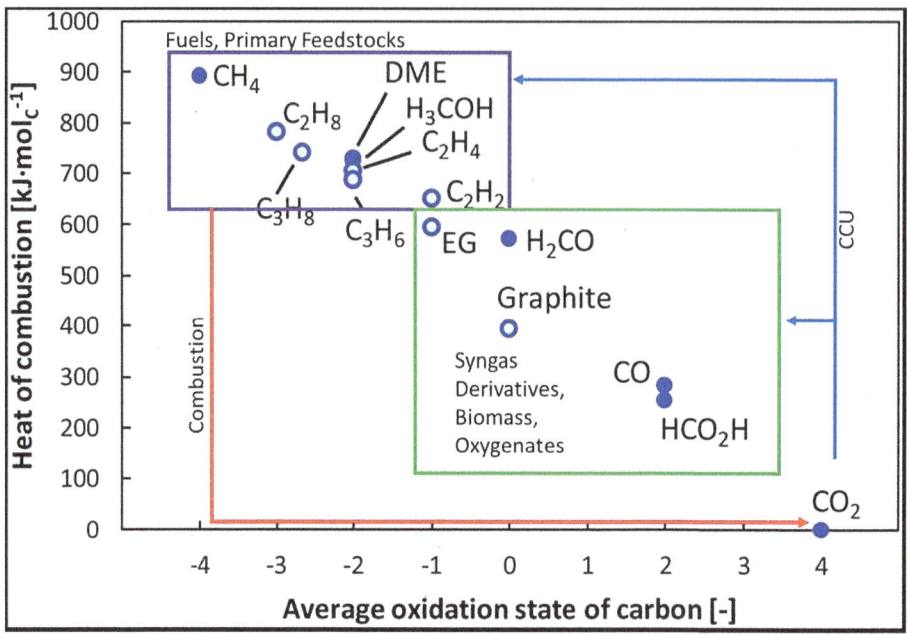

FIGURE 3-11 Heat of combustion versus oxidation state. SOURCES: Bender et al., 2018; Tomkins and Müller, 2019.

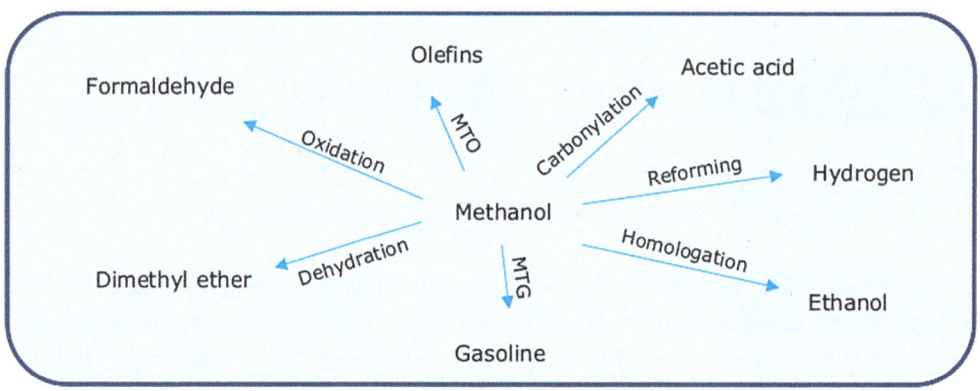

FIGURE 3-12 Methanol conversions. SOURCE: Alcasabas et al., 2021.

Biological processes to convert CO_2 are also showing significant progress. In addition to conversions of CO_2 via direct photosynthetic processes to generate biomass and/or crops, which can then be transformed into chemicals and fuels, efforts have been directed toward microbial syntheses using photosynthetic cyanobacteria or nonphotosynthetic chemolithotrophs. These microbial systems can be engineered to give a desired product slate (García-Granados et al., 2019), and a wide range of chemicals have been demonstrated (NASEM, 2019c). Currently, these gas fermentation routes are dominated by anaerobic processes. For example, LanzaTech has commercialized a 46-kiloton/year ethanol process with its proprietary acetogen, using steel mill waste gases as the carbon source. The LanzaTech technology utilizes a proprietary anaerobic acetogen, which is derived from a bacteria originally found in cat droppings. These bacteria produce ethanol from

CO-rich industrial gases (Daniell et al., 2012). The LanzaTech process can also make a variety of other chemical products. Through a combination of techniques including directed evolution and artificial intelligence, LanzaTech has demonstrated the possibility to form more than 50 different chemicals from industrial waste streams (Kobayashi-Solomon, 2021a,b,c). Most recently, researchers from LanzaTech, Northwestern University, and the Department of Energy's Oak Ridge National Laboratory reported newly developed acetogen strains capable of producing acetone and isopropanol from industrial waste gases (Liew et al., 2022). The acetone and isopropanol processes are now ready for commercialization (Kobayashi-Solomon, 2021b).

Another approach to CO_2 conversion that has made significant progress is electrochemical conversion. There are two ways of classifying electrochemical processes involving CO_2: indirect and direct. For indirect processes, H_2 is generated via water electrolysis and then used to reduce CO_2 to useful products through a variety of chemical processes. When this methodology is used for generating fuels, the fuels are sometimes referred to as e-fuels because the H_2 is electrochemically produced even if the actual generation of the final product is via a non-electrochemical route. An oft-cited e-fuel example is Carbon Recycling International's CO_2-to-methanol George Olah Facility (Burkart et al., 2019; NASEM, 2019c). Creation of hydrogen via water electrolysis is a commercial process and is the subject of ongoing research (Brauns and Turek, 2020; Shiva Kumar and Himabindu, 2019).

In direct electrochemical processes, CO_2 is directly reduced to yield products such as formic acid, CO, methane, ethane, and ethanol (Appel et al., 2013; Centi and Perathoner, 2009). In many cases, the direct route combines two steps into one where the hydrogen is generated in situ to generate the final product. These routes in turn can be broadly separated into two categories, low-temperature solvent processes and higher-temperature gaseous processes (Bonheure et al., 2021). The lower-temperature processes can utilize many of the same principles used in water electrolysis cell design, and several companies are commercializing technologies using these approaches. High-temperature electrolysis utilizes solid oxide electrolyzer cells (SOECs) to generate CO and syngas. The design of these systems is closely related to solid oxide fuel cell systems, which has aided in their development (Küngas, 2020). Although high-temperature electrolysis systems cannot directly synthesize other hydrocarbons and oxygenates from CO_2, they are being commercialized for the production of CO (Alcasabas et al., 2021). The efficiency and lifetime of SOECs will be improved through materials development and better understanding of nanoscale processes occurring in SOECs (Hauch et al., 2020).

While these commercial successes are encouraging, there is much more R&D to be done. A 2019 report from the National Academies recommends that "the U.S. government and the private sector should jointly implement a multifaceted, multiscale research agenda to create and improve technologies for waste gas utilization" (NASEM, 2019c). The report goes on to say, "Specifically the U.S. government and the private sector should support

- Research and development in carbon utilization technologies to develop pathways for making valuable products and to remove technical barriers to waste stream utilization;
- The development of new life-cycle assessment and technoeconomic tools and benchmark assessments that will enable consistent and transparent evaluation of carbon utilization technologies; and
- The development of enabling technologies and resources such as low- or zero-carbon hydrogen and electricity generation technologies to advance the development of carbon utilization technologies with a net life-cycle reduction of greenhouse gas emissions." (NASEM, 2019c).

In addition to the sequestration of CO_2 in deep geological formations and its reuse into products, there are other strategies for capturing CO_2 in natural carbon sinks. These include afforestation and reforestation and sequestration on agricultural land and in coastal and pelagic ocean waters. These strategies are thoroughly presented in three recent reports (NASEM, 2019a,e, 2021a) and ongoing work from the National Academies.[18]

3.4.4 Chemical Research to Improve Quality of Life

3.4.4.1 Air Quality

In 2013, a detailed characterization study of trapped air in Greenland snowpack reported a surprising conclusion. Mapping of the samples showed that atmospheric CO levels in the Arctic steadily increased from the 1950s through the 1970s and then began dropping, ultimately resulting in levels lower than what were found for samples corresponding to the 1950s. Prevailing models did not predict there to be a reduction of atmospheric CO levels of this nature. The results were even more unexpected because these reductions were occurring simultaneously with increasing fossil fuel use over that same period. The authors credited improvements in emission control technologies and the introduction of catalytic converters in the United States and Europe for lowering CO levels (Petrenko et al., 2013). This trend mirrors measurements reported for CO by the EPA.[19] Atmospheric levels for NO_2, SO_2, and particulates also showed significant declines since the 1980s and were all below established air standards (Figure 3-13). While a number of factors are undoubtedly responsible for the improved air quality, the introduction of catalytic convertor technologies plays a major role in these improvements (Farrauto et al., 2019).

Although emission control advances have significantly improved air quality, according to the WHO, air pollution caused an estimated 7 million deaths in 2016, with Africa and Southeast Asia being the most impacted by poor air quality (WHO, 2016). COVID-19 lockdowns provided direct evidence of the impact of many modern technologies on air quality (F. Liu et al., 2021) and highlighted the health benefits that reducing the dependence on fossil fuels could have if appropriately managed. However, some of the changes in air quality during the lockdowns did not fully follow the predicted behavior and demonstrated that this is an area where more basic research would be beneficial. For example, ground-level ozone in some locations rose because decreased urban NO_2 levels facilitated the production of ozone from free radical reactions with volatile organics (Bourzac, 2020). Projected climate change may make future air quality conditions worse, and thus there is a need to continue research in this topical area (Hong et al., 2019).

Even in a future when fossil energy is no longer used, there will be a need for emission control chemistry, though it will look different than the technologies of today. For example, if hydrogen combustion replaces fossil fuels, that will reduce CO_2, CO, and particulate emissions, but it could lead to increased NO_x emissions. These emissions of NO_x form because H_2 burns at a high temperature, and when done in the presence of air, a mixture of H_2, O_2, and N_2 all combust, which forms NO_x (Lewis, 2021). New energy technologies will create new problems to be solved. Solving them will require research in new areas of fundamental chemistry as well as building on what we know, such as our current knowledge of automotive and diesel emissions control (Lewis, 2021).

---

[18] See https://www.nationalacademies.org/our-work/carbon-utilization-infrastructure-markets-research-and-development.
[19] See https://www.epa.gov/air-trends/carbon-monoxide-trends.

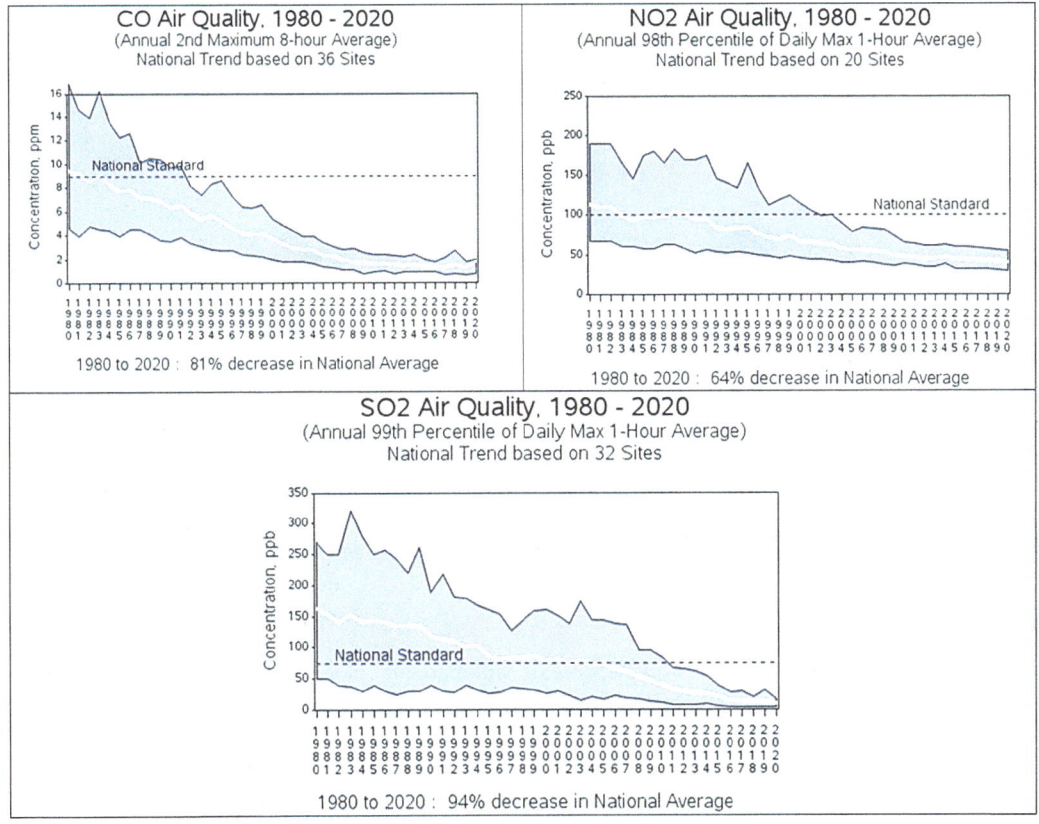

FIGURE 3-13 Emission concentrations of CO, NO_2, and SO_2 from 1980 to 2020. SOURCE: https://www.epa.gov/air-trends.

3.4.4.2 Water Safety

Without question, clean, safe drinking water is of the utmost importance to humanity, and chemistry has played a major role in ensuring that our water is safe to drink. There are three methods for disinfecting a water supply: ultraviolet treatment, ozonation, and chlorination. These treatments typically require significant infrastructure and are commonly deployed for population centers. Although most places with access to centralized treatment facilities have safe water, for some rural areas, particularly in geographic locations with fewer resources, access to potable water is quite limited (Figure 3-14). Developing treatment approaches for cost-effectively producing potable water at small scale is an area of need where fundamental research would have a significant global impact (Shinde and Apte, 2021).

Several new approaches are being explored, utilizing a combination of various chemical technologies such as electrochemistry, photochemistry, and nanomaterials (Amaro-Soriano et al., 2021; Bridle et al., 2015; Hand and Cusick, 2021). Novel materials and catalysts are also being probed to understand their efficacy and offer new avenues for materials development overall (Nasrollahzadeh et al., 2021). For example, the high biocidal activity, normally low human toxicity, and tunability of carbon nanomaterials have encouraged a number of research teams to start exploring how to employ them alone or in combination with approaches such as photochemistry as novel water disinfectant methods (Wang et al., 2019).

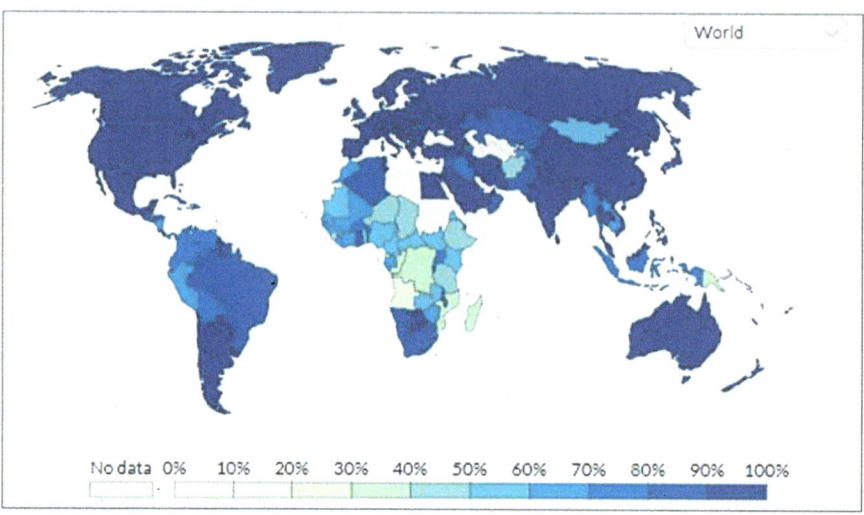

FIGURE 3-14 Share of rural populations globally in 2015 with access to improved water sources. SOURCE: Our World in Data, 2015.

Understanding the chemical interactions of water treatment methodologies with a range of chemical contaminants in water supplies is another rich area of research. While chlorination is used widely as a water treatment strategy, due in part to its broad ability to destroy pathogens, chlorine's strong oxidizing power and ability to directly halogenate other chemical molecules lead to a range of chemical transformations of other species present in a water supply. This can result in harmful by-products that are genotoxic, carcinogenic, and mutagenic. Several factors affect the formation of these harmful by-products, including pH, natural organic matter, ions, and the pipe material. A variety of chemical techniques have been applied to evaluate methods for eliminating pathogens while minimizing harmful disinfection by-products (Kali et al., 2021). Because new chemicals—related to the production and disposal of consumer products, agrichemicals, and pharmaceuticals—continue to enter our water supplies, studying these moieties under a range of chemical conditions and mapping their interactions will need to be an ongoing effort that requires buy-in from a number of different stakeholders and includes corporate responsibility and government oversight. This is an area of impactful science requiring new sampling, separation, and analytical techniques (Richardson and Ternes, 2014).

As our toxicological understanding advances, the need to develop faster and more sensitive detection capabilities continues to grow. These improved capabilities have identified previously unknown disinfection by-products, which in turn has led to safer drinking water supplies (Wawryk et al., 2021). Once advanced detection methodologies are adopted at water treatment facilities, opportunities to embed artificial intelligence and machine learning to provide better process control will engender other areas of research (Li et al., 2021).

3.4.4.3 Food Safety

A 2020 report from the Food and Agriculture Organization of the United Nations highlights how truly global the food supply has become. Their work estimates that about one-third of all agriculture and food exports are traded globally, crossing borders at least twice (FAO, 2020). While global food chains have increased access to a wide range of food choices for consumers, reducing seasonal changes in availability, they have also added complexity when tracking and tracing the source of the food. For

FIGURE 3-15 Countries of origin for various salad ingredients. SOURCE: IOM, 2012.

example, the 10 salad ingredients in Figure 3-15 originate from more than 37 countries. Global food chains greatly increase the possibility of food contamination and spoilage (IOM, 2012). Application of machine learning and related techniques in the service of food safety is a rapidly developing area, which will continue to be important (Deng et al., 2021). While there is no evidence of a link between SARS-CoV-2 infections and food ingestion, concerns around COVID-19 and the subsequent supply-chain disruptions have heightened interest in advancing food security monitoring, pathogen detection, and food safety–related applications. This will open a range of new avenues of scientific exploration (Lacombe et al., 2021).

Advances in analytical methods have led to significant improvements in the overall safety of the food supply, because these methods permit detection and monitoring of a wider range of potential contaminants and with increased sensitivity. Because of the variety of potential contaminants of concern, both traditional techniques of detection and new approaches are being explored.[20,21] Novel materials, such as magnetic nanoparticles (Yu et al., 2022) and metal-organic frameworks (Zhang et al., 2021) are being developed to help detect species of concern. This research space will be a rich area for further exploration for many years to come.

Food safety issues can also be created during routine food preparation. The application of advanced analytical approaches combined with chemical experimental probes has led to a detailed mechanistic understanding of reactions that can create potentially hazardous compounds during food processing and storage (Jackson, 2009). An example is our improved understanding of factors controlling formation of acrylamide, which the IARC classifies as a potential human carcinogen

---

[20] See https://www.fda.gov/food/science-research-food/laboratory-methods-food.
[21] See https://www.fda.gov/food/laboratory-methods-food/other-analytical-methods-interest-foods-program.

(Krishnakumar and Visvanathan, 2014). As new food types such as meat substitutes are added to the food supply, there will be even more need for these kinds of studies to build on fundamental knowledge required to ensure food safety (Sun et al., 2021).

Similarly, fundamental chemistry has led to improvements in food preservatives (Carocho et al., 2018) and packaging materials. Climate change could increase food spoilage rates and will require a multidisciplinary approach to ensure the long-term security of food supplies (Misiou and Koutsoumanis, 2021). Food spoilage and waste is already an area of great concern due to both its GHG impact as well as the financial cost to society. As much as 6% of global CO_2 emissions are linked to food waste (Our World in Data, 2020). This estimate includes both the fossil energy used throughout the food's lifetime and any emissions related to its decomposition. The EPA estimates that about 40% of food is lost or wasted, costing the United States about $218 billion/year, which is approximately 1.3% of the GDP. Food waste makes up approximately 24% of solid waste entering U.S. landfills and is estimated to be the third-largest U.S. source of methane associated with humans (EPA, 2021). Thus, this is an important area of focus and presents an opportunity for chemical sciences to have broader impact.

While the carbon and physical footprint of food packaging is a concern, such packaging plays a significant role in protecting food from contamination and spoilage as well as offering convenience to the consumer. Because of the range of food types and ingredients and the types of packaging materials and their composition, understanding potential interactions that might occur at or near packaging–food interfaces is a complex area of exploration. Migration of compounds within the packaging material to the food can result in contamination that causes potential safety and quality concerns, which are further impacted by storage conditions. However, such transference can also be beneficial and form the basis of many active packaging concepts (Alamri et al., 2021). There are also concerns that exist along the entire end-to-end supply chain related to food packaging, detection of spoiled food at retailers, and many other issues that can occur over the extended lifetime of food products (Chen et al., 2020). Thus, there is a complex interplay of factors to map out in this space. A particularly active area of research is in the development of sensor technologies that can be combined with active packaging concepts to offer the potential to improve food quality, safety, and shelf life (Han et al., 2018). Understanding how to co-design functionality of the packaging with improved recyclability will be a rich area of research for the foreseeable future (Bauer et al., 2021; López de Dicastillo et al., 2020). Similarly, detailed attention to potential safety issues arising from contaminants in recycled materials will spawn research needs in packaging design, recycling process technology, and plastic additives (Geueke et al., 2018)

3.5 CONCLUSIONS

Fundamental chemical research has historically led to significant advancements that improved quality of life for billions of people. These benefits have sometimes been accompanied by unintended adverse consequences. To sustainably meet the needs of society moving forward, the chemical economy faces both challenges and opportunities for innovation. To meet the challenges and to take full advantage of the opportunities, the chemical economy needs to dramatically rethink its approach to production and consumption. Unfortunately, markets and public policies have yet to fully reward sustainability and climate protection in ways that enhance profitability and incentivize companies to invest in a shift to sustainability. Notably, some companies that have embraced green chemistry and circular economy principles have become more competitive or, at least, remained competitive. With these factors in mind, the committee arrived at a number of general conclusions for this chapter.

Conclusion 3-1: To implement a circular economy, the future will require a paradigm shift in the way products are designed, manufactured, and used, and how the waste products are collected and reused. These new processes, and the use of clean energy and new feedstocks to enable these processes, will require novel chemistries, tools, and new fundamental research at every stage of design.

Conclusion 3-2: Transitioning the chemical economy into a new paradigm around sustainable manufacturing, in which environmental sustainability is balanced with the need for products that will improve quality of life, enhance security, and increase U.S. competitiveness, will require substantial investment and innovation from industry, government, and their academic partners to create and implement new chemical processes and practices.

Conclusion 3-3: As fundamental chemical research continues to evolve, the next generation of research directions will prioritize the future of environmental sustainability and new energy technologies. Keeping sustainability principles in mind during every stage of research and development will be critical to accomplishing this goal.

Conclusion 3-4: Chemical research will have the greatest impact addressing energy and environmental sustainability if researchers and practitioners develop and use tools to quantify and mitigate environmental and human health impacts of new discoveries and are aware of the societal implications of their work, and if the research is driven by policies that identify specific environmental sustainability outcomes.

In addition to these conclusions, the committee noted the importance of fundamental chemical research in addressing key steps in decarbonization. As the chemical enterprise continues to look for areas where chemical research could make the largest impact in sustainability, there are a number of concrete areas where initial steps have already been made, and further advancement is possible. The areas that are prime for chemical innovation include

- better measurements for life-cycle assessments;
- enhancement of recycling technologies and co-design of plastic products for recyclability;
- sustainable syntheses;
- sustainable feedstocks and energy sources;
- carbon capture, utilization, and storage;
- monitoring and improving air quality;
- monitoring and improving water safety; and
- monitoring and improving food safety;

Conclusion 3-5: As the world moves deeper into its current energy transition—including the switch to electric vehicles, the implementation of clean energy alternatives, and the use of new feedstock sources—coupled with an increasing focus on circularity, the committee expects that decarbonization, computation, measurement, and automation will significantly alter the operations and processes of current industries, creating new opportunities and challenges that will benefit from fundamental chemistry and chemical engineering advances.

4

Emerging Areas in the Chemical Sciences

Key Takeaways:

- Advances in fundamental chemical research rely on new tools and technologies, and some of the most prominent tools and technologies driving fundamental research are measurement, automation, computation, and catalysis.
- Advances in the areas of measurement, automation, computation, and catalysis will benefit from new discoveries in fundamental chemistry that will, in turn, drive forward technologies in these four areas.
- Access to data that are collected and presented in a standardized format will help advance these four key technologies and accelerate chemical research generally.
- Chemical measurement is becoming faster, smaller, and more accurate, which is driving new research with increasing accessibility to measurement capabilities and the subsequent measurement data.
- Automation, particularly in combination with flow chemistry, offers new avenues to researchers by enabling large numbers of chemicals or reactions to be tested, measured, and analyzed, and thus to more quickly determine new research questions to pursue.
- Fundamental research in computational chemistry is fundamental research in chemistry.
- Computational chemistry, especially when combined with data science (machine learning [ML], artificial intelligence [AI]), accelerates the chemical discovery process.
- Fundamental, multidisciplinary research in chemistry, physics, and engineering has played a critical role in the ongoing development of modern computer architecture.
- The future of catalysis requires new approaches that do not rely on energy-intensive processes with high-temperature or -pressure conditions. Some of these methods include photocatalysis, electrocatalysis, and biocatalysis, coupled with a strong push to synergize theory with experimentation.
- Catalysis research will be most successful if changes in scale related to manufacturing are considered from the beginning.

Success in the chemical sciences relies on our ability to observe, measure, and understand the fundamental chemical interactions that happen when we create or alter molecules and materials. Tools and techniques such as computation, automation, and advanced analytics are fueling advances in our fundamental understanding of chemistry. The fact that the discovery of such tools is also reliant on fundamental chemical principles emphasizes the continually evolving and interdependent nature of fundamental research. Although it is difficult to predict which experiment or project will lead to the next major breakthrough in chemistry, it is clear that the evolution of new tools and technologies is essential to advance our capabilities and further our understanding of basic chemistry.

Among the various tools and technologies that are available to scientists at present, a few are emerging that are particularly impactful to understanding the molecular world and promoting real-world discovery. These are measurement, automation, computation, and catalysis. Investments in these four areas will enable chemistry and related fields to advance more rapidly and facilitate research discoveries likely to help us solve global challenges in the fields of energy, human health, national security, and environmental stewardship. As the chemical sciences have matured, advances in these tools and technology are promoting exciting new opportunities for convergent research. The synergies among them are being harnessed to develop resource-efficient chemistry and processes that conserve feedstock and energy while minimizing adverse effects on human and planetary health. These advances in the chemical sciences are spawning interdisciplinary work both among various subdisciplines of chemistry, including resurgent areas such as photochemistry and electrochemistry, and with researchers in other disciplines in science and engineering such as biology, chemical engineering, and data sciences.

The remainder of this chapter looks more deeply at measurement, automation, computation, and catalysis.

4.1 MEASUREMENT

By enabling us to accurately and precisely ascertain the composition, structure, properties, and quantities of a variety of materials, measurement science fosters innovation across the chemical enterprise. State-of-the-art measurement technologies allow us to probe everything from subatomic particles to human health to ocean sustainability to our solar system, providing real-time or near-real-time data even when looking back to the beginnings of the universe. The development of all measurement technologies is an interdisciplinary feat that requires expertise in physics, engineering, chemistry, biology, data science, computation, and many other fields to create accurate and precise tools. We focus on the topic of measurement because it is critically important to every aspect of the chemical sciences, and advances in measurement are both driven by and help to drive chemical discovery.

There have been many reports and review articles that cover different aspects of measurement in the chemical sciences, and every area of chemical research benefits from frequent new advances in measurement. The committee chose to focus on advances in measurement and analytical chemistry that benefit all disciplines of the chemical sciences. These areas include improvements in the visualization of matter, the enhanced speed of taking and analyzing measurements, and the increased accessibility of measurement technologies. All of these improvements also play a role in making the practice of chemistry more sustainable and contribute to advancements that will help address grand challenges such as climate change.

4.1.1 Improvements in Visualization and Imaging

The ability to sensitively and accurately image chemical matter has increased dramatically over the past several years. To understand imaging and visualization, a Natioal Academies' report from 2006 on *Visualizing Chemistry: The Progress and Promise of Advanced Chemical Imaging* defined chemical imaging as "the spatial (and temporal) identification and characterization of the molecular chemical composition, structure, and dynamics of any given sample" (NRC, 2006b). This definition outlines the fact that chemical imaging has contributed to visualizing the very small (single atoms or molecules) to the very large (entire organs and ecosystem dynamics) by decoding the chemical and atomic composition of a sample.

Some of the most publicized improvements to measurement have been in the area of microscopy, for which Nobel Prizes were awarded in 2014 and 2017 for, respectively, the "development of super-resolved fluorescence microscopy" (Nobel Prize Outreach, 2022b) and "developing cryo-electron microscopy for the high-resolution structure determination of biomolecules in solution" (Nobel Prize Outreach, 2022c). Both of these techniques enabled measurement of different types of materials, reactions, chemical compounds, and biological molecules in unprecedented ways. With cryo-electron microscopy (cryo-EM), for example, computation-enabled improvements in resolution of structural images of biomolecules improved from around 15 angstroms (Å) before 2014 to as low as 1.2 Å in 2021 (Figure 4-1) (Peplow, 2020). Single-molecule fluorescence microscopy has not only allowed for a clearer spatial and temporal understanding of molecules but also provided an important technology for tracking reaction rates, including those driven by electrocatalysis (Li et al., 2018). This technique, along with other measurements of electrocatalysis such as surface plasmon resonance, will help drive the basic chemical research needed for the optimization of fuel cells.

Increased imaging sensitivity has allowed researchers to deconvolute the chemical composition of complex samples, including in environmental research. One example is identifying and quantifying different plastic polymers from water samples (Ivleva, 2021). It can be quite difficult to determine which plastic is being measured, due to structural similarities in their polymers. The most common method for interpreting the composition of these samples is by pyrolysis-gas

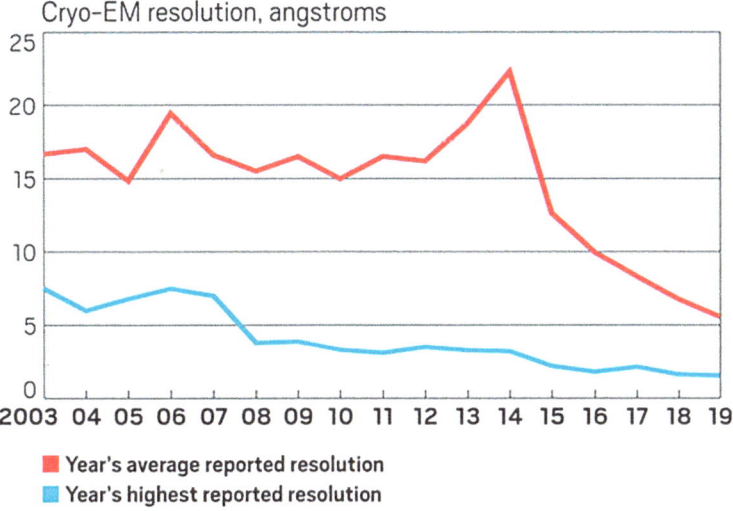

FIGURE 4-1 Resolutions of cryo-electron microscopy images submitted to the Protein Data Bank from 2003 to 2019 with resolution measured in angstroms (Å). SOURCE: Peplow, 2020. Reprinted from Chemical and Engineering News, copyright © 2020 by the American Chemical Society. This image was first published in C&EN on Sept. 27, 2020 and appeared in Vol. 98, Issue 37.

chromatography/mass spectrometry (Py-GC/MS), in which the sample is decomposed under heat and separated using gas-phase chromatography before the mass is analyzed. The full image developed from a Py-GC/MS data set helps researchers to understand the amount of different types of plastic in the water and can even lead to source identification. Similar types of MS methods have been applied to measuring the chemical interactions between different species within microbial communities in order to get a snapshot of community interactions (Dunham et al., 2017).

In addition to optical imaging, EM, and MS, there are a number of other imaging techniques that are rapidly improving and contributing to chemistry. These include atomic and molecular spectroscopies, other types of optical imaging, nuclear magnetic resonance (NMR), noninvasive imaging, and quantum imaging. Continued improvement in our understanding of basic chemical principles will enable analytical chemistry to better utilize all of these different techniques, and those insights will continue to improve imaging technologies for chemical research, as well as research in a variety of other fields.

One area on this list that has generated a lot of interest is the use of quantum principles to enhance sensing and imaging technologies. In a recent review, Yu and colleagues (2021) note that "a molecular approach to quantum sensing offers the unmatched combination of atomic structural control and tuanbility, enabling transformative discovery in fields spanning biology to astrophysics." The impacts that quantum physics can have on chemical discovery as well as the ways that chemistry can contribute to our understanding of quantum mechanics include a plethora of ideas and avenues, especially in a well-explored area such as quantum sensing. A National Academies' study is currently exploring this topic, specifically how quantum information science and measurement and modeling in chemistry can inform and advance one another.[1]

4.1.2 Real-Time Chemical Measurements

Real-time chemical measurements and analytics have become critical for everything from monitoring the time courses of reactions to ensuring acceptable air, food, and water quality (more detail on the latter is in Section 3.4.4). In this report, the concept of real-time chemical measurements includes near-real-time chemical measurements where a sampling phase is required in order to gather an appropriate amount of analyte to produce an accurate measurement. The process of near-real-time chemical measurement still happens quickly and provides time-dependent and actionable information. The speed and accuracy of real-time measurements regularly improve, and those improvements are reliant on advances in computation, automation, data analysis methods, engineering, and separation science (see Box 4-3 for more information on separation science). The discoveries in analytical chemistry that drive these processes create new tools and technologies that are used to identify differences in an established baseline, make experimental comparisons, and identify and intervene in dangerous or life-threatening situations.

Real-time chemical imaging and measurement can be applied to monitor changes in the environment such as monitoring volcanic activity (see Box 4-1) (Bi and Han, 2019) or surveying water sources (Yaroshenko et al., 2020) for pollutants and other unusual activity. One notable example is the Ocean Observatories Initiative, funded by the National Science Foundation (NSF), that takes data measurements on "physical, chemical, geological and biological properties and processes from the seafloor to the air-sea interface" and makes the data available to any researcher who wants to use them.[2] Another example is the mobile chemical analysis stations that have been deployed to identify pollution in rivers and are, in one particular case, able to take measurements of "temperature,

---

[1] See https://www.nationalacademies.org/our-work/identifying-opportunities-at-the-interface-of-chemistry-and-quantum-information-science.

[2] See https://oceanobservatories.org/.

> **BOX 4-1**
> **Real Time Measurements of Volcanic Eruptions**
>
> Volcanologists, geologists, and geophysical chemists have been building up the capacity to observe and measure properties of volcanic eruptions in an attempt to produce data for scientific study, as well as data for real-time information to ensure a prompt response that helps save lives. The recent eruption of Hunga Tonga Hunga Ha'apai, near the island nation of Tonga, devastated the country and also showed researchers that there is still much work to be done to understand and respond to volcanic eruptions. Observations of the volcano showed that it was unlike any previous eruption that scientists have seen, and satellite images showed that the cloud was higher than what volcanologists thought possible. Because of this discrepancy, researchers are trying to understand different aspects of the volcano by using "high resolution satellite imagery… to track how ash, gas and certain chemical species are drifting through the atmosphere" (Witze, 2022). Further in depth analysis of the ash showed that the chemical composition was different than what had previously been observed because the magma had not spent much time underground undergoing chemical changes (Witze, 2022). Previously collected samples of magma from the same volcano "had gone through some telltale chemical changes, almost like wine ageing in a barrel, before ultimately erupting onto the surface" (Witze, 2022). To further study this scenario, researchers have sent balloons carrying instrumentation into the eruption plume. The teams are looking to understand the chemical composition of volcanic particles that might help inform them of what happened during the blast and how to better predict eruptions in the future. The volcano was extraordinarily destructive, and researchers hope it will provide an opportunity to learn, observe, and have better real-time analytics to mitigate future harm and damage.

total phosphorous, pH and ammonium ions, dissolved oxygen, conductivity, nitrate ions, and total organic carbon" (Yaroshenko et al., 2020). These measurements rely on different types of electrochemical and optical sensors, and due to the extended duration of measurement (sometimes decades), there is a strong established baseline, meaning that abnormal levels of various pollutants are discovered instantly. Real-time measurements of the environment can also utilize other types of instrumentation such as MS (Zuth et al., 2018). Much research is still needed to improve sensitivity and deconvolute complex samples using rapid separation science (NASEM, 2019a). These types of systems are also critical for chemical manufacturing, enabling workers to monitor product quality and ensure that pollutants are not spilling into the surrounding environment (Schmitz, 2015).

Measurement speed is related to the instrumentation and how well analytical technologies can accurately quantify or image a subject. But it is also directly tied to the availability of rapid data analysis. High-speed data analysis requires both computing power and efficient algorithms that can quickly process a data input (see Section 4.3 for more detail on computation). Analysis of chemical measurements has benefited greatly from the use of data interpretation algorithms that use ML and AI principles. For example, a variety of ML algorithms can be paired with electrochemical sensors to "extract complex relationships between chemical structures and their electrochemical properties and to analyze complicated electrochemical data to improve calibration and analyte classification" (Puthongkham et al., 2021). In these cases, ML methods help to calibrate the sensors and enhance the sensitivity of measurement, creating faster measurement capacity without the need to adjust the setup by hand (Figure 4-2). The incorporation of advanced computational techniques, such as AI and ML, into real-time measurement tools will increase the speed with which researchers and other users can make decisions based on new measurement data.

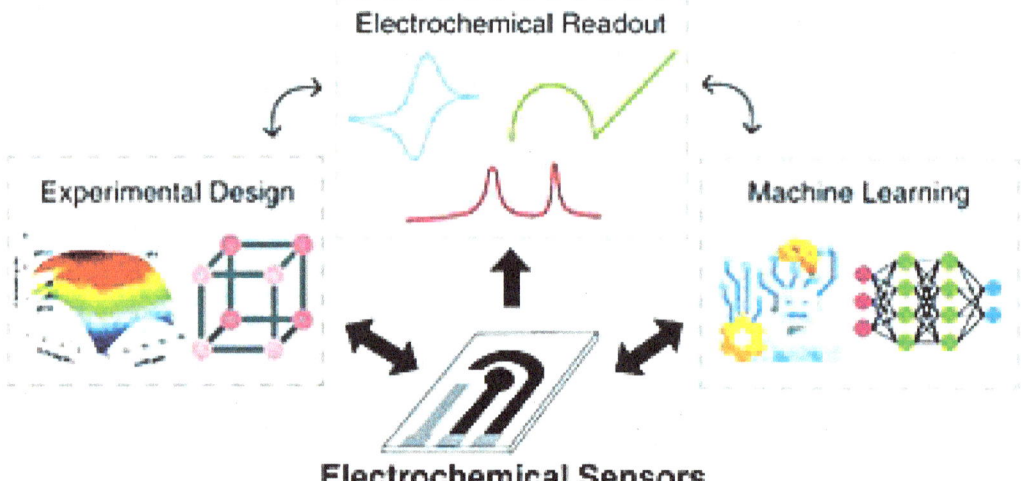

FIGURE 4-2 Interactions among sensors, sensor readouts, machine learning algorithms, and experimental design. SOURCE: Puthongkham et al., 2021.

4.1.3 Increased Accessibility of Chemical Measurement

Many chemical measurement tools are getting smaller, more portable, and cheaper. These developments contribute to the increased accessibility of chemistry. Although not every analytical technology has benefited from increased accessibility, this is a definitive trend in a number of different areas. In the 2017 Annual Review Issue of *Analytical Chemistry*, it was noted that, "[c]learly, analytical tools are getting cheaper and more available as seen in reviews on paper microfluidics, 3-D printing, digital assays, and point of care diagnostics" (see Box 4-2) (Kennedy, 2017). Miniaturization of chemical instrumentation is partly enabled by the silicon revolution, and in part because chemical knowledge of sample preparation, separation, and sensor technology has facilitated the engineering of smaller devices. Frequently, decreasing the size of an instrument comes at a cost of reduced sensitivity and accuracy, but some recent advances have shown that this is not always the case.

Some instrumentation such as MS has recently undergone a transformative miniaturization process. Mass spectrometers that used to require entire rooms and specialized setups now can fit in the palm of your hand. Of course, full-room MS setups still exist and provide state-of-the art analysis of chemical samples, but there is a critical need for MS capabilities in places where entire rooms cannot be set aside for instrumentation. For example, companies such as 1st Detect Corporation and 908 Devices are building MS instruments that weigh less than 4 lbs for use in quality control of food and water as well as many security and forensics applications (Perkel, 2014). The scientific community also needs analytical capabilities at places that are challenging to get to, such as the bottom of the ocean or outer space. MS technologies have already been deployed for use in space research, but their capabilities are limited and there is a need for continually miniaturizing MS systems, so that single instruments can be used for purposes such as geological, chemical, and biological testing in space (Wainerdi, 1970).

BOX 4-2
Accessibility of Chemical Measurements and At-Home Point-of-Care Testing

One of the most prominent examples of the increased accessibility of chemical measurement are at-home point-of-care tests, such as tests to measure blood-glucose levels, pregnancy tests, and rapid-antigen tests for infectious disease diagnostics, including the ones used for detecting SARS-CoV-2. These types of tests are low cost and can be used by the general public. They enable anyone to acquire information that was previously only available to chemists, medical professionals, or individuals who were specifically trained in instrumentation use (Figure 4-2-1) (Nayak et al., 2017). Getting to where point-of-care diagnostics are widely distributed, reliable, sensitive, and accurate required decades of fundamental chemical research to improve chemical sensors and understand how to identify the target chemical in each test. For example, the first modern pregnancy test was developed in 1927, when women's urine—which would contain the hormone human chorionic gonadotropin (hCG) if pregnant—was injected into immature rodents, and the tester would monitor if the animal entered an estrous cycle ("heat"), which would indicate a positive test (Romm, 2015; Tyssowski, 2018). Increased knowledge of hormones, antibody production, and decades of test iterations led to the current version of pregnancy tests incorporating lateral flow technology. Lateral flow tests were discovered in the 1950s and were fully developed in the 1970s after many of the enabling technologies were developed through work or co-design between chemists and chemical engineers (Wong and Tse, 2009). All of the point-of-care tests that are available at the pharmacy today required decades of research and development by chemists and chemical engineers to create products that are simple and accurate.

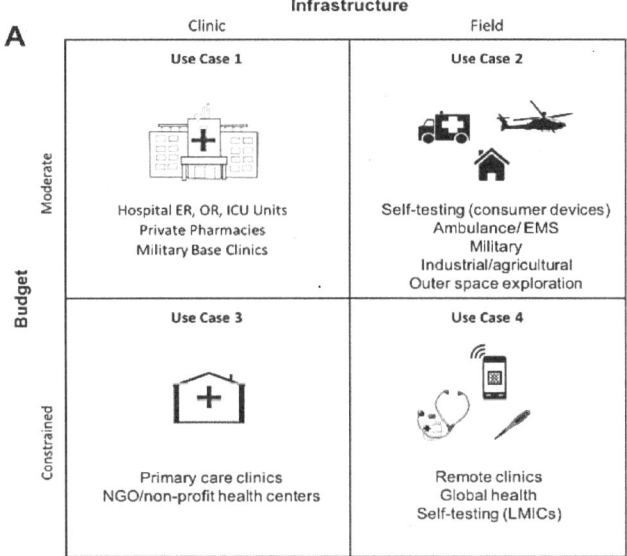

FIGURE 4-2-1 Understanding the advantages and disadvantages of different tests. SOURCE: Nayak et al., 2017.

continued

BOX 4-2 Continued

Infrastructure

Clinic	Field
Trained personnel	Untrained personnel
Portability- not important	Portability- very important
Minimal rough handling	Rough handling
Rapid analysis- moderately important	Rapid analysis- important
Ground electricity	No ground electricity
Controlled ambient conditions (temperature/humidity)	Ambient temperature/humidity fluctuations
Refrigeration	Limited Refrigeration
Cost of accessory equipment- high/low	Cost of accessory equipment- high/low
Cost of disposables- high/low	Cost of disposables- high/low
Moderately trained personnel	Untrained personnel
Portability- not important	Portability- very important
Moderate rough handling	Rough handling
Rapid analysis- moderately important	Rapid analysis- important
Ground electricity (supply fluctuations)	No ground electricity
Limited refrigeration	No refrigeration
Ambient temperature/humidity fluctuations	Ambient temperature/humidity fluctuations
Cost of accessory equipment- low	Cost of accessory equipment- low
Cost of disposables- low	Cost of disposables- low

FIGURE 4-2-1 Continued

4.1.4 Future of Measurement

Analytical chemistry and associated measurement technologies have made significant advances in the past couple of years, accompanied by advances in data acquisition, computation, sample preparation, microfluidics, and other enabling technologies. These tools can measure compounds in complex samples, create 3-D images of material by multiplexing different analytical methods, and trace material back to its original source with surprising accuracy. Two of the most important things that have facilitated these rapid advancements in analysis and measurement are the availability of data and the computational speed with which we can analyze data (Adams and Adriaens, 2020). As measurement continues to advance, so will the complexity of the data, which will require both better ways to integrate measurement and analysis and demand for new ways to store, retrieve, and use data. One increasingly popular concept in this regard is "democratic analytical chemistry" (Figure 4-3), which relates closely to the notion of accessible chemistry. The idea of democratic analytical chemistry is that measurements and instrumentation are available to any user directly (taking the measurements yourself) or indirectly (analyzing the measurement data from others) (de la Guardia and Garrigues, 2020). This concept is empowered by open-access software and readily available data repositories. As more systems like this develop, the network of chemists who can use and analyze data from measurements will expand significantly.

The concept of democratic analytical chemistry is also inherently a green chemistry concept. This is because instrumentation that already exists will be used to its full potential, mitigating the

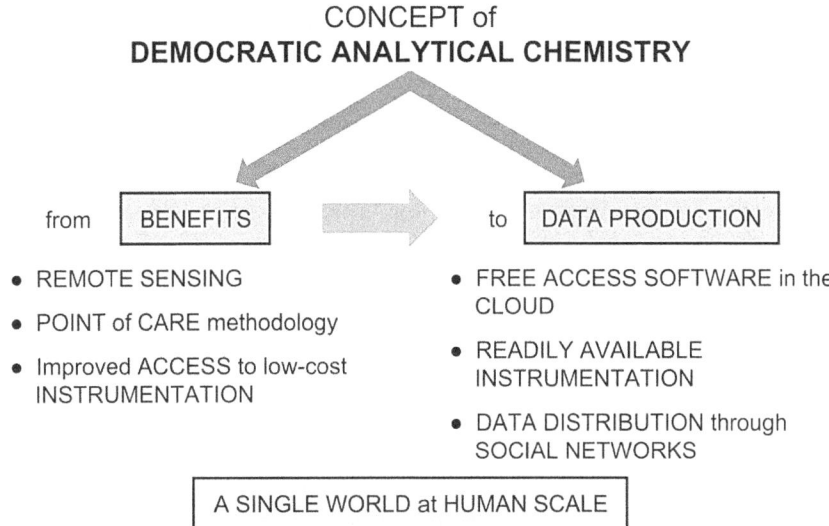

FIGURE 4-3 Descriptive model of democratic analytical chemistry. SOURCE: de la Guardia and Garrigues, 2020.

need to use materials and energy to make and operate new instrumentation. Additionally, in some cases, no use of instruments is required. Rather, existing data, analyzed using the latest analysis techniques, will provide new insights.

Another future area for chemical measurement is the ability to perform real-time remote analytics. As discussed above, most real-time analytics rely on sample collection at a particular site and the movement of that sample to an on-site or nearby measurement device. There are, however, some unique capabilities that would be afforded by completely remote sensing. Some remote sensing techniques are already a reality (Bogue, 2018), but there is much development still to pursue in this area. Real-time remote chemical sensing would afford extensive opportunities to environmental and forensic science, national security, manufacturing, and many other areas.

Elsewhere in this report, there are other examples of chemical measurement's role in advancing environmental stewardship and sustainability (see Sections 1.3.3 and 3.4.4). As noted earlier, many new tools and technologies are improving our ability to analyze air and water quality so as to be good stewards of the environment. Beyond the more direct impacts, the idea of "green analytical chemistry" is not a new one; it evolved simultaneously with the concept of green chemistry. These ideas are mostly related to reducing waste and preventing destructive interactions with the environment. As analytical chemistry increases in complexity, there is also an opportunity to be more innovative in the implementation of green and sustainable practices in chemical measurement.

4.2 AUTOMATION

The earliest mention of laboratory automation is from 1875 when chemists developed a device to "wash filtrates unattended" (Olsen, 2012). Over the next century, a number of unattended processes were developed for the chemical sciences. Laboratory automation in its current form, which utilizes robotics and computation, stems from the "demands in the life sciences in the 1980s for more productive means to aliquot and dilute biological samples for testing and analysis" (Selekman et al., 2017). This led to the development of automated liquid handlers capable of generating arrays of samples with minimal manual intervention. Further advances in equipment to automatically dose both liquids and solids led to its use in high-throughput screening assays for drug discovery and combinatorial chemistry for material discovery. The unifying element behind applications of automation and high-throughput experimentation (HTE) is the ability to rapidly generate large diverse arrays of experimental data where first principles and rational design are difficult, owing to the innate complexity of the systems and responses under study.

High-throughput techniques built upon laboratory automation technology, coupled to in-line analytical monitoring and measuring technologies, statistic experimental design, and parallel experimentation, are accelerating the discovery of new molecules (e.g., catalysts) and new reactions, and optimizing processes that generate an array of chemical products such as pharmaceuticals, agrichemicals, and materials (Selekman et al., 2017). The capabilities of automation increased significantly in the past decade with advances in software, computational power, robotics, analytical measurements, data science, and ML now incorporated into automated experimentation (Coley et al., 2019).

Advances in science are made by iterative cycles of experimentation, data analysis, and hypothesis formulation. These advances progress at a rate that is commensurate with the rates at which experiments can be performed and data can be analyzed. Thus, scientists have been motivated to automate the execution and analyses of experiments to accelerate discovery. The power of automation in research is readily apparent from recent breakthroughs at the interface of biology, chemistry, physics, and engineering. Some examples are the automated synthesis and analysis of DNA, RNA, and polypeptides. Automation technologies for these biopolymers have played a critical role in most of the transformative discoveries and advances in biology and medicine. Preparation of the mRNA vaccines for SARS-CoV-2 and sequencing of the human genome are just two recent landmark achievements in synthesis and analysis, respectively, that were enabled by automation.

While chemists can certainly claim partial ownership of the invention of automated methods for the synthesis and analysis of biopolymers in biological research, scientists are now using automation in new ways to advance research in the chemical sciences. This section highlights how chemistry is benefiting from automation technologies.

4.2.1 Automation for High-Throughput Experimentation

With automation, it is possible to execute millions of experiments on small molecules in parallel and in a short period of time. For example, pharmaceutical companies and academic and national research centers have facilities for HTE on libraries containing millions of small molecules. Robots are used to rapidly dispense molecules into 96-, 384-, and 1536-well plates in which assays of their binding to biomolecules of interest or their capacity to perturb enzymatic or cellular activity are measured (Mennen et al., 2019). Because of these robust automated systems, the synthesis of small molecules, rather than their evaluation in biological assays, has become the rate-limiting component in drug discovery. In addition to its use in drug discovery, chemical companies have recognized the value of automation, and understand its benefits in boosting the productivity of research and increasing worker safety (Bardin, 2021).

The technology for automating experimentation has been repurposed for many research areas in the chemical sciences. Robots enable parallel experiments in screens for catalysts for chemical transformations and for optimizing conditions for chemical reactions of interest. This is exemplified by the Merck Center for Catalysis at Princeton University, where a robotic system facilitates the setup, monitoring, and characterization of thousands of reactions in parallel.[3] It is equipped to dispense desired quantities of solid reagents and solutions in ambient atmospheres or in the inert atmosphere of a glovebox. Reaction analyses are enabled by automated, in-line GC/MS analysis.

---

[3] See https://chemistry.princeton.edu/research-facilities/merck-catalysis-center.

With this infrastructure, it is possible for skilled analysts to quickly evaluate a large number of reactants and catalysts such that data sufficient for model building can be generated. An early success of this technology was the discovery of a novel amino acid C—H arylation reaction (McNally et al., 2011). The research team noted that automation and rapid measurement and monitoring of successful reactions helps to "exploit [the] serendipity" present in scientific experimentation.

4.2.2 Automation for High-Throughput Synthesis

Historically, chemical synthesis was performed on an ad hoc basis focused on the optimal synthesis of a particular small molecule or its closely related analogs. Early deviations from this molecule-centric approach for synthesis was the preparation of polymers. In those cases, chemists recognized that the synthesis of polymers of defined structures from discrete building blocks was amenable to automation. Indeed, visionary chemists, such as Bruce Merrifield and Marvin Caruthers, developed methodologies for the automated synthesis of peptides (Kent, 2006) and nucleic acids (Caruthers, 2013), respectively. More recently, Peter Seeberger and others have implemented methods for the automated synthesis of carbohydrates, which are among the most complex of the polymers, due to the diversity of the monomeric building blocks, the complex branching of many carbohydrates, and the need to control stereospecificity (Plante et al., 2001).

Whereas polymers are particularly well suited for automated synthesis, it is a very different matter for molecules with molecular weights under 800 Daltons (Trobe and Burke, 2018). The structural diversity and complexity of small molecules is immense and there are no general solutions for their preparation. Conditions for chemical transformations in the synthesis of a small molecule are selected or screened for based on the peculiarities of the reactants, and approaches for the isolation of reaction products are either chosen by analogy or determined empirically. These facts have underscored arguments that chemical synthesis is tedious, challenging, and equally art and science. When there is an interest in the syntheses of closely related molecules, as in a medicinal chemistry program or ligand screening in a catalyst study, the ad hoc approach to small-molecule synthesis can slow progress. Accordingly, there has been growing interest in automating the syntheses of small molecules. Some of the earliest cases of small-molecule synthesis via automation date back to the 1970s and 1980s wherein volumetric transfer of solvents and solutions containing reagents to resin-bound reactants enabled parallel and combinatorial syntheses. The solid-phase methodology for chemical synthesis is highly amenable to automation but is limited to certain types of reactants and transformations (Leznoff, 1978). Accordingly, there has been much interest in and effort spent developing automated platforms for executing solution phase reactions.

4.2.2.1 Automated Chemical Synthesis Using Flow Chemistry

Some of the most promising platforms for automating solution-phase chemistry are based on a subdiscipline of chemistry called "flow chemistry," in which chemical reactions are effected by the controlled pumping of the input stream, and the mixture of solutions and reagents is developed and studied (Figure 4-4) (Hartman, 2020). Developing industrial-scale flow chemistry (i.e., continuous manufacturing) is under active investigation throughout the United States and the world, prominently at the Novartis-MIT Center for Continuous Manufacturing.[4] In flow chemistry, conventional bench operations of batch approaches can be automated (reagent dispensing, reaction mixing and work up, product purification and analysis, etc.). In this regime, reaction rates and productivity are enhanced by thorough mixing of reagents with reactants and by efficient heat and/or mass transfer. Moreover, reactions can be pressurized at superheated conditions wherein reactivity is greater

---

[4] See https://novartis-mit.mit.edu/.

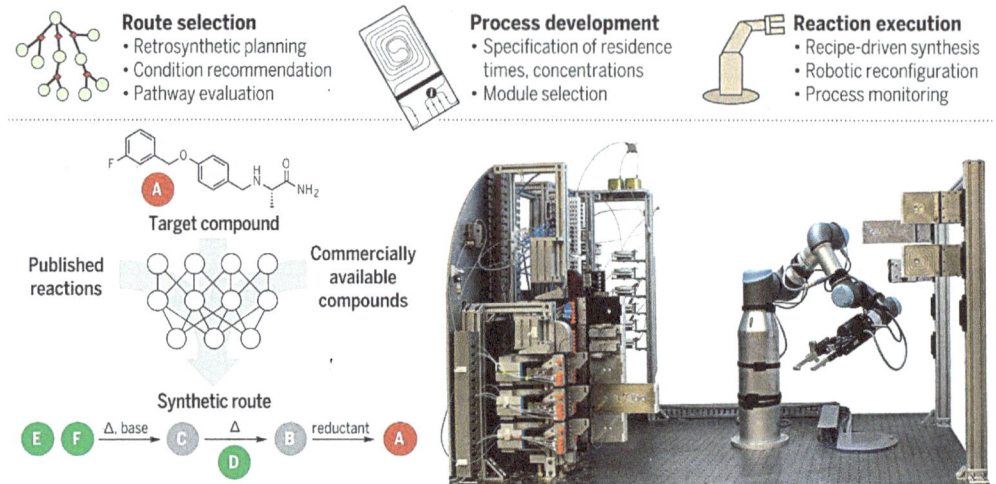

FIGURE 4-4 Planning and execution of a robotically reconfigurable flow chemistry platform that performs multistep synthesis using artificial intelligence. SOURCE: Coley et al., 2019.

(Hartman, 2020). Importantly, reactions performed in flow are often safer than those in round-bottom flasks or other conventional reaction vessels because hazardous substances and combustible reagents are contained within stable devices.

Reflecting the promise of the technology, there are now commercially available flow synthesizers in use by both academic and industrial groups. These devices have peristaltic pumps, mixing elements, residence-time loops, and separation units that are controlled by a computer (See Box 4-3 for more information on separation science). The hardware can be coupled to process analytical technology and can assess "in-line" multiple physicochemical properties of the reactants and products via thermocouple, spectroscopy (ultraviolet, infrared, and Raman), MS, NMR, crystallization monitoring, and particle size analysis (Browne et al., 2017). There are autonomous flow devices that utilize computational tools and software such as "design of experiments" and algorithms (e.g., evolutionary, self-optimizing, and ML) to monitor, manage, and fine-tune the operation of flow systems (Coley et al., 2019). Continuous-flow systems have proven to be an enabling technology for the efficient synthesis of small molecules of all kinds, including the manufacturing of active pharmaceutical ingredients. Indeed, the U.S. Food and Drug Administration has declared continuous manufacturing as one of the most important tools in the modernization of pharmaceutical chemistry (FDA, 2019). The combination of automated flow chemistry with comparably automated product and data analysis promises to change drug discovery and other areas within the chemical industry by giving us the ability to synthesize a wide range of organic matter (Li et al., 2015).

4.2.2.2 Automation and the Accessibility of Chemical Synthesis

Automation, and particularly flow chemistry, are changing the way in which academic chemistry functions. Martin D. Burke, a leader in the field, contends that automated flow chemistry will increase the accessibility of synthetic organic chemistry and make small molecules more readily available to all scientists (Burke, 2021). Dr. Burke cofounded the Molecule Maker Lab Institute, which is "an interdisciplinary initiative with leaders in AI and organic synthesis intensively collaborating to create frontier AI tools, dynamic open access databases, and fast and broadly accessible

> **BOX 4-3**
> **Separation Science to Enable Discovery**
>
> Separation techniques process mixtures to their constituent parts, with separation science focused on advancing our theoretical and practical knowledge of the chemistries that govern separations. Separation science is a well-established field that originated with many of the techniques that still exist today, such as distillation and chromatography (NASEM, 2019a). The range of modern separation techniques spans from metric ton–scale separation processes encountered in chemical manufacturing for the purposes of enriching a product stream into its pure components, all the way to micro- to nanoscale separations in flow-based devices. Separations play an important role in ensuring that quality products are made and accurate measurements are taken, and are critical to many aspects of the emerging tools and technologies for chemical sciences. Examples of some separations that are particularly relevant to large-scale products include distillation operations to separate ethane+ethylene and propane+propylene mixtures, and absorption with a suitable solvent to separate carbon dioxide (CO_2) from flue-gas streams for sequestration purposes. Some small-scale separations include the use of technologies in microfluidics and high-throughput flow devices used in point-of-care diagnostics, environmental monitoring, and forensic investigations. Fundamental research in the chemical sciences is contributing to the development of separation science, including the discovery of novel materials such as metal-organic frameworks, graphenes, and polymer membranes that effect chemical separation by selectively interacting with specific constituents in a mixture to manipulate their permeabilities and absorption or adsorption tendencies. Examples of such advances in energy-efficient separations and future challenges in the field are discussed in a recent National Academies' report titled *A Research Agenda for Transforming Separation Science* (NASEM, 2019a). According to this report, chemistry and other fields should focus advances in separation science on "two major themes: designing separation systems that have high selectivity, capacity, and throughput; and understanding temporal changes that occur in separation systems."

small molecule manufacturing and discovery platforms."[5] To make organic synthesis more accessible, they are making advances in modular synthesis by treating small molecules like Legos or building blocks (Figure 4-5). To reach a broader audience, the institute is also running "make-a-thons," where students develop ideas to solve health-related problems and design molecules that can be synthesized using the institute's infrastructure. This provides an in-depth opportunity to excite students about chemistry and its practical applications (Burke, 2021).

4.2.3 Future of Automation in Chemistry

Automation is transforming the chemical sciences and manufacturing, and there are undoubtedly more advances ahead that will push against conventional boundaries. Importantly, automation, HTE, and data capture can play an important role in sustainability, especially in processes where automation can help optimize green and sustainable reaction conditions. These technologies can also help with sustainability in areas such as early-stage drug discovery where microscale or nanoscale HTE uses minute amounts of chemicals to find the desired reaction condition, and data science will enable researchers to get to the end point with far fewer experiments. Additionally, automation is enabling small-molecule syntheses on scales where computational data can be stored in the form of small molecules (Cafferty et al., 2019). Because automation is a central part of a rapidly changing landscape in the chemical sciences, academic institutions may want to reconsider both their curricula and research avenues to minimize the gaps among basic science, translational research, and manufacturing. In this regard, Carnegie Mellon University is at the vanguard. The

---

[5] See https://moleculemaker.org/.

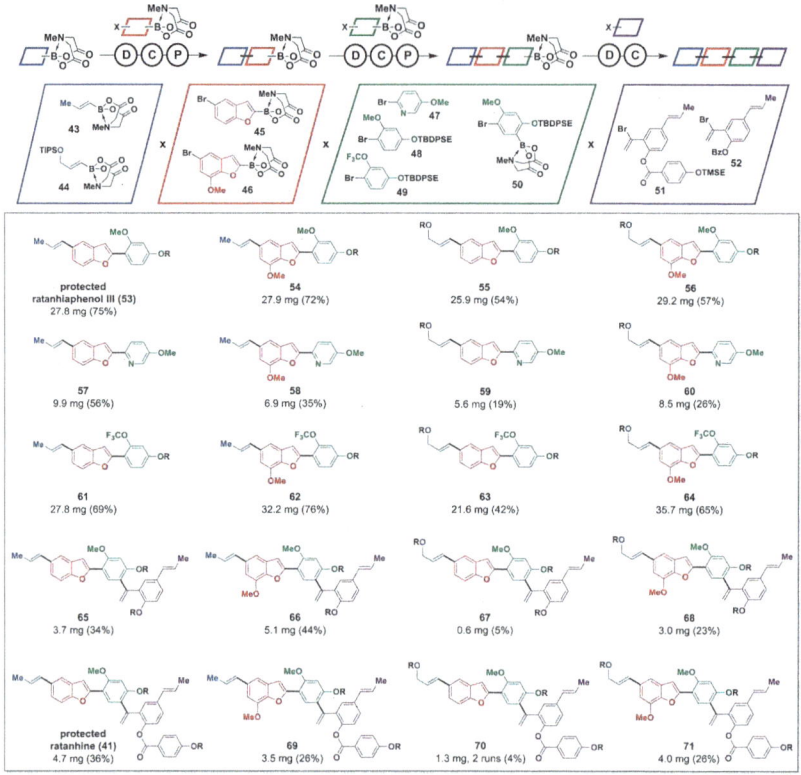

FIGURE 4-5 Automated synthesis of ratanhine derivatives showing the modular synthesis scheme to build a large suite of different molecules with different molecular building blocks. SOURCE: Li et al., 2015.

university has partnered with Emerald Cloud Lab to establish the Carnegie Mellon University Cloud Lab (CMU, 2021). Similar to the Molecule Maker Lab Institute, this remote-controlled lab is providing a universal platform for AI-driven experimentation in the biological and chemical sciences. It contains state-of-the-art instrumentation that can execute all aspects of daily work, from experiment design to data acquisition and analysis. The university envisions it as a resource not only for its faculty, students, and staff but for scientists around the world. It is easy to imagine that the successes of these laboratory platforms will inspire other institutions to break out of conventional ways of doing science and to embrace automation.

4.3 COMPUTATION

Building on concepts and theories from chemistry and physics developed in the late 18th, 19th, and early 20th centuries, scientists began turning theory into computable algorithms around the early 20th century, inaugurating the field that became computational chemistry. Early discoveries in computational chemistry include development of quantum chemistry and theories of chemical bonding of atoms in molecules (Esposito and Naddeo, 2014), the chemical ensembles of statistical thermodynamics that describe interactions between molecules (Gibbs, 2010), and elucidation and calculations of rates of transformation in chemical reactions (Eigen, 1961; Klippenstein et al., 2014). Even ML has an early precursor from chemistry in the work of Louis Hammett, who elucidated a mathematical relationship between equilibrium constants and descriptors of varying substitutions on a benzene core (Hammett, 1937).

Initially, solutions to these algorithms were calculated manually, using numerical methods for integration, differentiation, enumeration of ensembles of states, and curve fitting, with the result that algorithms were applied only to small, more tractable systems such as the hydrogen atom. Coming up with these solutions could take thousands of person-hours to calculate. The advent of mechanical, electromechanical, and vacuum-based computers vastly increased the speed of calculations and concomitantly expanded the size and complexity of the systems that could be studied. ENIAC, or Electronic Numerical Integrator and Computer, the first programmable, general-purpose computer, could perform thousands of computations per second (Kopplin, 2002; Levy, 2013; Mangalindan, 2021). Hand-in-hand with this computational speed, ENIAC's architecture of branching and stored memory enabled programmers to move beyond a linear sequence of operations and incorporate conditional execution (if-then-else for loops) in their codes. This innovation in computer architecture led directly to computational innovation: the first-ever computational Monte Carlo simulation, a calculation run on ENIAC on behalf of Los Alamos National Laboratory that simulated the movement of neutrons through the fissile material of a nuclear bomb (Haigh et al., 2014). Since this first calculation, Monte Carlo simulations have become a widespread and valuable tool across a wide range of computational chemistry applications (Amar, 2006; Doll and Freeman, 1994; Earl and Deem, 2008; Saito, 1997). This synergy among computational speed, computer architecture, and algorithmic innovation is a recurring theme throughout the development of computational chemistry as a discipline.

There have been many reports and review articles that cover the current state of the art for computational chemistry, and every area of chemical research benefits from frequent new advances in computational power and algorithm development. While this chapter and many other sections in the report point out some specific and important examples, the committee chose to focus on chemistry-enabled advances in computational hardware that benefit all disciplines of the chemical sciences and that have the potential for transformative impact on the types of chemistry challenges that can be addressed by computational chemistry algorithms.

4.3.1 Present-Day Computational Chemistry

As noted in Section 2.3.3.3 of this report, fundamental chemistry, along with advances from many other fields, have led to a steady increase in computing power and provided an improved architecture (Clancy, 2012; NRC, 2003). Research on photolithography, in particular, led to monumental advances in the miniaturization of microchips to allow more transistors per microprocessor (see Section 2.3.3.3).

Because of these steady increases in computational power and speed, and the decreases in environmental impact, computation can be applied to larger and more complex chemical systems. Computational chemistry has become a mature discipline that is used across all chemical subdisciplines to derive understanding, enable prediction, and drive experimentation. Quantum mechanics and statistical thermodynamics theories are regularly applied to biological macromolecules such as proteins, DNA, and RNA and to complex materials such as catalysts and batteries. ML and AI methods have gone far beyond Hammett's examinations of a few substituted benzoic acids to model building for data sets containing millions of molecules. These advances have become a driving reason for why the collection, storage, and standardization of chemical data are so critical. These more advanced computational processes only work with reliable data and training sets, which come from a wide breadth of information collected through experimentation and computational modeling. These new computational tools are also being used with supercomputers to analyze complex chemical models. In more complex simulations with many increasing atoms and variable space, supercomputers are the best option for computation, especially because of the modern-day increase in computational power associated with these machines (Figure 4-6).

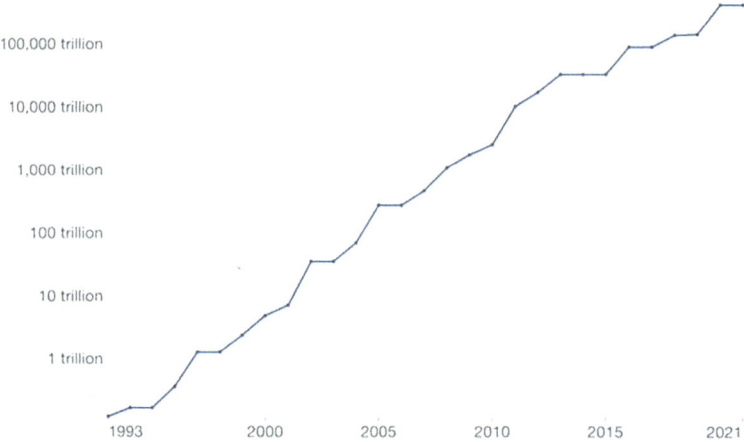

FIGURE 4-6 Growth of supercomputers as measured by the number of floating-point operations carried out per second by the largest supercomputer in any given year. SOURCE: Our World in Data, 2022.

Evidence of this breadth of application can be seen in usage statistics for supercomputing facilities available through the NSF's Extreme Science and Engineering Discovery Environment (XSEDE) computing environment and the Innovative and Novel Computational Impact on Theory and Experiment (INCITE) program for access to the Department of Energy's (DOE's) Leadership Computing Facilities. Chemists and materials scientists regularly account for approximately 20–35% of the principal investigators (PIs) who received INCITE allocations between 2017 and 2021, and chemists also made up approximately 22–24% of the PIs that used the XSEDE environment over the same time period (Figure 4-7). Within the field of chemistry, all subdisciplines are represented among XSEDE users, with materials scientists accounting for the biggest portion of users of this resource (Figure 4-7).

While computing is enormously helpful to solving complex chemistry questions, it comes at a price: Computing exacts a huge energy burden. While some programs, including one in DOE,[6] are working to solve this issue, the energy usage from computers and supercomputers is still an issue. One article noted that "electricity demand by data centers in 2018 was an estimated 198 terawatt hours, or almost 1% of demand for electricity in the world" (Ayanoglu, 2019), and the numbers will certainly keep increasing. Groups have started to measure the energy usage of supercomputers, and there is a push toward vastly increased energy efficiency for high-performance computing. The most efficient supercomputers are tracked using the Green500 List.[7] However, the energy efficiency seen in supercomputing has not been incorporated into all aspects of computing. This is especially true for data analytics of increasingly large data sets, which is a critical aspect of chemical research and will only grow in importance. As with advances in computing power and miniaturization, chemistry will be a vital contributor to improving energy efficiency for supercomputing, data centers, and other high-performance computing needs, both through the creation of new, more sustainable architectures such as neuromorphic computing (see Section 4.3.2.1) and in materials chemistry innovations that lead to greater sustainability for all computational systems.

---

[6] See https://www.energy.gov/eere/femp/energy-efficiency-data-centers.
[7] See https://www.top500.org/lists/green500/2021/11/.

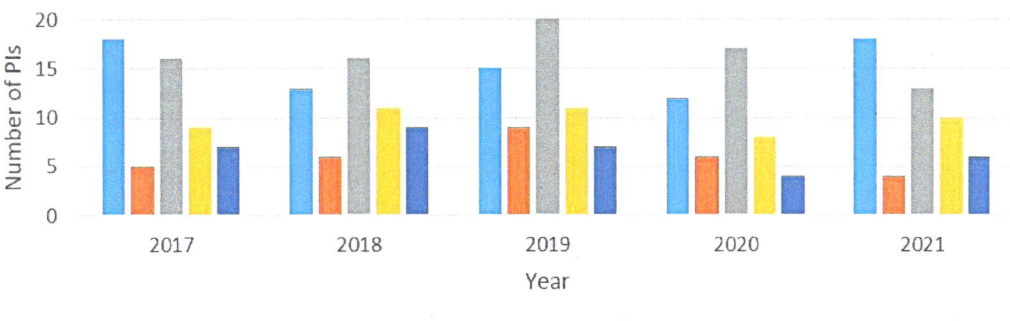

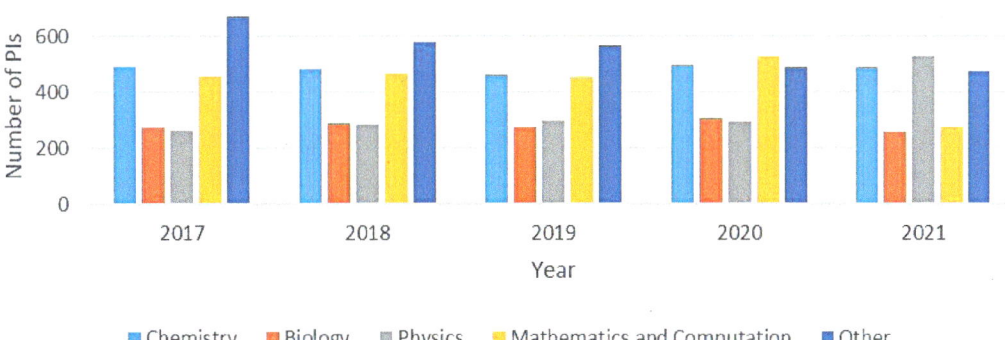

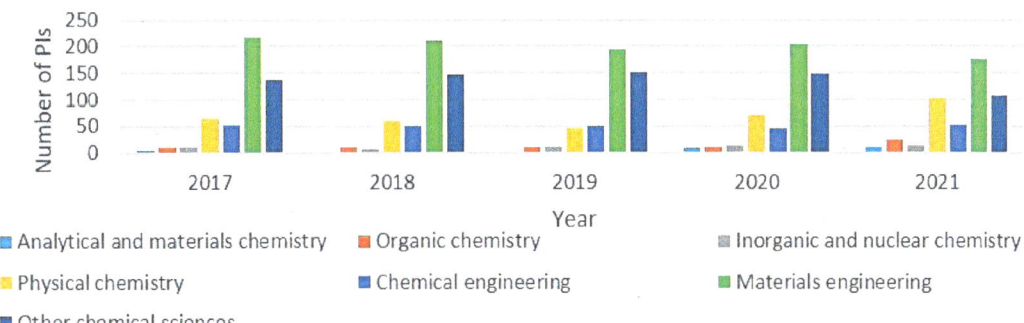

FIGURE 4-7 Chemistry usage of XSEDE and INCITE supercomputer systems by discipline of research. (a) Total number of PIs in different disciplines who use INCITE. (b) Total number of PIs in different disciplines who use XSEDE. (c) Total number of PIs in each chemistry subdiscipline who use XSEDE. SOURCES: Data from Palmer et al., 2015, and INCITE awards website, http://www.doeleadershipcomputing.org/awardees/.

4.3.2 Looking Toward the Future of Computational Chemistry

Miniaturization, increased computational power, and faster computational speeds have vastly increased the size and complexity of chemistry problems accessible to computation, but we are reaching the limits of miniaturization. As transistors are reaching scales at the size of atoms, we can go no smaller. In addition, as computation has become a regular and important component of chemical research, and data science becomes embedded across a wide range of disciplines, energy consumption becomes a limiting factor. Developments for the future will need to focus on improving sustainability and energy efficiency across all scales of computation—supercomputers, data centers, and personal computing devices—and on new computing paradigms that allow us to move beyond Moore's Law. As we explore new materials and computer architectures, chemistry and chemical engineering will continue to play a role in making fundamental discoveries to drive sustainability in computation, just as they drove miniaturization in the past. As was the case with development of our current semiconductor architectures, new materials and computer architectures will drive algorithmic innovation that enables us to ask new chemistry questions in entirely new ways.

4.3.2.1 Quantum Simulation with Quantum Computers for Chemistry

Nonclassical computing architectures such as quantum computers represent a new class of computational tools for chemistry simulations. These computer architectures will complement classical tools since performing certain calculations with quantum computers provides a computational resource advantage for specific problems in quantum chemistry. Quantum computers behave fundamentally differently from classical computers because they are made of different types of materials than semiconductors and store and encode information based on the principles of quantum mechanics (Figure 4-8). Though the development of quantum computers is an active area of research, several classes of them have begun to emerge. At a high level, these include ion trapping, transmon-based technologies involving superconductors, and theoretically topological quantum computers. Quantum bits, "qubits," are being produced today that leverage ion-trapping materials and superconductors. (At the time of this report, topological quantum computers had not been actualized.)

Since quantum computers operate under fundamentally different principles than do classical computers, they are best suited to tackle certain types of problems. As a concrete example of the differences in problem solving, there is no advantage (speed up) to using a quantum computer for addition or multiplication operations, but there is a theoretical advantage for number factoring (e.g., Shor's algorithm). Through the construction of quantum circuits, programmers leverage quantum mechanical properties of the qubits—such as superposition, entanglement, and interference—for computation. Thus, quantum computers are best suited to handle complex, interconnected information.

Chemists may intuitively associate complex, interconnected system problems with electron correlations, and indeed, the behavior of electrons in a molecular system is well described by quantum algorithms. Quantum chemistry calculations using Schrödinger's equation are typically exponential in resource requirements. As the problem size gets larger, the calculation quickly becomes intractable on classical devices. So, "exact" calculations on molecules, where the entire set of particle interactions in a system is simulated, become impossible. Furthermore, the Pauli exclusion principle adds to the cost because two electrons cannot sit at the same energy with the same spin, which is referred to as the "sign problem." State-of-the-art full-configuration interaction calculations are intractable past approximately 20 orbitals and approximately 20 electrons, even with today's most powerful supercomputers. The calculations can take more than a week to run, and calculations for reactions with popular approximate computational chemistry methods such as density functional theory (DFT) can be inaccurate. To mitigate these problems, simulations involving DFT invoke approximations to avoid high computational cost—for example, modeling the distinct electrons in an "electron field" compared to calculating all of the distinct electron–electron interactions. Determining the energy landscape for a reaction mechanism can take months or years of trial and error.

Although all of these issues with classical computers might be solved using quantum computers, there is still a large overhead associated with quantum chemistry calculations on quantum computers. A quantum chemistry circuit can require millions of gates to do a full configuration interaction calculation of a molecule. For quantum algorithms, such as quantum phase estimation for chemistry simulations, error-corrected devices (e.g., universal fault-tolerant quantum computers) will be necessary in order to carry out the operations (Gambetta, 2020). Quantum computing in a fault-tolerant regime promises more accurate chemistry calculations (possibly leading to *predictive* chemistry) (Motta and Rice, 2022). However, while research continues in quantum hardware, software, and theory, researchers can test small examples on devices available today. Researchers will often use a simulator that runs on a classical computer to test resources for calculations before trying the calculation on a noisy real device. The quantum computers all have differences in connectivity, number of qubits, and noise profiles, which enables customization for a calculation of interest. How to best leverage the hardware for chemistry calculations remains an open research question.

Beyond quantum chemistry, quantum computing might have advantages for chemistry broadly, through quantum ML, optimization, and Monte Carlo calculations. Applied to chemistry examples, it is possible to imagine exploring quantum computers for classification, anomaly detection, conformation optimization, solvation, and kinetics, to name a few. Finally, it is feasible that due to hardware performance differences, different types of quantum computers may be better suited for different classes of problem solving. Thus, incorporating quantum computers into computational chemistry workflows is a promising use of this emerging technology.

4.3.2.2 Biology-Inspired Computer Architecture

Neuromorphic computers, also referred to as "brain-inspired computing," aim to maximize the energy efficiency of computational processing and communication, thereby offering a low-energy computing platform, especially when combined with accelerators designed with the same principles (Schuman et al., 2017). New and emerging hardware for neuromorphic computing can operate with energy input on the order of milliWcm^{-2}, whereas complementary metal oxide semiconductor (CMOS) technologies operate in the 50–100 Wcm^{-2} range (D. Liu et al., 2021). Given the significant energy input required to run the world's silicon-based supercomputers, offloading key computations to low-energy requirement architectures would improve the sustainability of computation. Additionally, since the neuromorphic architecture handles data input differently than conventional computers, neuromorphic computers may excel at certain tasks for chemistry applications. A key research question will be to determine which tasks should be managed by neuromorphic, quantum, and conventional architectures.

Neuromorphic computing is a new architecture made up of artificial neurons and synapses and inspired by the current understanding of how brains work (Schuman et al., 2017). These architectures aim to capture properties of how brains function as learning machines that operate efficiently. Current semiconductor architectures are made up of large numbers of separate computational (CPU) and memory components, and current computational algorithms aim to maximize the simultaneous usage of all CPU and memory components to attain computational complexity and speed. In contrast, neuromorphic architectures make use of a vast number of relatively simple processing components that contain both memory and computational properties in one device. Algorithms for these architectures sparsely activate only a small number of the processing components at any time.

FIGURE 4-8 IBM Quantum scientist Dr. Maika Takita in the Thomas J. Watson Research Center IBM Quantum Lab. SOURCE: https://www.flickr.com/photos/ibm_research_zurich/51098680334/.

Neuromorphic computers are incarnated as neurons and synapses on traditional CMOS chips, in mixed analog and digital systems known as MEMristors,[8] and, increasingly, in future devices that are modeled after actual chemistry-driven synapses. Examples of the latter include water–lipid mixtures (DOE Office of Science, 2020b) and electron transport through ion channels (Lee et al., 2021; Tang et al., 2019). These shifts in architecture hold promise for the creation of computers that can use energy more efficiently by orders of magnitude.

A key challenge for making neuromorphic computing a reality is the feedback between architecture and algorithm. All aspects of computing are being rebuilt at the same time, and neuromorphic hardware needs to exist to motivate creation of new algorithms, while an understanding of neuromorphic-appropriate algorithms is necessary to drive the design of appropriately matched algorithms. Solving this challenge will require true multidisciplinary research at the intersection of chemistry, biology, neuroscience, and computer science.

4.4 CATALYSIS

Catalysts are used in most of the chemical reactions that make commonplace products, such as fuels, food, pharmaceuticals, synthetic fibers, and plastics. They drive chemical processes by facilitating the rearrangement of atoms and molecules. It is estimated that more than 95% (by volume) of everyday products contain ingredients that are made using catalysts, accounting for >80% of added value in the chemical industry (Hagen, 2015). Despite the well-established importance of catalysis, feedstock changes, energy transitions, and other environmental, health, and safety concerns will bring about fundamental changes in the chemical economy requiring new approaches for catalytic systems as well as a reexamination of previously explored technologies. In particular, the fields of electrocatalysis, photocatalysis, and biocatalysis are expected to play significant roles in the resource-efficient conversion of emerging feedstocks.

---

[8] More information on MEMreistors available online at https://www.nanowerk.com/memristor.php.

4.4.1 Homogeneous Versus Heterogeneous Catalysts

Two main types of catalysts drive reactions: homogeneous and heterogeneous. Homogeneous catalysts are in the same phase as the reaction they are catalyzing, usually meaning they are in the liquid phase, while heterogeneous catalysts and reactants are in different phases. Both types of catalysts are employed in the chemical economy; however, because homogeneous catalysts are less thermally stable at the elevated temperatures used in many high-volume petrochemical applications, heterogeneous catalysts are more commonly used. More than 80% of industrial catalytic processes are based on heterogeneous catalysis where the catalyst exists as a solid and the reactants are present in a gas and/or a liquid phase surrounding the catalyst (Wacławek et al., 2018).

Even in systems where heterogeneous catalysts dominate commercially, there are often homogeneous catalyst systems as well. For example, while heterogeneous Ziegler–Natta catalysts produce the majority of polyolefins, the commercialization of single-site catalysts in the 1980s has allowed exceptional control of polymer molecular weight and weight distributions, branching, stereochemistry, and a wide range of other properties not possible using heterogeneous catalytic systems (Stürzel et al., 2016). Continued progress in ligand design has expanded the thermal stability of some homogeneous polyethylene catalysts to 160°C and greatly extended the potential application space (Klosin et al., 2015). Design of new homogeneous polymerization catalyst systems remains a robust area of research (Chen, 2018).

Homogeneous catalysts typically have superior selectivity over heterogeneous catalysts because ligand selection can be used to tune the electronic and steric properties at the metal site. As such, homogeneous catalysts are used in a number of fine chemical and pharmaceutical processes (Howard et al., 2006; Zecchina and Califano, 2017). The desire to improve the sustainability of pharmaceutical preparations is leading to the design of novel metal catalytic systems (Hayler et al., 2019). Application of techniques such as NMR and infrared spectroscopy enables direct investigation of the catalytic reaction (Howard et al., 2006) which when combined with computational methodologies promises to deliver even more detailed mechanistic understanding with the potential for increased selectivity (Durand and Fey, 2021).

One downside of homogeneous catalysts is that the recovery of the catalyst often requires additional materials and energy that could have adverse environmental impacts, while the separation of heterogeneous catalysts from the reaction mixture is usually quite simple. Because homogeneous catalysts are present at low levels, many of the standard separations techniques are not well suited for this application (Schnoor et al., 2019). Furthermore, near-quantitative catalyst recovery is often demanded by economic and/or product purity considerations, especially when precious metals are used. As a result, developing new approaches for catalyst separation is an active area of research (Cole-Hamilton, 2003; Vural Gürsel et al., 2015). Use of membrane separation (Janssen et al., 2011; Schnoor et al., 2019; Xie et al., 2020), soluble polymer supports (Bergbreiter et al., 2009; Xie et al., 2015), magnetically separable catalysts (Kazemi and Mohammadi, 2020), and mimicking homogeneous metal sites through metal organic frameworks (Drake et al., 2018) are a few of the many approaches under exploration to address this challenge. These advances in homogeneous catalysis may have applications in processing biorenewable sources of carbon, such as conversion of lignocellulosics to chemicals, where low-volume high-value products are required for economic viability of biorefineries (Bender et al., 2018).

The importance of catalysis has long been recognized. Beginning with F. Wilhem Ostwald's 1909 Nobel Prize for the discovery of fundamental principles of equilibria and reaction rates, the scientific contributions of both homogeneous (including enzymatic catalysis) and heterogeneous catalysis have been recognized with Nobel Prizes more than 15 times (Thayer, 2013). Probably one of the more cited examples is the 1918 Nobel Prize awarded to Fritz Haber for "the synthesis of ammonia from its elements" (Nobel Prize Outreach, 2022a).

The conversion of atmospheric nitrogen into ammonia via the Haber-Bosch process, used today primarily to produce synthetic fertilizers (see Section 1.3.1), is a well-known example of a heterogeneous catalytic reaction. While the natural process of nitrogen fixation uses soluble molecules and enzymes as homogeneous catalysts, the Haber-Bosch process uses powdered iron particles as a heterogeneous catalyst. The key reaction steps between nitrogen and hydrogen occur on the surface of the iron particles.

The evolution of the field of heterogeneous catalysis is shown in Figure 4-9. Most heterogeneous catalysts comprise metal particles anchored to high-surface-area oxide or carbon supports. To maximize the number of active sites per mass of metal, the metal particle diameter is ideally on the order of a few nanometers. However, industrial catalyst synthesis and development are still largely based on trial-and-error approaches; accumulated practical knowledge; and the basic principles of physical chemistry, solid-state chemistry, and advanced experimental tools for characterizing surface reactions. New methods and tools in surface science are helping to address some of these gaps (see Box 4-4), but if catalysts can be improved with precision design to maximize activity, selectivity, and durability, the projected cost savings could be between $3 billion and $6 billion/year, with corresponding energy savings of 300–600 trillion BTU/year (Thayer et al., 2006).

Bridging homogeneous and heterogeneous catalysis continues to be a grand challenge (Cui et al., 2018). In heterogeneous single-metal-site catalysts (also referred as single-atom catalysts), the electronic properties of the atomically dispersed metal centers and their catalytic activity are tuned by the interaction between the metal and the neighboring surface atoms such as nitrogen, oxygen, or sulfur. This is similar to how the activity of metal centers in homogeneous catalysis is tuned by the choice of ligands. In this way, heterogeneous single-metal-site catalysis introduces new opportunities for bridging homogeneous and heterogeneous catalysis. Supported ionic liquid phases (SILPs) represent another advance in attempts to "heterogenize" homogeneous catalysts (Selvam et al., 2012). In SILP catalyst materials, a thin film of ionic liquid (IL), containing the dissolved (homogeneous) transition metal catalyst complex, is deposited within a porous solid, such as silica. Given that ILs have vanishingly low vapor pressure at typical reaction conditions, the IL film

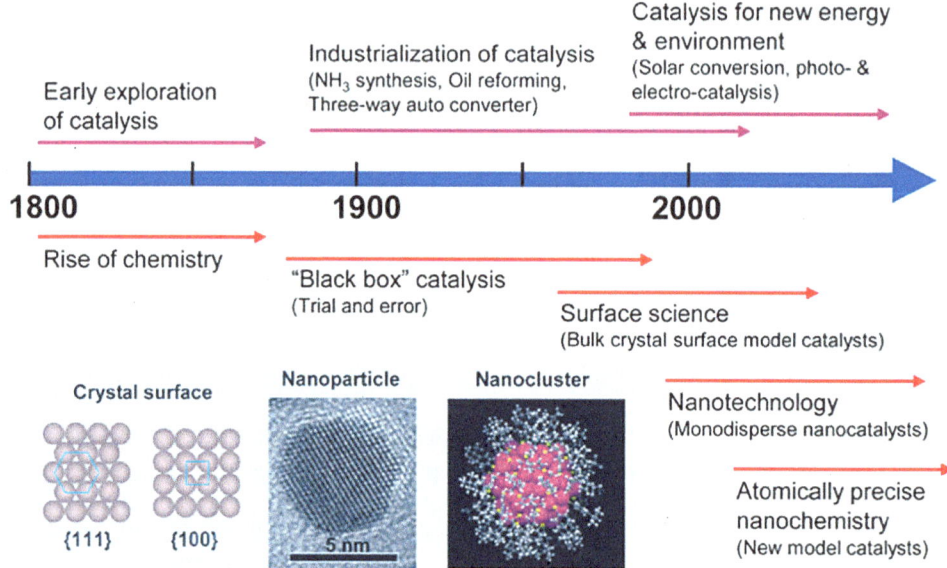

FIGURE 4-9 Evolution of heterogeneous catalysis. SOURCE: Jin, 2021.

BOX 4-4
Surface Science Contributions to Catalysis

It is well known that heterogeneous catalytic activity and selectivity is dependent on the geometric structure of surface metal atoms. To better understand how these catalysts work, measurements are taken to understand their surfaces. Surface science experiments are typically performed under ultrahigh vacuum (UHV) to avoid contamination issues and allow for the detection of electrons or ions from the surface being studied. While such studies have been invaluable to understand the thermodynamics, kinetics, and structure sensitivity of reactions on model surfaces, industrial catalytic processes occur over several orders of magnitude of pressure, at either atmospheric (~0.1 MPa) or higher pressures. The difference in pressure introduces the so-called "pressure gap" in the knowledge accrued from traditional surface science experiments and practical heterogeneous catalysis. To bridge this gap, surface science techniques that do not require UHV conditions are being developed. In addition, to model catalyst surfaces, some advanced techniques have also been deployed to study the structure and composition of practical heterogeneous catalysts in situ and operando in functional environments involving multiple phases including combinations of solid, gas, and liquid. Recent examples include the application of x-ray absorption spectroscopy to study the structure, composition, and dynamics of heterogeneous catalysts (Timoshenko and Roldan Cuenya, 2021) and scanning tunneling microscopy to probe surface structures of crystalline materials such as transition metals, oxides, and alloys exposed to gases in the milliTorr to atmospheric pressure range (Salmeron and Eren, 2021). These studies enable information on the atomic, chemical, and electronic structures of heterogeneous catalysts to be correlated with catalyst performance at relevant reaction conditions. Such tools are accelerating the field toward the goal of rationally designing and synthesizing advanced catalysts for specific function.

Advances in experimental surface science are being complemented by the development of microkinetic models and powerful quantum-chemical computational methods that allow simulation of catalyst surfaces and their interactions with the reaction mixture. A review from Motagamwala and Dumesic (2021) notes, "Microkinetic modeling is used to identify critical reaction intermediates and rate-determining elementary reactions, thereby providing vital information for designing an improved catalyst." Computational methods combined with spectroscopic techniques enable visualization of atomic-scale interactions at real reaction conditions. Such advances also help bridge the aforementioned material and pressure gaps between ideal and real reaction conditions and models, and thereby lead to a better understanding of the underlying fundamental catalytic phenomena (Xia et al., 2021).

Helped by advances in surface science measurement and computational tools, significant progress is being made in nanoscience that is relevant to the development of advanced catalytic materials. Metal nanoparticles of ultrasmall sizes (i.e., subnanometer to 2 or 3 nm) have shown "promise in heterogeneous catalysis because of their high surface-to-volume ratio, rich active surface atoms, and unique electronic structures as compared with their bulk counterparts" (Figure 4-4-1) (Gao et al., 2020). The geometric shape of a metal nanocrystal determines the atomic structure manifested on its surface. This tunability has been harnessed to enhance the performance of metal nanocrystals for a variety of industrially significant catalytic reactions including ammonia synthesis, methanol synthesis, Fischer-Trøpsch reaction, ethylene epoxidation, and C–C coupling reactions (Shi et al., 2021). For example, gold nanoparticles display high CO oxidation activity, copper-based nanoparticles promote efficient conversion of syngas to methanol, and plasmonic nanoparticles of silver and gold are effective for photocatalytic water splitting and CO_2 reduction.

continued

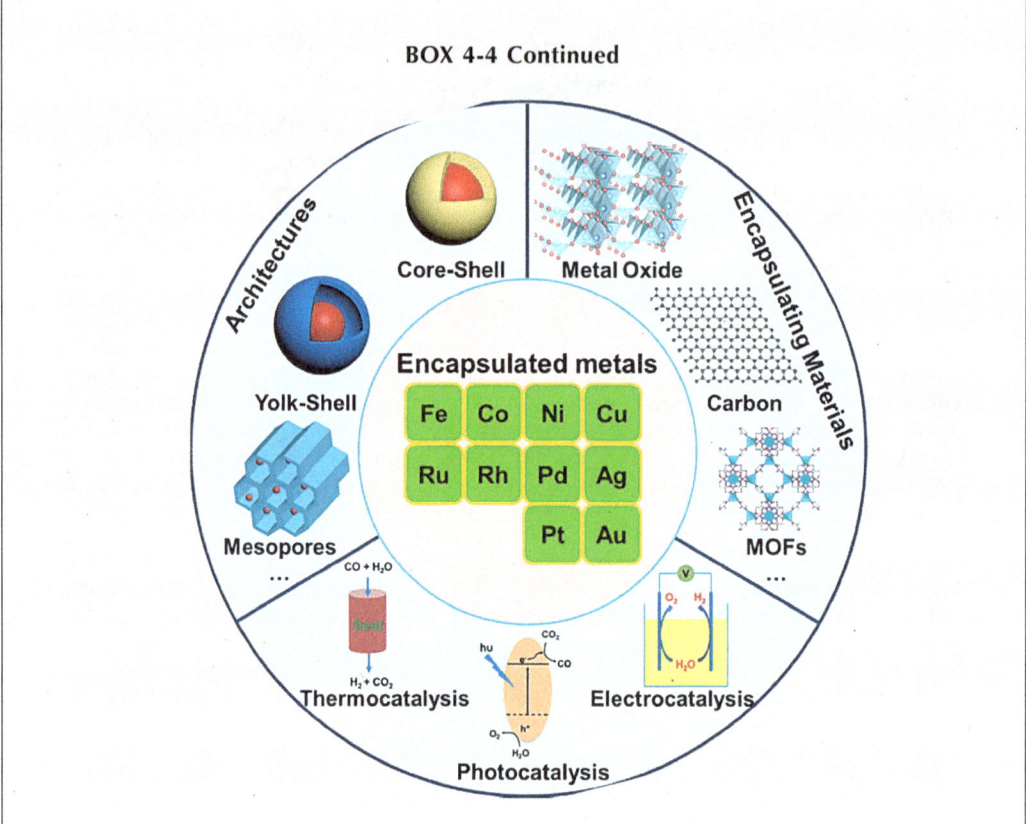

FIGURE 4-4-1 Encapsulation of metal nanoparticles with different materials and architectures for catalytic applications. NOTE: MOFs = metal-organic frameworks. SOURCE: Gao et al., 2020.

remains relatively stable during continuous fixed-bed reactor operation (Riisager et al., 2005). SILP catalysis has been demonstrated for industrially important reactions, such as hydroformylation, hydrogenation, carbonylation, hydroaminomethylation, metathesis, water-gas shift reaction, and oxidations (Fehrmann et al., 2014). The SILP catalytic materials have been demonstrated to surpass heterogeneous counterparts in terms of activity and selectivity. A better fundamental understanding of the physicochemical processes underlying SILP catalysis as well as its stability in multiphase environments is key to broader practical application of this uniquely tunable system.

4.4.2 Reemergence of Photocatalysis, Electrocatalysis, and Biocatalysis

While alternative methods to classical metal-based catalysis, such as photocatalysis, electrocatalysis, and biocatalysis, have been under consideration for decades as useful methods to drive chemical reactions, they are recently experiencing a resurgence in popularity as alternative methods to classical metal-based catalysis (Figure 4-10). The literature describing new transformations via electrochemistry, photochemistry, and biocatalysis portends their potential for vast applications in materials synthesis, chemical production, and new pharmaceutical and agrochemical molecules. Though there are many challenges to be overcome, these methods provide an alternative to

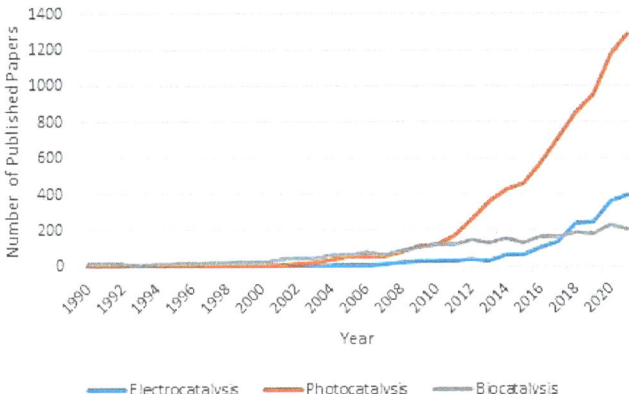

FIGURE 4-10 Number of published papers in the SCOPUS database that mention "electrocatalysis," "photocatalysis," or "biocatalysis," and "organic synthesis." Each search was done with "electrocatalys*," "photocatalys*," or "biocataly*," and "organic synthesis." Please note that the number of publications related to "biocatalysi," is likely not as accurate as the other publication numbers due to the fact that the study of biocatalysis is frequently described using other terms such as "synthetic biology" or "metabolic engineering."
SOURCE: Data from https://www.scopus.com/search/form.uri?display=basic&zone=header&origin=resultsl ist#basic.

energy-intensive synthesis methods that use high heat, high pressure, or both to perform chemical transformations in the production of some industrial chemicals.

4.4.2.1 Photocatalysis

Photochemistry is the use of light (i.e., radiant energy) to excite the electrons of a molecule to drive chemical change. The most ubiquitous photochemical reaction is photosynthesis, performed by phototrophic microorganisms and plants that use solar energy to convert carbon dioxide (CO_2) and water into glucose and oxygen.

The emergence of photochemistry as a sustainable technology for the synthesis of complex organic molecules can be attributed to developments over the past decade in light sources, photocatalysts, and flow reactors. Advancements in light-source technology, specifically light-emitting diodes (LEDs), fabricated from layered crystalline semiconductor material, offer advantages over compact fluorescent light sources. LEDs are smaller, more energy efficient, and able to generate light with very narrow emission spectra. Though the energy content of visible light is often insufficient to directly cleave bonds and induce chemical reactions, this can be addressed by adding a catalyst molecule that can more effectively adsorb the radiation, and then exchange the energy, electrons, or hydrogen atoms with other organic molecules. In the past decade, a large library of photocatalysts has been developed that can catalyze a wide array of reactions (Djurišić et al., 2020). These include photochemical transformations like C–H activation, C–H arylation, C–H alkylation, C–O bond formation, C–N bond formation, fluorination, dehalogenation, isomerization, rearrangements, cycloadditions, and many more (Djurišić et al., 2020). Some other specific advantages of photocatalysis include the ability to access highly saturated molecules or to modify chemically obstructed areas of molecules, which are common needs in many pharmaceutical syntheses.

Photochemical reaction rates are dependent on the local light intensity; thus, the reactor configuration and light source placement are critical to reaction performance. Reactors with shallow dimensions provide efficient irradiation of the entire reaction medium and would be favored for photochemical reactions. These requirements can be met by using continuous flow reactors as discussed in Section 4.2.2.1. Custom flow microreactors for performing photocatalytic organic chemistry have proliferated thanks to 3-D printing and inexpensive LEDs and materials. Commercial flow reactors for reaction screening are now available as well. With these advances in light sources, catalysts, and flow technology, photocatalytic reactions can be performed with precise control over the reaction progress (Buglioni et al., 2022). Although there are a few industrial processes using photochemistry (e.g., see Box 4-5), large-scale photocatalysis of complex organic molecules is still an emerging area.

BOX 4-5
Photocatalysis in Trifluoromethylation Reactions

Photocatalytic chemistry in organic synthesis is providing pathways for incorporation of functional groups under greener and more efficient conditions. Straathof et al. (2014) note that "S–CF$_3$ bonds are important structural motifs in various pharmaceutical and agrochemical compounds." Construction of these bonds via direct trifluoromethylation of aryl and alkyl thiols requires harsh conditions, long reaction times, and expensive reagents. A mild and fast photocatalytic trifluoromethylation of thiols was reported recently by the Noël Group (Straathof et al., 2014).

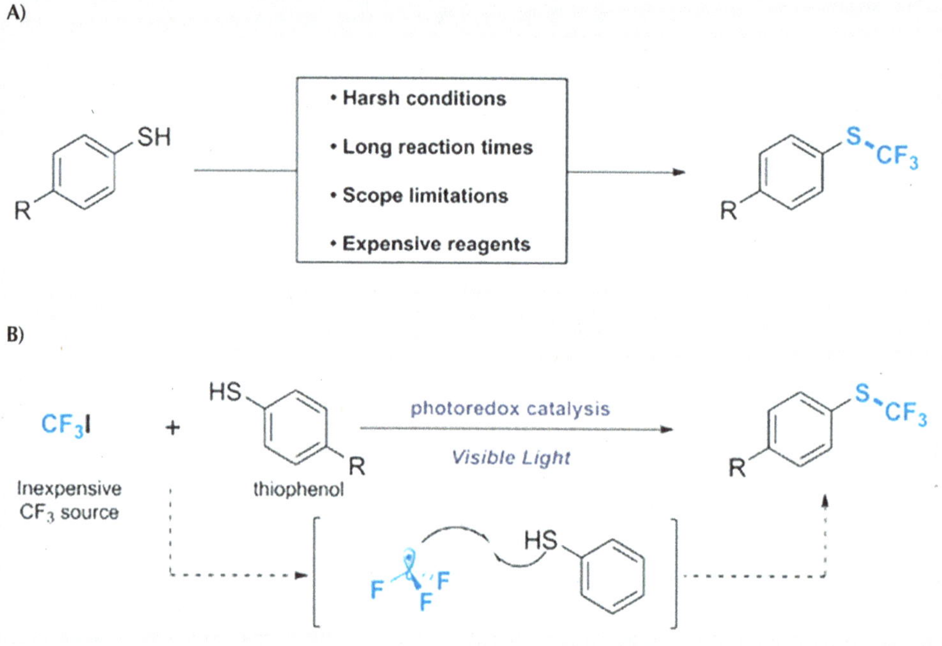

A)

[Reaction scheme: R–C$_6$H$_4$–SH → R–C$_6$H$_4$–S–CF$_3$ with issues: Harsh conditions, Long reaction times, Scope limitations, Expensive reagents]

B)

[Reaction scheme: CF$_3$I (Inexpensive CF$_3$ source) + HS–C$_6$H$_4$–R (thiophenol) → R–C$_6$H$_4$–S–CF$_3$ via photoredox catalysis, Visible Light; intermediate shows ·CF$_3$ radical and HS–C$_6$H$_5$]

FIGURE 4-5-1 (A) Reported direct trifluoromethylation of benzenethiols and (B) direct trifluoromethylation of thiols via photocatalytic CF$_3$* generation. SOURCE: Straathof et al., 2014.

This chemistry was recently demonstrated at commercial scale in a plug flow reactor using LEDs as the light source. The reaction generated high yields of an early intermediate molecule used in the pathway for a pharmaceutical drug candidate (Harper et al., 2022).

4.4.2.2 Electrocatalysis

Electrochemistry uses electrical potential to drive the arrangement of electrons to create a chemical change. Within an electrolytic cell, electricity is used to drive an oxidation-reduction (redox) reaction to generate a flow of electrons from the anode to the cathode. One of the first electrochemical reactions was the decomposition of water into hydrogen and oxygen by electrolysis, discovered in 1800 (Smolinka, 2009). This soon led to the discovery of electroplating (addition of metal to surfaces) and electropolishing (removal of metals from surfaces). These are examples of direct electrolysis, in which molecules undergo electron transfer directly at the electrode surface. Electrochemistry also includes indirect electrolysis, in which a catalyst serves as a mediator of the electron transfer. Indirect electrolysis as well as advances in laboratory equipment for these reactions are now enabling complex molecular transformations. These transformations include carboxylation, C–N formation, functionalization of C–O, methoxylation, oxidative intermolecular coupling, and oxidative intramolecular cyclization. Electrochemical reactions using common feedstocks such as CO_2, ammonia, and water to supply carbon, nitrogen, and oxygen, respectively, can form the basis to build other chemical feedstock molecules. These are essential transformations as we move toward electrification and decarbonization of the chemical industry (Schiffer and Manthiram, 2017).

Until this century, electrochemistry had not been broadly investigated by organic chemists as a reaction for high chemoselectivity to create new molecules. The setup for these reactions was believed to be too complex, with too many reaction variables to consider, such as electrolytes, electrode composition, cell type, and potentiostat. In addition, there was a lack of standard instrumentation and protocols to run these reactions. The fundamental research work of several academic organic chemistry groups, starting around 2000, showed that these types of electrochemical transformations were feasible (Moeller, 2000). The discovery of new electrocatalysts that can lower the overpotential of electron transfer and impart chemo-, regio- and stereoselectivity during the transformation of reactive intermediates is another significant advancement in the reemergence of electrochemistry. Strategies to provide precise control over the reaction progress now include externally controlling the electrical input, establishing new reactivity by coupling multiple redox events, and using new electroanalytical tools to understand the mechanisms of redox transformation. The growing interest in electrochemistry by organic chemists is driving the integration of electrochemical reactions into flow systems and the development of commercially available reactor systems to screen electrochemical reactions, such as the IKA ElectraSyn 2.0.[9] Electrochemistry is poised to expand from the academic laboratory to industrial R&D, as industrial organic chemists adopt electrocatalysis for the development of new molecules.

4.4.2.3 Biocatalysis

As nature's catalysts, driving biochemical reactions for billions of years, it may seem peculiar to highlight enzymes as an emerging area of chemical research leading to new and impactful applications. Furthermore, biologically driven catalytic processes used for food and beverages date back to the earliest known civilizations. However, it is only quite recently that biocatalysis has proven to be useful for large-scale, targeted synthesis of a wide array of chemical compounds.

At first glance, the scope of reactions catalyzed by enzymes may seem too narrow to be suitable for broad use in the chemical industry. These specialized proteins are classified by the Enzyme Commission of the International Union of Biochemistry and Molecular Biology into six classes: oxidoreductases, transferases, hydrolases, lyases, isomerases, and ligases. Each class is further delineated based on the functional groups on which the enzyme acts, the reaction being performed,

---

[9] See https://www.ika.com/en/Products-Lab-Eq/Electrochemistry-Kit-csp-516/ElectraSyn-20-Package-cpdt-20008980/.

and the nature of the substrate. It is true that unmodified enzymes have been used for many decades in industries that range from food ingredients (e.g., glucose isomerase to produce high-fructose corn syrup) to pharmaceuticals (e.g., lipase-catalyzed hydrolysis to resolve enantiomers), proving that biocatalytic production systems are scalable.

But a significant breakthrough, based on fundamental chemical and biological research, has been the development of methods for precisely modifying the structure, and hence the function, of enzymes. These tools allow researchers to engineer new enzymes. In one example, researchers at Merck and Codexis engineered a transaminase to replace a rhodium-catalyzed reaction step in the production of sitagliptin (see Section 2.3.3.2 for more details). The resulting process increased productivity and yield and decreased waste. Additional innovations in enzyme design, including using computational approaches, have produced biocatalysts that perform new reactions (Siegel et al., 2010), create new molecules (Coelho et al., 2013), and incorporate new atoms (Kan et al., 2016).

The advantages of biocatalysis are well known, including moderate reaction temperatures and pressures, minimal organic solvent waste due to the use of largely aqueous solutions, and exquisite regio-, stereo-, and substrate specificity. Naturally, there are disadvantages, too, though these vary widely and are enzyme specific. They include low catalytic rate constants, poor enzyme stability, low total turnover numbers, and a limited substrate range, with respect to both the molecular structure surrounding the target functional group and the types of atomic bonds that can be formed. Recent advances in addressing these limitations, however, suggest that biocatalysis is primed to play a much larger role in chemical transformations, especially as work moves toward more sustainable chemical production.

4.4.2.4 Scaling Up Photocatalysis, Electrocatalysis, and Biocatalysis

Advances in electrocatalysis, photocatalysis, and biocatalysis have created new reactions with promising applications in many different industries. However, these chemistries differ from traditional organic chemistry in that scale-up equipment and strategies are more complex. For traditional organic reactions, the general reaction parameters (e.g., solvent, concentration, additives, catalysts, and temperature) and reactor-related parameters (mixing conditions, heat transfer, chemical compatibility, and residence time) are well understood and quantifiable. However, for electrocatalysis, photocatalysis, and biocatalysis, industrial-scale conditions and processes have to be developed, tested, and optimized.

Scaling up electrocatalytic reactions requires consideration of the standard potential, overpotentials, faradaic efficiency, electrolyte choice and concentration, resistance of the electrolyte, and variations in voltage and current. Existing large-scale electrochemical reactor units are either optimized for a specific reaction type or built for aqueous systems and not compatible with organic solvents.

In the case of photocatalysis, there are radiation-related reaction parameters to optimize, such as wavelength-dependent quantum yield, molar extinction coefficient of the reaction species, and wavelength selectivity of the desired reaction. Additionally, there are reactor-related parameters to work out, such as the wavelength and intensity of the light source, the reactor geometry (which impacts light distribution), the photon flux on the reactor walls, and the reactor material absorbance. A range of scale-up units for photochemistry have been custom developed using capillary and tubular plug flow reactors, immobilized photocatalysts, thin-film reactors, rotor-stator reactors, and continuous stirred tank reactors (Cambié et al., 2016). Customization has been aided by using 3-D printing to rapidly prototype designs.

Many of the reaction, energy source, and reactor parameters are coupled in electrochemistry and photochemistry. The challenge will be in decoupling these parameters to ensure that electrochemical and photochemical processes can be transferred across different platforms and scales.

To facilitate that, fundamental research in electrochemistry and photochemistry would be tied to research and development on the platform and reactor equipment that will permit scaling up these chemistries. Standardization of equipment for photochemistry and electrochemistry will enable the manufacturing supply chain to produce the new products and chemicals enabled by the chemistry.

Although there have been many successful translations of biological enzyme-mediated reactions into commercial processes, significant gaps remain that currently limit the ability of biocatalysis to play a larger role in sustainable chemical synthesis. The scope of known enzymatic reactions is far less than the full slate of reactions accessible through nonbiological catalysis. While multiple reaction schemes are possible for any specified target, it stands to reason that broader replacement of thermochemical catalysis necessarily requires an expansion of the chemistry that can be performed by enzymes. However, it is also the case that these new biological catalysts must perform well outside of academic laboratories. New insights are needed to enable not just proof-of-concept synthesis, but large-scale production. Directed evolution, for example, has proven to be a powerful tool to improve the function of both naturally occurring and de novo designed enzymes. A significant limitation to its use, though, is the need to generate extremely large libraries (typically on the order of 10^9 variants) to search for the best-performing enzyme. Advances in measurement could launch new screening methods, but an alternative approach is reliable computational design. New knowledge leading to greater understanding of sequence–structure–function relations could lead to predictable enzyme engineering, requiring tractable library sizes on the order of 10^2 rather than 10^9.

4.4.3 Future Challenges in Catalysis

Fundamental challenges in catalysis and surface science are articulated in two DOE reports (De Yoreo et al., 2016; DOE Office of Science, 2017). To address these challenges, DOE established five research priorities: considering both the binding site and allosteric effects when designing catalysts, understanding the dynamic evolution of catalysts, manipulating reaction networks in complex environments to selectively steer catalytic transformations, designing electrocatalysts that are highly selective and energy efficient, and driving new catalyst discoveries by coupling data science, theory, and experiment. Research programs and centers funded by the DOE Office of Basic Energy Sciences are addressing several of these challenges. The goal is that fundamental science advances and tools made available by this research will lead to precision design of catalysts that maximize yields of desired products under mild conditions, thus conserving both feedstocks and energy. While catalysis has been instrumental in major industrial success stories such as ammonia synthesis (Section 1.3.1) and the catalytic converter (Section 2.3.3.5), there are processes that have eluded a catalytic solution for decades. An example is the steam cracking of naphtha to produce ethylene, a major platform chemical with an annual capacity of 160 million tons. Among petrochemical processes, this one consumes the largest amount of energy with the endothermic cracking reaction accounting for roughly 13–22% of overall energy and the downstream separation of ethylene from ethane accounting for the rest (Wong and van Drill, 2020). Ongoing efforts to power the steam cracking furnaces with renewable electricity (the so-called electrification of steam crackers) are expected to significantly reduce the carbon footprint of this energy-intensive process (Hydrocarbon Processing, 2021). Efforts to develop a catalyst that further reduces energy requirements for this endothermic process have been challenged by rapid catalyst deactivation due to coking and catalyst stability issues. Advances in heterogeneous catalysis, materials science, and process engineering are needed to effectively decarbonize current ethylene manufacturing (Gao et al., 2019).

Many industrial heterogeneous catalytic reactions, including the most relevant reactions related to energy and decarbonization, are limited by the Sabatier principle (Elnabawy et al., 2020; Montoya et al., 2015). The activity of a catalytic site is dictated by one or more rate-determining steps in the catalytic sequence of adsorption, surface reaction, and desorption. Overcoming this limitation is a fundamental research challenge in heterogeneous catalysis. Recently, Ardagh and colleagues (2020) put forward the catalytic resonance theory to understand and potentially enhance rate and selectivity in heterogeneous catalysis beyond the Sabatier limit. The theory recognizes that surface binding energy and transition state energies oscillate over time such that at a certain point, it may be easier for a reactant to adsorb on, or a product to desorb from, the catalyst surface. By superimposing an external wave at the catalyst surface that resonates with these oscillations, the simulation work shows that it may be possible to accelerate a surface-catalyzed reaction by several orders of magnitude compared to static conditions. The resonance theory was demonstrated experimentally for intensified formic acid formation via dynamic electrocatalysis (Gopeesingh et al., 2020). However, creating the required frequency of temperature or concentration oscillations in thermally driven conversions may be more challenging. This is an example of fundamental chemical research where experimentation and theory intersect to produce some groundbreaking possibilities for catalytic processes.

Plasma-driven heterogeneous catalysis is another area that shows promise for the chemical transformation of hard-to-activate molecules such as N_2, CO_2, and CH_4, all relevant in ammonia synthesis and in mitigating greenhouse gas emissions (Mehta et al., 2019). Mehta et al. (2019) note that there is a challenge in discovering catalytic materials

> that can take full advantage of a given plasma operating environment to selectively produce desired products. Key to the discovery of such materials is the development of relationships between plasma configurations, catalytic materials, and plasma-generated species by carefully and controllably combining plasmas with catalysts and isolating contributions of the catalyst from those of the plasma."

For example, using ozone, a powerful oxidant, at mild conditions is being tried in organic syntheses in flow reactors (Polterauer et al., 2021) and light alkane activation (Zhu et al., 2021). Electrification of the chemical industry using affordable renewable energy could spur increased use of plasmas and ozone in chemical transformation schemes.

Multiphase catalysis, involving gas, liquid, and solid phases, may be key to profitably processing emerging feedstocks such as biomass, CO_2, recycled plastics, and natural gas liquids. For example, catalytic hydroprocessing of biomass to produce fuel and chemical precursors is often performed in a liquid phase (Bagnato et al., 2021). There is growing interest in performing catalytic CO_2 hydrogenation in the liquid phase to produce fuels and chemicals (Mitchell et al., 2019). Similarly, chemical upcycling of plastic waste involves catalytic hydroprocessing in a melt phase (Celik et al., 2019). To understand how the multiphase environment around a catalytic site influences its elementary reaction steps, advanced experimental operando/in situ tools are needed to probe catalytic surfaces in condensed media under pressurized environments. Recent examples include the development of high-pressure operando x-ray absorption spectroscopy and NMR techniques for probing heterogeneous catalytic reaction mechanisms involving conversion of biomass model compounds (Walter et al., 2018).

The broad field of catalysis, encompassing heterogeneous, homogeneous, bio-, electro-, and photocatalysis, will be key to solving many of the challenges related to climate change and ensuring a sustainable supply of energy and materials. Success will require building a bridge between theory and experiments to understand how intrinsic material properties determine catalyst performance: what some people call the quest for the "catalyst genome" (Nørskov and Bligaard, 2013). This will require fundamental chemical research that integrates expertise in materials, measurement, and computation.

Research on catalysis covers broad length and time scales (Figure 4-11). Ideal catalytic conditions on a small scale may not produce the desired product at larger scale. To bridge this "scale gap," it is vital to consider reaction and process engineering aspects along with life-cycle analysis

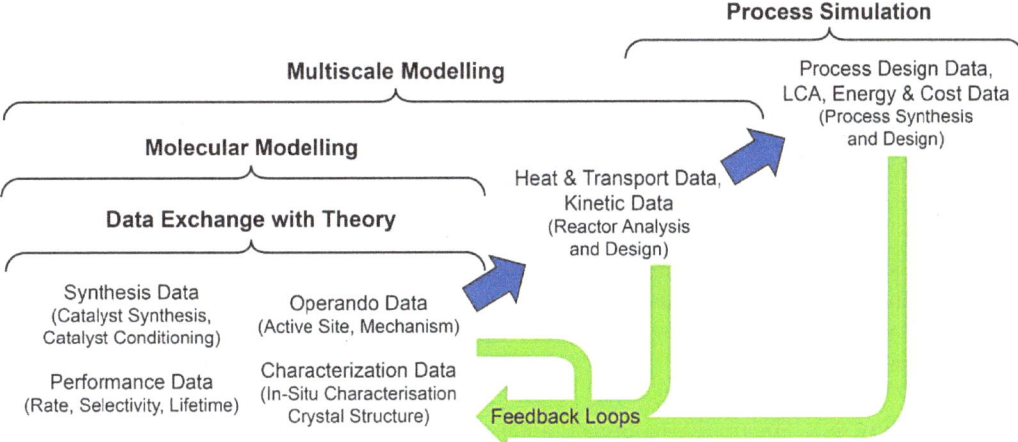

FIGURE 4-11 Data value chain for catalysis sciences spanning multiple time and length scales. SOURCE: Wulf et al., 2021.

during early stages of catalyst and process development (Figure 4-11). This will ensure that energy efficiency is considered across all manufacturing steps, including separation processes.

The urgent need to tackle sustainability-related challenges is driving the need to improve the pace and efficiency of discovering new catalysts and developing new catalytic processes. Toward this end, available materials, adsorption, and reaction data are being harnessed using high-throughput computation and ML to gather further insights and predict new materials (Bo et al., 2018). However, this knowledge extraction approach is challenged by a data deficit, or non-uniform reporting of materials-related data that are not computer readable (Himanen et al., 2019). Wulf et al. (2021) note that "in order to make data widely useful, rather advanced and well-coordinated approaches are needed that are beyond what a single group or institution can develop and sustain." Digitalization of the catalysis field is essential to "enable efficient data-driven interdisciplinary development of catalysts and catalytic processes" (Wulf et al., 2021). The NFDI4Cat (National Research Data Infrastructure for Catalysis-Related Sciences), supported by the German government, is an example of such a large-scale effort.[10] Establishment of an "internet of catalysis" will guide research along the development chain from molecules to chemical processes. The creation of digital workflows that bridge theory and experimental studies in catalyst design, characterization, kinetics, and related engineering aspects will accelerate discovery and innovation in the catalysis sciences.

4.5 CONCLUSIONS

This chapter explored state-of-the-art tools and technologies in the chemical sciences that are enabling, and enabled by, fundamental research. Advances in measurement, automation, computation, and catalysis have already driven chemical research forward, and there are many examples where advances in these tools have created subsequent advances in chemistry. Though there are a large number of enabling technologies that move chemistry forward, the committee identified these four because they are considered to have the biggest immediate impact and offer enormous potential for the future. The committee also noted some similar needs in these four emerging areas, especially around the use of chemical data. The committee reached a number of conclusions about

---

[10] See http://gecats.org/NFDI4Cat.html.

the importance of measurement, automation, computation, and catalysis for enhancing the chemical economy.

> **Conclusion 4-1:** Chemistry is an enabling scientific discipline that will continue to have the largest impact on society when chemists collaborate with experts from other areas such as engineering, biology, physics, computation, and data science to generate new fundamental knowledge and create translational impact at larger scales.

> **Conclusion 4-2:** Measurement, automation, computation, and catalysis are the enabling tools and technologies of fundamental chemical research that will have a substantial impact on both the adoption of novel methodologies and future discoveries in the chemical economy.

> **Conclusion 4-3:** The ability to collect, document, store, share, and use chemistry-related data is needed to advance the use of new tools, such as computation and automation in fundamental chemical research, and increase the accessibility of chemical research to a larger community of practitioners. This information architecture will produce an indispensable tool for the chemical sciences research community to increase the pace and efficiency of innovation by fully harnessing advances made with previous research investments.

In considering each of the four areas outlined in this chapter, important conclusions emerged related to measurement, automation, computation, and catalysis.

> **Conclusion 4-4:** Analytical chemistry will continue to play a substantial role in driving fundamental chemistry and having an important impact on numerous sectors including medicine, environmental and forensic monitoring, and national security, especially as analytical tools increase in speed and accuracy and instrumentation miniaturizes.

> **Conclusion 4-5:** Automation has the potential to change the way chemistry research is designed by increasing the number of syntheses, measurements, or process steps that can be done in rapid succession and by producing data that can be analyzed to help move research forward.

> **Conclusion 4-6:** Fundamental research in computational chemistry is fundamental research in chemistry. A synergistic relationship among advances in computer architectures, computational chemistry algorithms, and the application of computational chemistry enables innovation in all chemistry disciplines.

> **Conclusion 4-7:** Fundamental multidisciplinary research in chemistry, physics, and engineering has played a critical role in the ongoing development of modern computer architecture, and will continue to do so with the continued miniaturization of computers and the emergence of new architectures such as quantum and neuromorphic computing.

> **Conclusion 4-8:** Advances in catalysis remain critically important for driving new products and processes, and the subfields of catalysis covering heterogeneous, homogeneous, biocatalysis, electrocatalysis, and photocatalysis will all play key roles in advancing the fundamental science and technologies needed for making renewable energy, decarbonizing the chemical industry, and promoting a circular economy.

5

Preparing and Empowering the Next-Generation Chemical Workforce

Key Takeaways:

- Diversity in the chemical workforce is essential for the productivity of the chemical economy and is supported through inclusive and equitable practices in the classroom and workplace.
- Effective mentorship at all career levels is important for the success of individual members of the workforce.
- Minority-serving institutions excel at recruiting and retaining diverse chemistry cohorts.
- Flexibility in chemical sciences and engineering education is vital for the exposure of the workforce to new tools and emerging concepts and can be implemented through the support of basic education research.

Chemical advances are most impactful when research and development (R&D) are guided by diverse teams, and individuals are adequately supported. Empowering the chemical workforce to innovate requires equitable and inclusive practices, effective mentorship, and exposure to new knowledge and tools at every developmental stage, beginning at the K–12 level and continuing throughout the entirety of one's career. As noted throughout this report, innovations originating in the chemical sciences will be essential for addressing the significant challenges facing our planet today and in the future. The complex solutions to these challenges will benefit from a diverse and well-trained workforce with a deep knowledge of the chemical sciences and the ability to use cross-disciplinary tools and technologies. Whereas Chapter 2 of this report considers the impact of the entire chemical workforce on the U.S. economy, this chapter focuses on the preparation and support of the members of the workforce that drive research and innovation.

To remain within the scope of the committee's Statement of Task and best reflect the committee members' lived experiences and expertise, the discussion concentrates on those chemical scientists and engineers within or preparing to join academia, industry, or government.

5.1 A DIVERSE AND EQUITABLE CHEMICAL WORKFORCE

A talented chemical workforce is comprised of individuals whose intellectual curiosity in the chemical sciences was nurtured and reinforced through supportive pedagogy and opportunities to engage in impactful research. Cultivating interest in chemistry from grade school to graduate school and beyond ensures the continuation of a highly skilled workforce in the chemical sciences. It is critical that all community members are welcome in the exploration of chemistry. Thus diversity, equity, and inclusion (DEI) have become fundamental tenets of the sciences. The U.S. Department of Housing and Urban Development defines diversity as encompassing "the range of similarities and differences each individual brings to the workplace, including but not limited to national origin, language, race, color, disability, ethnicity, gender, age, religion, sexual orientation, gender identity, socioeconomic status, veteran status, and family structures."[1] Although many definitions of equity and inclusion exist, the committee has chosen to adopt the following:

> Equity is about the fair treatment and equal opportunity for success and advancement for all people, irrespective of their identities. Inclusion refers to an organization's active efforts to invite and nurture the participation of its diverse members. (Reisman et al., 2020)

While efforts have been made to cultivate and maintain diverse, inclusive, and equitable work and educational spaces, there continue to be strong barriers that prevent talented individuals from entering or staying in the chemical sciences. The Open Chemistry Collaborative in Diversity Equity (OXIDE) initiative has worked alongside *Chemical & Engineering News* since 2012 to collect data from Ph.D.-granting chemistry departments and publish demographic assessments of research-active tenured or tenure-track faculty.[2] From the 75 departments surveyed for the 2015–2016 academic year, only 19.4% of chemistry faculty were female, and only 5.9% of the faculty identified as underrepresented minorities (JHU, 2017). These numbers are reflective of the many significant barriers to entering and staying in the chemical workforce. According to a 2011 National Academies' report, underrepresented minorities have lower retention rates as science, technology, engineering, mathematics, and medicine (STEMM) majors at undergraduate institutions because they face more barriers to persistence and completion (NAS, NAE, and IOM, 2011). Beyond the better-known policy and structural barriers, there are also significant psychological barriers that marginalized populations face, such as discrimination and lack of social connections (Urbina-Blanco et al., 2020). As stated in a recent National Academies' report on mentorship in STEMM,

> Talent is equally distributed across all sociocultural groups; access and opportunity are not. . . . Individual STEMM professionals identifying as African American, Latinx, American Indian, first generation, or sexual or gender minority individuals and individuals with disabilities continue to be less likely to be successfully integrated in STEMM environments. These individuals may be questioned about their competence, challenged in their science, and simultaneously invisible as scientists, yet under the microscope as members of underrepresented groups in STEMM. (NASEM, 2019f)

The chemical enterprise does not reflect the potentially available talent, as true diversity in the chemical workforce is lacking and not representative of our nation's demographics. According to

---

[1] See https://www.hud.gov/program_offices/administration/admabout/diversity_inclusion/definitions.
[2] See http://oxide.jhu.edu/2/about.

the American Chemical Society *Diversity Data Report 2021* (ACS, 2021), far fewer women and nonbinary individuals submitted manuscripts for publication to ACS journals compared to men, and men made up 73% of publication reviewers. Additionally, white- and East Asian–identifying authors comprise almost 70% of all ACS published authors and experience higher publication acceptance rates than those from other identifying groups. While publications are only one way to measure scientific contribution, these numbers show that the chemical workforce still disproportionally centers on white men to stimulate growth and innovation in the chemical sciences, leaving behind experts from marginalized communities.

Interventions and support mechanisms are crucial to improve the diversity of the chemical workforce, as they work to improve equity and inclusion while breaking down systemic barriers so that inclusive excellence can be achieved. Such mechanisms are varied in scope, design, and application. At the level of the individual, actions that chemists can take to support workforce diversity include professional advocacy, value-based leadership, meaningful mentorship, bias training, and bystander intervention (Sanford, 2020). A 2020 *Journal of Organic Chemistry* editorial argues that "individual actions will not be sufficient. . . . It follows, then, that combating institutional bias requires us to hold our departments, journals, and scientific societies accountable to the principles of diversity, equity and inclusion that they proclaim as central values" (Reisman et al., 2020). There are many programs sponsored by such institutions across the nation that aim to enhance DEI in the chemical sciences and STEMM more broadly, and some examples of these programs are highlighted in Box 5-1. Other programs exist to support individuals from underrepresented communities who are pursuing an advanced degree. Table 5-1 presents an illustrative, rather than exhaustive, list of opportunities currently available at multiple career levels and sectors to support a diverse chemical workforce. For a more comprehensive list, the Institute for Broadening Participation[3] maintains a hub of opportunities for a range of education and career stages. Such programs were also highlighted at a recent National Academies' workshop hosted by the Chemical Sciences Roundtable and are summarized in a proceedings document (NASEM, 2021c).

While support mechanisms at every level of an individual's career are important, a diverse and equitable chemical workforce cannot exist without diverse and equitable recruitment and retention practices. In higher education, systemic issues exist broadly that prevent diversity, equity, inclusion, fairness, and justice, according to a National Academies' workshop proceedings (NASEM, 2022b). The proceedings notes that recruitment and retention processes in academia are "inequitable at every step," and a lack of diversity in these practices cannot be attributed in full to the lack of diverse Ph.D. candidates. Other resources, including a recent National Academies' consensus study, *Promising Practices for Addressing the Underrepresentation of Women in Science, Engineering, and Medicine: Opening Doors*, notes the inequities and major issues with hiring, recruitment, retention, tenure, and promotion structures within STEMM fields, especially those practices related to women of color (NASEM, 2020a). The report lays out a comprehensive set of recommendations and action items that universities and departments can take to assess their own recruitment and retention practices and to determine solutions for improvement. These recommendations include extensive quantitative and qualitative analyses of an institution or department to "understand the nature of their unit's particular challenge with the recruitment, retention, and advancement" (NASEM, 2020a). Another important facet of the strategy in that report is the emphasis on actions by leadership. One recommendation notes, "Leaders in academia and scientific societies should put policies and practices in place to prioritize, reward, recognize, and resource equity, diversity, and inclusion efforts appropriately" (NASEM, 2020a). Following this recommendation are several action items that include permanent positions in DEI, compensation for related efforts, and equitable promotion practices.

---

[3] See https://www.pathwaystoscience.org/urm.aspx.

BOX 5-1
Examples of Diversity, Equity, and Inclusion Programs in Chemistry and Chemical Engineering from a National Academies' Workshop

The Chemical Sciences Roundtable of the National Academies held a virtual workshop on Diversity, Equity, and Inclusion in Chemistry and Chemical Engineering on May 25–26, 2021, to provide a space for academic, government, and industrial professionals to discuss the barriers to and best practices for creating more diverse, equitable, and inclusive environments in chemistry and chemical engineering classrooms and workplaces.

Freeman Hrabowski, president of the University of Maryland, Baltimore County and co-founder of the Meyerhoff Scholars Program,[a] and Geraldine Richmond, Presidential Chair in Science and professor of chemistry at the University of Oregon and founding director of the Committee on the Advancement of Women Chemists (COACh),[b] served as the event's keynote speakers. Hrabowski emphasized that it is important to recognize the progress that has been made in increasing the number of underrepresented minority students in STEM fields and to build off of that progress by replicating successful initiatives. Hrabowski stated that the success of the Meyerhoff Scholars Program is largely due to the commitment of leadership and their focus on the advancement of undergraduate students. Richmond shared factors that contribute to the low retention of women and underrepresented minority students in U.S. chemistry departments, including lack of peer support, inadequate funding, and lack of diverse representation among faculty mentors.

The 2-day workshop highlighted a wide range of other programs and initiatives that work to increase diversity, equity, and inclusion in the chemical sciences and engineering. Programs featured included the Organization for Cultural Diversity in Science[c] at the University of California, Los Angeles; the BUILDing Scholars program[d] (Building Infrastructure Leading to Diversity); the Open Chemistry Collaborative in Diversity Equity (OXIDE);[e] the Inclusive STEMM Ecosystems for Equity & Diversity (ISEED)[f] and the STEMM Equity Achievement (SEA) Change program[g] within the American Association for the Advancement of Science; the Sloan University Center of Exemplary Mentoring;[h] the All for Good: Engineering for Inclusion Program[i] and the Future of STEM Scholars Initiative (FOSSI)[j] within the American Institute of Chemical Engineers (AIChE); the ACS Bridge Program;[k] and the Gilliam Graduate Fellowship Initiative.[l] Also featured were industrial efforts within Merck and GlaxoSmithKline, a nonprofit called Accessible Science,[m] and an overview and discussion of universal design for chemistry and chemical engineering learning environments. All of the programs mentioned during this workshop have helped to spur change in increasing diversity and inclusion within the population of STEM practitioners, and can serve as helpful models for new programs at universities, companies, and professional societies.

---

[a] https://meyerhoff.umbc.edu/.
[b] https://coach.uoregon.edu/.
[c] http://uclaocds.weebly.com/.
[d] https://buildingscholars.utep.edu/.
[e] http://oxide.jhu.edu/2/about.
[f] https://www.aaas.org/programs/inclusive-stemm-ecosystems-equity-diversity-iseed.
[g] https://seachange.aaas.org/.
[h] https://sloan.org/programs/higher-education/diversity-equity-inclusion/minority-phd-program.
[i] https://www.aiche.org/giving/impact/stories/all-good-engineering-inclusion.
[j] https://futureofstemscholars.org/fossi.
[k] https://www.acs.org/content/acs/en/education/students/graduate/bridge-project/about-bridge-program.html.
[l] https://www.hhmi.org/science-education/programs/gilliam-fellowships-advanced-study.
[m] https://www.accessiblescience.org/.

As chemistry departments strive to recruit and retain diverse talent, topical research and reports can be a helpful resource when used in coordination with the consulting of knowledgeable individuals at each institution who understand the culture and nuance of the individual program under consideration. Taking actions based on best-practice guidance and recommendations from relevant experts and community members can start to fix the inequities that are prevalent at every stage of recruitment and retention in STEMM and move toward a more diverse workforce in the chemical sciences.

Creating and maintaining equity-minded opportunities can ensure that the chemical workforce best reflects our nation's communities and talents, thus expanding the potential for discovery and innovation. Additionally, supporting diversity and inclusivity within educational and work environments lends support to the recruitment and retention of global talent, which are vital for the United States to maintain a competitive advantage in the chemical sciences.

5.2 MENTORSHIP AND SUPPORT FOR SUCCESS

Productive and supportive mentorship is essential to success throughout one's career, from recruitment and training to retention and advancement. Mentorship can be defined as "a professional, working alliance in which individuals work together over time to support the personal and professional growth, development, and success of the relational partners through the provision of career and psychosocial support" (NASEM, 2019f). Mentoring relationships can take several different forms, including formal or informal, dyadic or multiple-mentor structures, or group or peer mentorship (NASEM, 2019f). Mentorship is a critical aspect of an individual's personal and professional development that is not limited to periods of formal training and education. In the chemical sciences, mentorship is historically associated with a dyadic or apprentice model, in which aspiring practitioners receive expert guidance to develop a particular technical skill set (Lagowski and Stewart, 2003). Whether actively or passively, mentors serve as models for professional behaviors that are often emulated at later career stages once the apprentice has transitioned into the role of mentor. Today, it is widely accepted that the role of mentoring encompasses not just guidance on the acquisition of specialized scientific knowledge and understanding but also development of professional skills that help the protégé navigate the path toward securing their first post-training position and then operating effectively within it.

5.2.1 Challenges in Effective Mentorship

Students pursuing chemical sciences degrees can have vastly different educational, research, and professional growth experiences, depending on their access to, and relationships with, faculty mentors. Mentorship at the undergraduate level, while dependent upon strong dyadic interactions, often occurs at the departmental level and therefore naturally includes engagement with a broader range of instructors and advisors or counselors. Graduate-level trainees generally have far more localized interactions. One's doctoral advisor typically serves as the primary faculty mentor for a student. Thus, the quality of one's graduate school experience is highly dependent on the approach to mentorship taken by a single faculty member. Given that a student's graduate school success and professional outcomes are influenced strongly by faculty members and the quality of their mentorship, it is paramount that professors cultivate strong mentorship skills early in their careers.

While faculty jobs require mentorship, many academic institutions do not afford their faculty adequate mentorship training with appropriate levels of support and recognition. Traditional prefaculty experiences focus nearly exclusively on the science and grant making related to research. Once hired, faculty often then resort to a "do-it-yourself" approach to mentorship, gleaning what they can from their own mentors, independently finding resources on best practices in mentorship

TABLE 5-1 Programs and Initiatives That Support DEI and Diverse Individuals in the Chemical Sciences

Career Level	Programs and Initiatives
Pre-college	<ul><li>Upward Bound Math-Science Program[a]<ul><li>A U.S. federal TRIO program that aims to help pre-college students from low-income families, or who would be first-generation college students, to recognize and develop their potential to excel in math and science and encourage them to pursue postsecondary degrees in those fields.</li></ul></li><li>American Chemical Society Project SEED Program[b]<ul><li>Provides opportunities for high school students with diverse identities and socioeconomic backgrounds to participate in summer research experiences at institutions of higher education across the United States.</li></ul></li><li>Native American Science & Engineering Program (NASEP)[c]<ul><li>Offers Native American, Alaskan Native, and Hawaiian Native high school students experiences for 1 year aimed at developing their understanding of STEMM career opportunities.</li></ul></li><li>Future of STEM Scholars Initiative (FOSSI)[d]<ul><li>Supports high school students who have an interest in pursuing careers in chemical manufacturing, engineering, environmental health and sustainability, or other related chemical-industry fields.</li></ul></li></ul>
Undergraduate	<ul><li>Ronald E. McNair Postbaccalaureate Achievement Program[e]<ul><li>A U.S. federal TRIO program that provides competitive funding to institutions of higher education to prepare students from underrepresented segments of society for later doctoral studies to increase the number of diverse Ph.D. holders.</li></ul></li><li>Building Infrastructure Leading to Diversity (BUILD)[f]<ul><li>Housed by the National Institutes of Health (NIH), the initiative provides awards to undergraduate institutions to help them engage and retain students from diverse backgrounds while supporting institutional transformation.</li></ul></li><li>Meyerhoff Scholars Program[g]<ul><li>Supports prospective undergraduate students who are interested in later pursuing graduate studies with the goal of increasing diversity of future leaders in STEMM fields.</li></ul></li></ul>
Graduate	<ul><li>University Centers of Exemplary Mentoring (UCEM)[h]<ul><li>Provides scholarships and other forms of support to help underrepresented minority students obtain a doctoral degree in engineering, the physical and natural sciences, or mathematics at a partner university.</li></ul></li><li>GEM Scholars Program[i]<ul><li>Works to increase the participation of underrepresented groups at the master's and doctoral levels in applied engineering and science by matching the interests of individuals with the needs of GEM employer members.</li></ul></li><li>WiscProf: Future Faculty in Engineering Workshop[j]<ul><li>A multiday event for Ph.D. students and postdoctoral scholars from underrepresented groups in STEM engineering fields to learn about academic career paths and the process of securing and succeeding within a faculty position.</li></ul></li><li>American Chemical Society Bridge Program[k]<ul><li>Works to increase the number of chemical science Ph.D.s awarded to underrepresented students through the creation of transition (bridge) programs and a national network of doctoral-granting institutions.</li></ul></li></ul>
Postdoctoral	<ul><li>MLK Visiting Scholars Program[l]<ul><li>Hosted by the Massachusetts Institute of Technology (MIT), this program works to increase the presence of minority scholars at MIT. Scholars participate in research and academic programs.</li></ul></li><li>Presidential Postdoctoral Fellowship[m]<ul><li>Hosted by Brown University, the fellowship supports Ph.D. graduates from underrepresented groups.</li></ul></li></ul>

TABLE 5-1 Continued

Career Level	Programs and Initiatives

Multilevel and post-academia	• National Science Foundation (NSF) includes Planning Grants[n]
	   ○ Includes planning grants to support the development of infrastructure to increase the capacity for collaborative innovation to broaden participation in STEMM fields.
	• The Committee on the Advancement of Women Chemists (COACh)[o]
	   ○ Provides career-building and -development opportunities for women in chemistry and assists institutions in developing their diversity and inclusion efforts.
	• Inclusive STEMM Ecosystems for Equity & Diversity (ISEED)[p]
	   ○ Leads initiatives to support underrepresented individuals in STEMM by evolving and reconstructing systems and structures to increase inclusivity.

[a] https://www2.ed.gov/programs/triomathsci/index.html.
[b] https://www.acs.org/content/acs/en/education/students/highschool/seed/locations.html.
[c] https://nasep.arizona.edu/about-nasep.
[d] See Box 5-1.
[e] https://www2.ed.gov/programs/triomcnair/index.html.
[f] https://www.nigms.nih.gov/training/dpc/pages/build.aspx.
[g] See Box 5-1.
[h] See Box 5-1.
[i] https://www.gemfellowship.org/.
[j] https://wiscprof.engr.wisc.edu/.
[k] See Box 5-1.
[l] https://mlkscholars.mit.edu/about/about-program.
[m] https://www.brown.edu/about/administration/institutional-diversity/initiatives/presidential-postdoctoral-fellowship.
[n] https://www.nsf.gov/pubs/2019/nsf19600/nsf19600.htm#:~:text=NSF%20INCLUDES%20Planning%20Grants%20are,broadening%20participation%20challenge%20at%20scale.
[o] See Box 5-1.
[p] See Box 5-1.

in the form of reading materials and workshops, or learning from a senior faculty member who can share their personal experiences with mentorship. Placing the responsibility on the mentor to teach themselves the skills of effective mentorship is increasingly challenging for early-career professors who face enormous pressure to achieve certain research and publication metrics that will determine their own career success. The ability to secure grants to support their research, publish a sufficient quantity of papers on the path to tenure, and establish a robust and effective teaching program are priorities that strongly compete with the prioritization of student mentorship.

These realities highlight a potential conflict of interest between the simultaneous roles of research supervisor, mentor, and independent professional that are embodied in the personage of a single faculty member. Research advisors, in their roles as mentors, may at times need to act against their own professional interests in favor of those of their students. Career exploration is recognized and emphasized by several federal funding agencies, such as the NSF and NIH, as an important element of professional development for trainees. Experiences within companies and corporations are particularly valuable for students of the chemical sciences, as many will eventually take on one of the large number of industrial roles that make up the chemical workforce. However, many graduate students are supported by federally funded grants with strict timelines that could be affected by a student's temporary departure from the laboratory to pursue an external internship. Delays can affect the overall productivity of an advisor's lab, which in turn can affect subsequent grant applications or even consideration for promotion and tenure. If the grant involves milestones that determine future funding, students or trainees themselves could be affected as declined grants can equate to lower levels of funding support for students or trainees.

Faculty are faced with weighing their individual professional needs, the needs of their research program, and the needs of their students, which can be in direct conflict. It is no wonder then that some advisors are unwilling to support the decision of a student to take leave from the research group even if it is to the benefit of the individual student. The natural power imbalance that exists between students and advisors may lead some students to forgo requesting temporary leaves, even if doing so would result in significant personal and professional benefits.

Beyond these challenges, many faculty are unprepared to help students navigate nonacademic career options. A recent survey published in *Nature* found that the majority of students arrived at their career decisions by doing their own research; only 28% reported that an advisor's advice helped to guide their decisions (Woolston, 2019) (Figure 5-1). Some advisors are simply unable to assist students with an interest in these paths, but others may display outright hostility to such "nontraditional" careers. As a result, students at all levels often suffer from a lack of sufficient support and career guidance. The 2019 *Nature* survey also showed that nearly a quarter of graduate students would change their doctoral advisor if they could start over, with a fifth indicating outright dissatisfaction with their relationship with their supervisor. Additionally, the data revealed major difficulties in the ability of doctoral students to spend meaningful time with supervisors—nearly half report that they connect with advisors for less than 1 hour per week—contributing to a notable deficit in the ability to obtain important career guidance (Woolston, 2019).

While students of all backgrounds are experiencing deficits in their mentoring relationships, the careers of women, nonbinary individuals, and other underrepresented students (Black, indigenous, people of color, LGBTQIA+, and others) are more likely to be affected by poor mentorship. Strong mentors are often more essential for the success of individuals from historically marginalized communities, as their other support systems via social, cultural, or family networks may not be sufficiently empowered to support their career development (NASEM, 2019f). A larger conversation about the importance of mentorship for underrepresented students in STEMM can be found in Chapter 3 of the National Academies' report *The Science of Effective Mentorship in STEMM* (NASEM, 2019f).

Beyond academia, mentorship in the workplace is also imperative for success. Workplace mentors can help to identify existing barriers encountered by underrepresented individuals as well as approaches to tearing them down within a given institution. Mentors provide guidance and support to their mentees, which can be invaluable for someone early in their career or unfamiliar with the culture of the institution or company, particularly if their voice is less likely to be heard due to their identity. A positive mentorship experience can help an individual overcome challenges when faced with existing organizational hierarchy, power dynamics, and issues related to implicit bias, stereotypes, and outright discrimination. In large companies, there are typically multiple programs in place that provide opportunities for interoffice mentorship, including employee resource groups, human resources programs, functional organizations, and informal channels. In contrast to other sectors in the chemical enterprise, mentorship is a familiar and expected action for employee development in many large industrial settings. Because industrial workers tend to have more experience with, and training in, productive mentorship, there is value for those working in the chemical industry to collaborate with academic institutions to provide mentorship to students. There are opportunities for these mentorship relationships to be formally built into programs that encourage cross-sector collaboration, such as NSF's Industry-University Cooperative Research Centers program.[4]

Other reports have noted the opportunities to alleviate the various challenges of mentorship, and these recommendations could be implemented within the chemical workforce through alternative funding structures or alternative mentorship models that would allow both research advisors

---

[4] See https://iucrc.nsf.gov/.

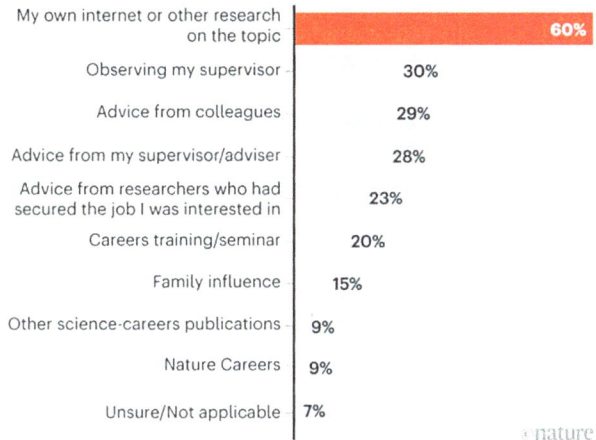

FIGURE 5-1 Results of a 2019 survey, in which Ph.D. students shared how they determined their future career paths. SOURCE: Woolston, 2019.

and trainees to flourish (NASEM, 2019f). Another option that has been explored in other reports that might decrease the risk of a potential conflict of interest for faculty mentors and improve the overall institutional climate is the development of mechanisms to provide research funding directly to the trainee rather than provide the funding via a principal investigator (NASEM, 2018). For the support and success of the individuals that comprise the chemical workforce, it is also worth considering a transition away from the traditional dyadic mentorship model toward a model that prioritizes a mentorship network.[5] A more expansive mentorship model recognizes that mentees are multifaceted individuals with a variety of interests and needs, while a single mentor can only impart the views reflective of a single individual's lived experience. Such a mentorship network would enable the mentee to contact different individuals at different career and/or educational stages and more effectively access and utilize a diverse range of voices and views to benefit their own professional and personal development. Creating a larger network of mentors also allows mentees to have more opportunities to learn from professionals that work in a variety of fields, both within and tangential to the chemical sciences. This cross-disciplinary exposure would better prepare students for a diverse range of careers and better support those already in their careers to be nimble in the ever-evolving landscape of the chemical sciences. Additionally, such a model would provide mentees who wish to pursue chemical research with greater knowledge and assistance in navigating challenging grant processes and funding structures.

5.2.2 Networks and Communities for Mentorship in Chemistry

Professional societies play a significant role in facilitating the creation and promotion of mentorship networks. Large professional organizations such as ACS and AIChE provide a variety of opportunities for interaction among students, academic faculty, and industrial chemists and

---

[5] A further discussion of the concept of a "constellation of mentoring relationships" can be found in *The Science of Effective Mentorship in STEMM* (NASEM, 2019f, Recommendation 5.3).

chemical engineers. Some programs currently exist within these societies that work to seed mentoring pairs. One such example is MentorNet, promoted by ACS, which is a "mentoring platform that combines the technology of social networks with the social science of mentoring" to provide effective virtual mentoring relationships within the STEM community.[6] AIChE's Future Faculty Mentoring Program[7] formally pairs senior graduate students and postdoctoral scholars with senior faculty members for 1 year to cultivate lifelong relationships and broaden the mentees' perspective of the chemical enterprise.

Programs also exist outside the professional society framework, such as the Chemistry Women Mentorship Network (ChemWMN), which aims to provide guidance to graduate and postdoctoral-level women by matching them to women faculty members.[8] Another example is Empowering Women in Organic Chemistry, established as a forum to engage women in organic chemistry at multiple career stages, including graduate students, faculty, and industrial practitioners, with the goal of providing an "environment for them to feel a true sense of belonging, develop powerful networks, and know the opportunities available to them."[9] Opportunities to partner with societies comprising a more targeted professional constituency, such as the National Organization for the Professional Advancement of Black Chemists and Chemical Engineers[10] or the Society for Advancement of Chicanos/Hispanics and Native Americans in Science,[11] help to enable the successful formation of effective mentoring networks for underrepresented students and professionals. Such programs and mechanisms promote the concept of a broader mentorship network, as mentees are often matched with mentors outside of their home institutions.

5.2.3 Supporting Gender Equity

Several other ways exist to support individuals for successful careers beyond strong mentorship and unbiased funding mechanisms. While many will not be discussed in this report, it is important to highlight the importance of attracting and retaining members of the chemical workforce of all genders through equitable support and compensation. A study by Huang et al. (2020) found that career-long publishing productivity in STEMM is significantly lower for women than for men, while the annual publishing productivity is essentially the same between the two studied genders. The study concluded that while men and women are equally productive when actively publishing, men have 19.2% longer careers on average, amplifying their total impact on the fields in which they work. The Royal Society of Chemistry found that women are leaving their careers in the chemical sciences "before reaching their full potential" due to the current academic funding structures, academic culture, and lack of support for balancing personal and professional lives (RSC, 2018). Not only does this diminish the available talent within the workforce, lower retention rates of women in chemistry have direct financial implications for the field. According to the findings from a workshop (ACS, NSF, and NIH, 2006), it costs approximately $500,000 to produce a Ph.D. chemist, where most of the funds come directly or indirectly from federal grants. As women leave the chemical sciences post-Ph.D. attainment, the return on investment by the federal government decreases. Nonbinary and transgender individuals also exhibit lower retention rates in STEMM, as they frequently experience discrimination and hostility in the classroom and workplace (Powell et al., 2020). Women of color also face a "double bind" while navigating their STEMM careers, which is related to experiencing both racism and sexism (Reardon, 2019).

---

[6] See https://greatmindsinstem.org/mentornet/#.
[7] See https://www.aiche.org/community/sites/divisions-forums/education-division/future-faculty-programs.
[8] See https://brandicossairt.wixsite.com/chemwmn.
[9] See https://ewochem.org/.
[10] See https://www.nobcche.org/.
[11] See www.sacnas.org.

Supporting underrepresented individuals in the chemical workforce is essential for the productivity of the chemical economy. An *ACS Axial* article (Miller, 2021) states that individuals can support women in chemistry by citing women-authored papers and supporting their work, nominating women for awards and leadership positions, and overall being an active bystander. Institutions can support all genders by implementing unconscious bias training, adopting climates and compensation structures to support those taking care of family members, and rethinking and redesigning their goals and management structures to be more inclusive and collaborative (Duncan, 2020). Groups exist to support gender equity in the chemical sciences, such as the ACS Women Chemists Committee,[12] which is committed to advocating for the promotion of women in chemistry-related fields.

5.3 DEVELOPMENT OPPORTUNITIES FOR ACADEMIC INSTITUTIONS

The majority of those who currently or will comprise the chemical workforce are educated in chemistry, chemical engineering, or other closely related science or engineering fields. The preparation that individuals receive to enter the workforce is the direct product of the institutions that they attend. Formal academic preparation is vital for the successful integration of individuals into and their contributions to the chemical workforce. Without basic knowledge of the foundations of chemistry and/or chemical engineering, it would be difficult for the workforce to contribute to the solutions of our world's greatest problems. However, this formal academic preparation will look different depending on the school(s) one attends, their geographic location and socioeconomic status, and the development opportunities available and afforded to that individual. Note that this section will primarily focus on 2- and 4-year colleges and universities. Information on K–12 STEMM education is provided in a 2012 National Research Council report (NRC, 2012).

Regardless of one's academic path, it is important that students learn to apply the foundational knowledge of the chemical sciences to work in emerging fields, use new tools and methodologies as they are rapidly developed, learn new techniques, and be able to apply them creatively throughout their careers. Equally important is removing barriers to accessing quality education and research experiences so that all students who wish to pursue a career in the chemical sciences and support the chemical economy are given the opportunities, tools, and support they need to succeed.

Chemistry and chemical engineering degrees at all levels remain popular. There were 22,156 undergraduate chemistry degrees awarded in 2019,[13] and 14,406 undergraduate chemical engineering degrees were awarded[14] in the United States. However, while the current academic system is working for some, there are changes that can be made so that all students are supported. Furthermore, opportunities exist to introduce novel ways of teaching chemistry and chemical engineering and for evaluating the teaching institutions themselves.

5.3.1 Academic Institutions to Model and Support

To travel a better path, one that prepares all students to support the chemical economy, it is valuable to look at institutions and programs that are uniquely successful and innovative at producing chemistry and chemical engineering graduates. Chemistry and chemical engineering departments have historically underproduced a diverse group of graduates, with numbers still well below their potential. In chemical engineering, the number of degrees awarded to white women and individuals of color of all genders have remained essentially unchanged since 2008 and are well

---

[12] See https://acswcc.org/.
[13] See https://datausa.io/profile/cip/chemistry.
[14] See https://datausa.io/profile/cip/chemical-engineering.

below the national average (NASEM, 2022b). For both chemistry and chemical engineering at all levels, students are overwhelmingly white (Figure 5-2).

Historically Black Colleges and Universities (HBCUs) and other minority-serving institutions (MSIs), such as Hispanic-Serving Institutions (HSIs), are unequivocally more successful at educating individuals who identify as persons of color, first-generation students, and/or students from low-income backgrounds for the chemical workforce. There are currently 101 HBCUs in the United States, and while 9% of Black undergraduate students attended an HBCU in 2018, these institutions were responsible for awarding 29% of the undergraduate degrees that Black students received. In 2018, 32% of the chemistry Ph.D.s awarded to Black students went to those who graduated from HBCUs (Widener, 2020). These data show that not only are HBCUs more successful at recruiting Black students, they are significantly more successful at supporting these students and nurturing their chemical careers.

Although HBCUs and other MSIs are better able to contribute diverse talent to the chemical workforce, the institutions themselves are not being supported adequately. For example, the number of Hispanic and Latinx students, as well as the number of HSIs, has grown significantly in the past 20 years. However, the federal funding to support these students and the institutions that

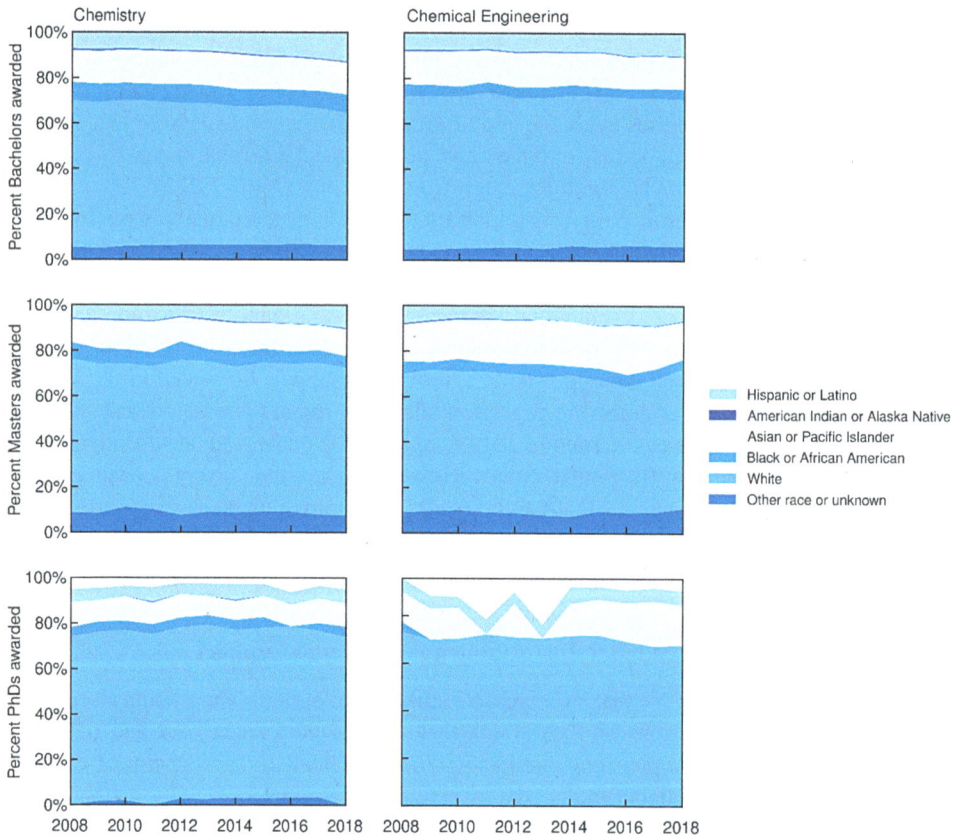

FIGURE 5-2 Demographics of students who received degrees in chemistry and chemical engineering programs at the bachelor's, master's, and doctoral levels. For Ph.D.s awarded, values do not add to 100% in some cases because data were redacted for privacy purposes, and Native Hawaiians and other Pacific Islanders were disaggregated from Asians, but those data were similarly redacted or zero. Vertical order of entries in the legend matches the order of bands in the figure. SOURCE: https://ncsesdata.nsf.gov/sere/2018/.

support them has not changed, as the federal appropriations have remained stagnant for the past two decades (NASEM, 2019d). That report looked specifically at the return on investment in MSIs and concluded that MSIs contribute significantly to the STEM workforce; with greater investment, the positive impact that they impart would only increase (NASEM, 2019d).

Community colleges and other 2-year degree programs are also uniquely positioned to support diverse chemical sciences and engineering students, because they allow for greater flexibility in taking courses (full time, part time, and evenings), focus on educating their local population, and are significantly less expensive than 4-year degree programs. These programs provide accessible education opportunities for students, and are more successful at attracting and retaining racially diverse students (NASEM, 2019d). Additionally, the majority of tribal colleges and universities in the United States are 2-year institutions (NASEM, 2022b). Although the success of these institutions is critical for supporting the populations of students that attend them, they are massively underfunded (NAE and NRC, 2012). STEMM programs at community colleges have strikingly limited resources, such as overloaded faculty who are undercompensated and overworked, limited to no professional development opportunities for faculty and staff, and a counselor-to-student ratio of 1 to 1,000, at best (NAE and NRC, 2012). Increased financial support is one mechanism to increase the number of students that such institutions could support, as well as increase the number of students that might then be able and be excited to transfer to 4-year institutions after graduation.

One great strength of chemical sciences education is the hands-on experiential learning provided through courses, modules, experiments, and research opportunities in the laboratory. Being able to apply what is learned in the classroom to a hands-on experience creates a richer and more impactful learning experience for students and teaches them the fundamental skill of applying concepts to solve real problems. It is important that lab experiences and research experiences are equitable across all levels of education, including undergraduate and community colleges, and are available to all students regardless of identity. Such opportunities are valuable for all students to gain workforce-applicable skills and to help them develop a passion for the chemical sciences, but these opportunities can be even more valuable for underrepresented students. A National Academies' report found that undergraduate research experiences improve the persistence of historically underrepresented groups in STEMM and can help to validate their disciplinary identity (NASEM, 2017b).

5.3.2 Flexibility in Chemical Sciences Education

The many critical courses included in the undergraduate chemistry and chemical engineering curricula have remained unchanged for a long time. Both disciplines are held to the standards of an external body: the Accreditation Board for Engineering and Technology for chemical engineering programs and the ACS Approval Program for chemistry programs. The large number of core courses in each of the curricula can make it difficult to add new required or elective courses without adjusting current requirements. However, as technology evolves, weather events become more severe, and new global threats emerge, such as the COVID-19 pandemic and the climate crisis, flexibility in education is becoming more important than ever.

The potential for chemistry and chemical engineering foundations to be applied to innovative solutions is ever expanding. While foundational chemical knowledge is essential, and altogether remains the same, the methods and tools used to apply that knowledge and the ways in which we are able to access chemical information change over time (Chapter 4 discusses some of the most important emerging tools and technologies in the chemical sciences). These changes are occurring more rapidly and more frequently, requiring those in the chemical workforce to innovate at warp speed. Operating at the boundaries of innovation requires that the chemical workforce be ever more flexible in their approaches and increasingly able to learn and use new skills and applications.

The flexibility required of the chemical workforce may not be mirrored by the education that the workforce receives. Ideally, chemistry and chemical engineering curricula would provide educators with ample opportunities to introduce new content, or simply adapt content, to match the current landscape of their respective fields. Providing education that is reflective of the state of the discipline engages and empowers students to obtain the degrees necessary for their desired careers in the chemical sciences and better enables those students to address the global challenges at hand once they enter the workforce. Such an education can only be provided if the courses and curricula remain nimble and are able to be adapted as fields evolve over time.

A key way that institutions can provide students with an education that is current with the state of the discipline is to provide opportunities for exposure to the new tools and technologies as they emerge. Some prevailing tools in the chemical sciences include computation and computer science, robotics, data science, and scholastic output. Impactful exposure to these tools does not necessarily require the present way that students are being taught chemistry and chemical engineering to be uprooted. Opportunities for introducing students to new tools and applications and reinforcing that knowledge can range from 1-hour laboratory modules or experiments to entire elective courses. Other opportunities for exposure can include research experiences and internships, student workshops and conferences, and online courses and tutorials.

However, current curricular models do not leave much room for the flexibility that is needed. At most institutions, if an educator wants to provide exposure to tools and methods that are not included in the set curriculum, their only option is to use their own incredibly limited time and resources to develop and incorporate such alterations. This places a heavy burden on the faculty, who are already stretched to provide meaningful mentorship and research experiences to their students, and requires them to do work for which they are not being adequately compensated or recognized. Additionally, this system of ad hoc injections of innovative content into courses does not promote the sharing of information among faculty and across institutions. When one professor takes the time to create new and innovative course content, there are limited methods of sharing that model with other faculty who are interested in making similar changes. Although such content can be published in the *Journal of Chemical Education*[15] or *Chemistry Education Research and Practice*,[16] increased mechanisms for dissemination of these innovative ideas would enable greater collaboration and implementation.

Although there are many challenges with changing chemistry curricula, there are specific topical areas and teaching approaches where the chemistry education community has already made important progress. These areas are exciting for the future of chemistry education but still require extensive work before they can be universally implemented. Two examples of this include the following:

- **Computation and data analysis in chemistry education**: With the increased use of computational and data analysis tools in chemistry research, this topical area could benefit individuals entering the chemical workforce. Computational interfaces and programs are already being developed to help incorporate these skills into chemistry education (Sendlinger et al., 2008; Snyder and Kucukkal, 2021), but there are still many challenges to overcome in effectively and seamlessly implementing computation and data analysis into the curricula.
- **Sustainable chemistry and systems thinking in chemistry**: With the chemical enterprise continually increasing its focus on environmental sustainability, it will be helpful to introduce the concepts of green and sustainable chemistry, as well as systems-level thinking

---

[15] See https://pubs.acs.org/journal/jceda8.
[16] See https://www.rsc.org/journals-books-databases/about-journals/chemistry-education-research-practice/.

related to chemistry, in earlier stages of chemical education. Groups such as the Green Chemistry Institute of the ACS are developing green and sustainable chemistry modules that can be incorporated into the material that is normally covered in general and organic chemistry classes at the undergraduate level.[17] Additionally, there are groups, such as Systems Thinking in Chemistry Education, who argue for the importance of systems thinking as a complementary approach to reductionist teaching styles that are typically used in chemistry classrooms (Orgill et al., 2019). This group argues that a mix of approaches will produce members of the chemical workforce who are environmentally conscious and socially aware.

To fully explore the best ways to universally and equitably implement topics such as these into the chemistry curricula, it would be critical to convene a group of chemistry education experts who could assess the educational landscape of chemistry, decide what is important for the next-generation chemical workforce, and chart a path toward implementation. An in-depth discussion of reforming chemistry education goes beyond the scope of this report, but the study committee does note the importance of chemistry education being flexible and adaptable to changes in the chemistry enterprise.

Introducing greater flexibility into chemistry and chemical engineering curricula without placing undue burden on faculty will require communication and collaboration among faculty, departmental leadership, university leadership, accreditation bodies, and funding entities. Conversations around how to best support excellence in chemical education are long overdue, and curricular development needs and opportunities must be considered by all parties. Discussions pertaining to continuous evaluation of the tools and applications utilized in the workforce would allow all educators to keep their courses current. The time needed to bring such parties together and have the conversations needed to enact change and create action could be substantial, and the agencies that support chemical research could consider financially supporting that time.

Furthermore, creating more equitable access to chemical education research and providing students with the tools and innovative spirit necessary to contribute to the future of the chemical workforce could be a substantially easier task if more advances were made in chemical education research. It is imperative to better understand the foundations of education and the optimal ways to impart knowledge with the current state of technological development. Basic chemical research can only flourish if chemical education research is also prioritized, because it is the key tool to develop the workforce.

5.4 WORKFORCE DEVELOPMENT

Professional development of individuals is vital for their success and fulfillment as well as their ability to contribute to the chemical economy. Professional development opportunities available to students and professionals at all levels facilitate transitions during a person's career, provide beneficial knowledge, create valuable networks, and help increase retention in the field. Students studying chemistry and chemical engineering have a wide array of career paths available, including ones outside of the realm of the chemical economy (e.g., medicine, law, financial services, business consulting, government, and more). Those who do enter the chemical workforce can work in sectors and industries involving R&D, production, or business and focus on numerous fields, such as energy, chemicals, materials, consumer products, food, health, pharmaceuticals, or information. Regardless of a student's focus or the path of an individual's career, attaining hands-on experiences is key to beneficial professional development.

---

[17] See https://www.acs.org/content/acs/en/greenchemistry/students-educators/module-development.html.

5.4.1 Undergraduate Student Development

Professional development at the undergraduate level helps students develop interest in and understanding of a field within their chosen major, as well as the role they might play in such a field. At the undergraduate level, students are tasked with determining whether to seek full-time employment after graduation and begin their professional career or pursue further education in a graduate program to eventually obtain a Ph.D., M.D., J.D., M.B.A., or other advanced degree.

Exploring career options is typically not an activity that is included in coursework requirements. Insights into available career paths can come from a variety of places: presentations by industry representatives, discussion in class about specific industries and jobs, participation in conferences (e.g., AIChE student conferences), and mentorship from someone in the field. One of the most important ways to learn about the working world is through professional development experiences in which students can apply their scientific and technical knowledge to real-world problems. These experiences allow students to participate in defining an approach to a problem and generating knowledge toward solutions. Opportunities for development occur through a range of direct experiences, many of which are discussed in Box 5-2. Although these experiences are vital for the student, there are many barriers to securing an undergraduate professional development opportunity. These include identifying opportunities, successfully competing for them, and balancing financial needs and personal priorities with potential opportunities. Professional development experiences can be especially influential for the recruitment and retention of underrepresented undergraduate students. Several organizations have created programs to help increase diversity in the chemical sciences by providing scholarships, and mentorship, and assisting with placement in summer internships (see Table 5-1).

Guidance and mentorship are important to successfully navigate the process of obtaining professional development experiences. In many academic institutions, a centralized office of career advising and services is a primary source of help for students. However, there is a lot of variability among academic institutions on how useful these offices are for chemical sciences students, as the focus of a career services office may be primarily on managing the interface between employers and students for full-time employment or assisting students in programs with the largest enrollments.

Some companies take an active role in preparing students to work in industry through programs that engage students beyond internships. For example, company-organized activities and conferences allow students to get to know a company and perhaps attract those students as future employees. They can be particularly helpful at engaging groups traditionally underrepresented in the field. Several universities, including the MIT ACCESS Program,[18] have developed similar programs to expose students underrepresented in research to the opportunities afforded by pursuing advanced graduate degrees. Professional societies also have a role in helping students with career exploration. ACS provides resources on its website[19] for students to learn about different career options with a chemistry degree, and AIChE launched an Institute for Learning & Innovation with a pilot on career discovery to help students "gain clarity on optimal job choices and receive direction on how to acquire the necessary skills and training" for a job.[20]

5.4.2 Graduate Student and Postdoctoral Development

Professional development at the graduate level serves to prepare students for their future careers after graduation. While some graduate students in chemistry and chemical engineering will become faculty members and continue to pursue research, many masters and Ph.D. graduates will

---

[18] See https://access.mit.edu/.
[19] See https://www.acs.org/content/acs/en/careers/chemical-sciences.html.
[20] See https://www.aiche.org/ili.

BOX 5-2
Undergraduate Professional Development Opportunities

Internships

Industrial internships expose students to a job role within a company, such as R&D, manufacturing operations, data analytics, and business processes. Students often first learn about internship opportunities when company representatives visit school career service offices or are at career fairs. Typically, company representatives that visit schools are from large Fortune 1000 companies. Any individual company only visits a finite number of schools based on proximity to company locations, majors at a school, number of open positions, and prestige and reputation of the institution. The opportunities at these companies, however, are only a subset of all the industrial internships available. Information about internship jobs at a company is often provided on the company's website through its careers portal, in theory making these internships available to all. Many internship opportunities can also be found on the ACS platform, "Get Experience."[a] The onus is on the individual students to seek out interactions or initiate applications with companies who do not come to their school. An additional challenge is that the pool of potential candidates for an internship is often much larger than the number of openings, so they are competitive. Many U.S. national laboratories also offer internships. For example, the U.S. Department of Energy (DOE) offers summer and semester-long internships, including their Science Undergraduate Laboratory Internships,[b] at its 17 national laboratories. The National Aeronautics and Space Administration hosts a similar set of internship programs at its research centers and other facilities.[c]

Research Experiences

Undergraduate research experiences provide students with a valuable opportunity to learn about different areas of research within their field of study. They can occur at one's academic institution throughout a semester as an enrolled student or over the summer at a different institution. A primary source of funding for undergraduate research is the National Science Foundation Research Experience for Undergraduates (REU) programs.[d] As of 2022, 75 universities offer REU programs in chemistry, and approximately 20% of the 141 Engineering REU programs are offered in chemical engineering or have projects in which chemical engineers would be able to participate.[e] REU programs typically prioritize giving opportunities to early-stage students and students that are underrepresented in the chemical sciences and engineering. In addition to REUs, faculty members may have grant funding to support undergraduate research assistants. Some universities offer competitive grants to undergraduate students hoping to conduct research at their institution. In some cases, companies have partnered with universities to fund undergraduate summer research experiences. For example, the Amgen Scholars Program,[f] which funds science and biotechnology research, has supported ~4,000 students since its inception in 2007. Additional opportunities are presented in the National Academies' report *Undergraduate Research Experiences for STEM Students* (NASEM, 2017b).

Cooperative Programs

Cooperative programs provide work experience at a company as part of the curricular sequence for degree completion. The academic institution assists in matching students to a company. The course offerings and scheduling are designed to allow students to have full-time work experience during one or two academic semesters. In schools with co-op programs, a bachelor's degree is typically completed in 5 years, with the additional year devoted to time spent in the co-op assignment. During a co-op assignment, students are paid at prevailing internship salaries, and tuition is not charged during the assignment. Cooperative programs for chemical engineering majors are more common than for chemistry majors; however, the number of schools that offer co-op programs is relatively small. A few schools (i.e., Northeastern University, Drexel University, and Kettering University) structure their undergraduate programs with co-op assignments for all majors and all students, and thus make them available for both chemical engineering and chemical science majors.

---

[a] https://getexperience.acs.org/.
[b] https://science.osti.gov/wdts/suli.
[c] https://intern.nasa.gov/#info-intern-0.
[d] https://www.nsf.gov/crssprgm/reu/.
[e] https://www.nsf.gov/crssprgm/reu/reu_search.jsp.
[f] https://amgenscholars.com/.

enter the industrial or government workforce. Opportunities for employment also exist in science policy, nonprofits or nongovernmental organizations, professional societies, and more. But, the two main paths for graduate students in the chemical sciences include preparation to become an academic professor and preparation to enter either industry or government. Note that industry is defined broadly in this context and includes established companies, start-ups, and entrepreneurial opportunities to create new companies.

Graduate students interested in academic careers will overwhelmingly transition from a doctoral program to a postdoctoral position. Accordingly, many students who choose such a path often delay deeper exploration of academic careers until they are already in their postdoc position. For those who do seek earlier engagement in understanding academic careers and the process of securing a faculty position, opportunities such as AIChE's Future Faculty Mentoring Program as well as many of the NSF ADVANCE programs[21] provide the means to do so. These programs also help students to prepare for their roles as mentors and teachers, in addition to their research responsibilities, once they become faculty members. These programs are typically open to participants across educational institutions, though some universities also offer programs specifically tailored to their populations.

For graduate students preparing to enter the industrial workforce, they will first need to determine what industry to consider and what role to pursue, because their specific area of graduate research does not necessarily define the role that they could have within a company. There is an expectation that many Ph.D. students will have an interest in R&D roles at a company, but other opportunities such as production, operations, consulting, data analytics, or business roles could be attractive. The Accelerate to Industry program[22] helps students explore opportunities in industry while developing the key skills needed to enter the workforce. Entrepreneurship is also a growing area of focus of universities and colleges to foster innovation from research conducted by students and faculty. Some schools have established programs in entrepreneurship to provide students with information and advice, workshops and courses, and alumni mentors and networks, including those in the chemistry departments at Case Western Reserve University[23] and Northwestern University.[24]

While there are some opportunities for graduate students to explore industries through a summer internship, they are more limited than those for undergraduates. Additionally, working at an internship means time not working on dissertation research, which can extend the time a student spends in graduate school. However, direct industrial experience is often very impactful and useful for graduate students as they seek to define their industrial career interests. One avenue that graduate students can take to connect to the chemical industry is through graduate student symposiums. At these symposiums, hosted by a variety of institutions, graduate students present their work to industrial representatives and are provided a forum to interact with professionals and learn about different companies. Companies with strong R&D foci, who seek to hire Ph.D.s for roles in R&D, often engage students through these research symposiums.

Postdoctoral researchers occupy a unique role within the academic research system. As neither students nor faculty, the opportunities for postdocs are both more varied and more limited. A postdoc who is hired directly by a research group is classified as an employee of the university. Therefore, opportunities such as internships are typically not available because work above a full-time appointment at the university is not permissible. A postdoc whose appointment occurs through a fellowship is subject to the rules of the granting agency, which may defer to the guidelines in place for nonfellows at the university. The net result of the unique status of postdocs is that the

---

[21] See https://www.nsf.gov/crssprgm/advance/.
[22] See https://grad.ncsu.edu/professional-development/career-support/accelerate-to-industry/.
[23] See https://case.edu/step/chemistry.
[24] See https://chemistry.northwestern.edu/graduate/current/professional-development1/management%20and%20entrepreneurship.html.

vast majority of professional development opportunities available to them are offered through either the home institution or professional societies. However, most programs that target senior graduate students interested in academic careers are also open to postdocs. Opportunities for direct employment in industry are typically prohibited based on employment agreements; however, "fly-in" or other introductory programs offered by companies to senior graduate students may be available for postdocs as well.

Federal funding agencies have started to recognize that postdoctoral researchers need quality mentoring for the betterment of their personal and professional development. NSF now requires that a mentoring plan be part of any proposal that includes funding for postdocs, while other agencies, such as NIH and DOE, have emphasized the importance of mentoring for postdocs. As discussed in Section 5.2, better and more consistent mentoring practices for postdocs contribute toward their professional development and lead to a higher retention rate of those researchers within the chemical workforce.

5.4.3 Continuing Professional Development

Prioritizing professional development throughout the entirety of an individual's career within the chemical workforce is essential for the health of the chemical economy. Supporting lifelong learning through professional development sets up the workforce for success, gives people the opportunity to learn new skills, and allows them to seek new opportunities, which together support the long-term health of the chemical economy. Too often people place professional development on the back burner as their lives becomes more demanding. But it is important that professional development opportunities for members of the chemical workforce be continuously prioritized, so that the workforce is prepared for and empowered by new challenges, tools, and opportunities.

Workforce professional development can be driven by workplace expectations set by the employer, whether it be a company, an academic institution, or the government. Professional development at this stage is varied, because the needs of early-career, mid-career, and experienced workers differ and expectations for continuing education vary across sectors. Broadly, development may consist of expanding an employee's technical and nontechnical knowledge and skills, preparing them for future roles in the organization or field, and developing their leadership and mentorship skills.

Individuals working in scientific and technical roles in industry learn on the job how to apply their formal education to their profession. But new tools and techniques are always being developed, so continuing technical education is an important component of professional development in the workplace. This can occur through part-time study for advanced degrees, topic-specific short courses, and training sessions. Opportunities are offered at technical conferences, through the employer, or accessed online. Many companies cover the cost of these activities as an educational benefit provided directly to the employee or through an employer's training budget. Nontechnical professional development tends to focus on the skills that can enhance an individual's ability to contribute to innovation. These include communication and presentation skills, working as a team, unconscious bias training, and continuous improvement methodologies. There appear to be fewer professional development opportunities for those working in academia and fewer requirements by schools for faculty to engage in professional development. Faculty who do so are generally driven by their own ambitions. In addition, professional development opportunities are sometimes not supported by universities, further increasing the barrier to access.

As was noted for the previous career stages, there are challenges for professionals in the chemical workforce to identify and access development opportunities. It is important that an institution's leadership set the expectations for, provide information about, and help people gain access to opportunities, training, and programs to further their education and understanding of their role within their field. Individuals can also access professional development through professional societies that focus on the chemical sciences and engineering. These professional societies organize conferences, offer courses and training, and provide networking opportunities.

Networking is a key skill that can be used to support one's professional development. For some, networking may suggest a social mixer to meet others who work in similar fields, and to cultivate relationships with people to help further one's career. However, networking can be viewed more broadly as interactions with colleagues (present, past, and future) in which ideas are exchanged and questions are asked with three goals in mind: learning, generating ideas, and problem solving. Many technical environments, such as labs, naturally provide a venue for networking, as they promote collaboration and lead to productive interactions among colleagues. Networking beyond one's project team and organization is typically more beneficial for someone's professional development, since those interactions contribute to both institutional innovation and employee development through sharing knowledge and skills.

5.5 CONCLUSIONS

As highlighted in this chapter, the U.S. chemical economy is best supported by a well-trained workforce. This chapter covered many different components of chemical training, including some that are subject-matter based, such as the need for nimble and flexible academic training programs, and some that are not specific to the chemical sciences, but equally important, such as the need for adequate mentorship and professional development. Based on the information gathered and presented, the committee came up with several conclusions related to building and maintaining the skilled and diverse chemical workforce that is foundational to support and enhance the U.S. chemical economy.

Conclusion 5-1: A skilled science and engineering workforce paired with a diverse, inclusive, and equitable science and engineering research enterprise is central to a thriving, nimble chemical economy equipped to respond to emerging challenges and maintain U.S. competitiveness.

Conclusion 5-2: The current structures and systems governing funding, promotion, retention, and professional development are in conflict and can stymie holistic career advancement for students, faculty, and research staff.

Conclusion 5-3: MSIs, including HBCUs and HSIs, and community colleges excel at supporting the academic preparation of diverse populations of students that will enter the chemical workforce upon graduation.

Conclusion 5-4: Effective mentorship is essential for the success of the chemical workforce at all levels and across sectors. Positive mentoring relationships can increase equity and inclusion in the classroom and the workplace, but mentors often have insufficient access to the resources and training needed to enable such relationships.

Conclusion 5-5: Creating an equitable and inclusive learning environment that exposes trainees of the future chemical workforce to new and innovative chemical tools, technologies, and instrumentation, as well as interdisciplinary knowledge and critical collaboration skills, will require a serious and sustained investment from funding agencies, universities, industry partnerships, and accreditation programs. This investment is critical

because the tools and practices that enable chemical research are constantly evolving, and training programs must be able to adapt to best facilitate the learning of basic-to-advanced chemical principles that will help students succeed.

Conclusion 5-6: Professional development is a key factor for the success of undergraduate students studying the chemical sciences and can determine their future role in the chemical economy. While there are many opportunities for professional development for undergraduates in the chemical sciences, barriers exist that reduce the equity in competing for these opportunities.

Conclusion 5-7: Professional development is vital for the chemical workforce at all career stages, but it can be difficult to prioritize and navigate. Institutions and professional societies can aid in promoting lifelong learning among those who contribute to the chemical economy.

6

Funding Chemical Research

Key Takeaways:

- Much of fundamental chemical research is done in universities.
- Academic institutions depend on federal, corporate, and philanthropic funding to support this research.
- Small Business Innovation Research and Small Business Technology Transfer (SBIR/STTR) programs have been critically important in translating fundamental chemical research discoveries to final products and processes.
- The landscape of funding is evolving, with opportunities for corporate and philanthropic support to increase as new relationships are forged.
- Despite spending much less on fundamental research than it used to, the chemical industry remains an innovative and vital player in the chemical economy.
- Funding of research infrastructure at universities remains a challenge and continues to be complicated by the shifting landscape of research funders.

In the United States, there is a wide variety of options for funding chemical research. Numerous federal agencies distribute funds in the form of contracts, grants, and cooperative agreements to support basic and applied research, start-up endeavors, and education and training. In addition, the United States supports 17 Department of Energy (DOE) national laboratories, along with numerous other large collaborative laboratory facilities. These laboratories include a dozen that engage directly in chemical research and others that provide facilities, such as supercomputers and advanced analytical instrumentation, used extensively by chemists. In addition to federal support for chemical research, there has been an exemplary tradition of industrial labs developing breakthrough research, from Bell Labs to DuPont Central Research, among many others. Today, chemical companies have expanded collaborations with academia and start-ups on fundamental research while continuing to invest in-house in applied research and development (R&D). Philanthropy makes significant targeted contributions supporting academia through gifts that are used to build centers for chemical research and training, fund endowed faculty positions, and support programs for training the next generation of chemists. One area in which philanthropy excels is by providing the money to create interdisciplinary centers of research that tackle many of the world's most pressing problems, such as climate change. In this chapter, the report examines funding sources for chemical research, who receives these funds, and, broadly, how the funds are used. It also considers some of the difficulties in funding fundamental chemical research under the current system and highlights opportunities for supporting fundamental chemical research as the funding landscape evolves. This chapter is not meant to serve as an exhaustive list of funding opportunities for chemistry. Rather, it highlights some key funding mechanisms that enable fundamental chemical research to contribute to the chemical economy.

6.1 FEDERAL INVESTMENTS IN CHEMICAL RESEARCH AND EDUCATION

Chemistry and related areas of research are funded by a diverse set of federal agencies, from the National Science Foundation (NSF) to the various mission-oriented agencies including the DOE, the National Institutes of Health (NIH), the Environmental Protection Agency (EPA), the National Institute of Standards and Technology (NIST), the U.S. Department of Agriculture (USDA), and the Department of Defense (DoD). This diversity of sources is regarded as a strength of the U.S. system for funding fundamental science, because the differing missions mean that the different agencies will have different criteria for what will be funded. Regardless of the agency, proposals for grants and contracts in the U.S. system are usually reviewed by experts, and awards are made on the basis of merit.

The United States has been making robust investments in chemical research for decades. In 2019, more than $3.4 billion was committed to basic and applied chemical research (Fleming and Basco, 2021), and from 1999 through 2019, federal obligations for basic research roughly doubled to $2 billion (Figure 6-1). Largely through its Divisions of Chemistry and Materials Research in the Math and Physical Sciences Directorate, NSF supports basic research and education in the chemical sciences. Other, more "mission-oriented" agencies of the federal government are charged with ensuring our national security (DoD), improving human health (NIH), preserving the environment (EPA), advancing measurement science (NIST), protecting food and agriculture (USDA), and providing the energy we need (DOE). Fulfilling these missions depends on advances in chemistry, providing a compelling rationale for sustained, large investment in chemistry research. But the ultimate missions of the federal agencies are likely to be executed, at least in part, by companies, including defense contractors, pharmaceutical companies, and energy companies, among others. These companies will employ chemists who are prepared to address the problems associated with mission-oriented agencies of the federal government. Thus, coupling advanced education and research is a critical part of the federal investment in fundamental research at academic institutions, which provide the talent pool needed by companies.

NSF's FY 2021 budget was $8.5 billion, of which $589.5 million went to the Divisions of Chemistry and Material Sciences. NSF also provides significant support to STEM education from kindergarten through graduate school. It participates in the SBIR program funding approximately 400 start-up companies per year, and it supports collaborative research centers, such as the Centers for Chemical Innovation (CCI).

As with basic research, the federal government has appropriated substantial funds for STEM education. The federal government published a STEM education strategic plan in 2018, which set out a 5-year strategy based on a vision to provide high-quality STEM education to all Americans

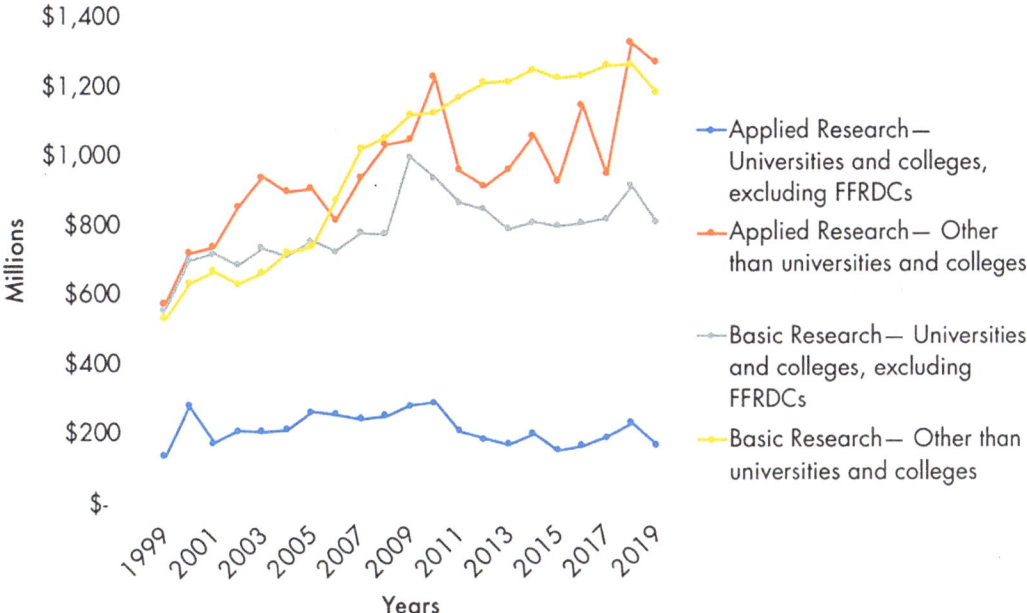

FIGURE 6-1 Federal obligations for research in the chemical sciences. NOTE: FFRDCs = federally funded research and development centers. SOURCE: Fleming and Basco, 2021.

throughout their lifetimes (NSTC, 2018). The responsibilities for supporting STEM education in the United States is spread over 17 departments and institutions, from DoD and DOE to NSF and the Smithsonian Institution. In FY 2021, the estimated federal budget for STEM education was $3.91 billion (Figure 6-2) (OSTP, 2021). This investment in STEM education is for 22 disciplines, of which about one-third are wholly or partially related to chemistry (OSTP, 2021).

6.1.1 Targeted Funding and Investigator-Initiated Research Funding

In general, federal agencies fund scientific research that is either investigator initiated or targeted to address specific research priorities of the agencies (Myers, 2020; NIMHD, 2021). NIH and NSF award the majority of research grants through investigator-initiated (also called unsolicited) awards, but the reverse may be true for other agencies, such as DOE's ARPA-E (Advanced Research Projects Agency-Energy) program,[1] that have research objectives that are of a more applied nature (Myers, 2020).

Targeted research is funded under "one-time competitions, which request proposals on specific" agency priorities (Myers, 2020). Chemistry-related targeted research sponsored by federal agencies, for example, might include research on materials for photovoltaic devices, or hydrogen and fuel systems at DOE; in situ resource utilization for missions to the Moon or Mars for the National Aeronautics and Space Administration (NASA); or research on chemical toxicity for EPA. Although it is not a mission agency, NSF's Division of Chemistry funds multidisciplinary targeted research in conjunction with NSF's other units, such as the Sustainable Chemistry, Engineering, and Materials (SusChEM) program, which supports basic research in green chemistry.

---

[1] See https://arpa-e-foa.energy.gov/Default.aspx#FoaId2b1605fb-a156-4d55-aa5b-b4c1a213c736.

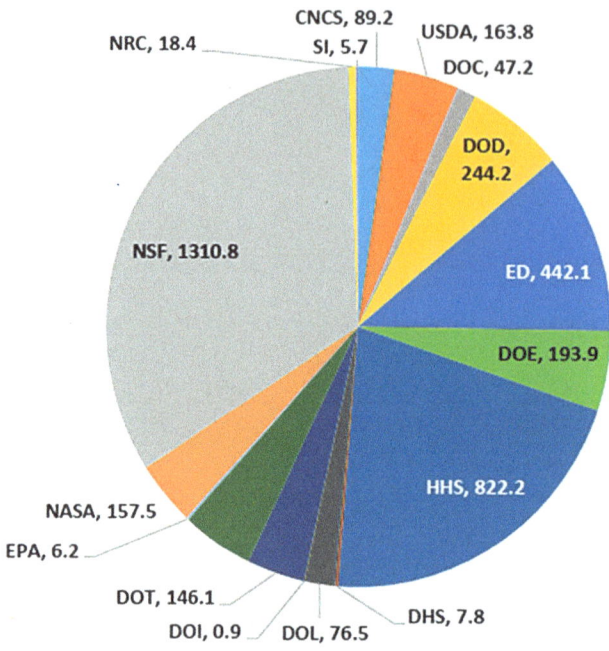

FIGURE 6-2 FY 2021 federal estimated STEM education budget. NOTE: DOD = Department of Defense, ED = Department of Education, DOE = Department of Energy, HHS = Department of Health and Human Services, DHS = Department of Homeland Security, DOL = Department of Labor, DOI = Department of the Interior, DOT = Department of Transportation, EPA = Environmental Protection Agency, NASA = National Aeronautics and Space Administration, NSF = National Science Foundation, NRC = Nuclear Regulatory Commission, SI = Smithsonian Institute, CNCS = Corporation for National Community Service, USDA = United States Department of Agriculture, DOC = Department of Commerce. SOURCE: OSTP, 2021.

6.1.2 Different Scales of Federally Funded Research

With the variety of sources available for federal funding, there are also a number of different opportunities for varying scales in funding opportunities. While federal funding discussions in the United States tend to center on individual labs, the past couple of decades have seen a growing discussion around mid- and large-scale collaborative opportunities. All of these different scales of research have played an important role in moving research forward, and chemical research is present at every level to help achieve the goals of all science-funding federal agencies.

6.1.2.1 Single-Investigator Research

Some of the examples in this report have focused on individual laboratories whose chemical research endeavors have led to Nobel Prizes or fundamental discoveries that were critically important to the economy, the environment, and society. Support for individual laboratories makes up a large portion of chemical research in the United States, and these individual laboratories take on the primary responsibility of mentoring and training the future chemical workforce, which was discussed in detail in Chapter 5. Based on the current system of research in the United States, in which primary training of scientists happens at universities at the undergraduate and graduate levels, support of individual, principal investigator laboratories is critically important. Critical breakthroughs in the chemical sciences are frequently attributed to individual labs or small-team collaborations.

It is important to note that many studies have shown the inequities of research funding distributions in the United States. Recently, extensive work has focused on funding from NIH but reveals important data about how women (when compared to men) and all individuals of color (when compared to White individuals) are receiving proportionally less funding (Lauer and Roychowdhury, 2021). There is a call in the research community to fund Black scientists as there is an increase in hiring of Black individuals but not the proportional increase in funding (Stevens et al., 2021). Recently, NIH announced a program to fund "at-risk investigators from diverse backgrounds," which is an important step to remedying the current discrepancy.[2] While many programs are starting to be put in place, further action and policy considerations would be a helpful step from the research funding community.

6.1.2.2 Mid- and Large-Scale Research Infrastructure and Collaboration

In 2016, NSF held two workshops to explore mid-scale research infrastructure needs in chemistry (NSF, 2016a,b; NSB, 2018). The main conclusions from the workshops noted the increasing complexity of the questions in the chemical sciences, and the need for broader approaches to measurement and synthesis. They noted the need for "large-scale, coordinated efforts of equipment and personnel of complementary capabilities and skills, which can only be made possible with mid-scale investment" (NSF, 2016b). The workshop participants also identified a wide range of chemistry questions that would benefit from mid-scale infrastructure, including work around understanding the dynamics of interactions and interfaces, and instrumentation for parallel operations such as synthesis and measurement (NSF, 2016b). The workshop was particularly encouraging about the need for mid-scale infrastructure for facilities and instrumentation that would exist to provide collaborative experiences for chemical researchers exploring complex questions. Additionally, the workshop participants emphasized the fact that mid-scale facilities "should not replace single-investigator grants" (NSF, 2016b).

Since this workshop in 2016, mid-scale research collaborations and facilities have been growing in popularity. In 2017, NSF made "mid-scale research infrastructure" one of its "10 Big Ideas" that would help move research forward. Since then, in response to a study from the National Science Board in 2018 that called mid-scale research infrastructure "underrepresented" in the NSF research portfolio, there have been several calls for projects in the $6 million to $70 million range. In looking through the awarded opportunities from NSF, many of them focus on chemical principles or require collaborations with chemists and chemical engineers to be successful. Although not an exhaustive list, some of the titles of chemistry-specific mid-scale infrastructure projects include

- NSF National EXtreme Ultrafast Science (NEXUS) Facility,
- Consortium: Biogeochemical-Agro: A global robotic network to observe changing ocean chemistry and biology,
- Atmospheric Science and Chemistry mEasurement NeTwork (ASCENT), and
- Grid-Connected Testing Infrastructure for Networked Control of Distributed Energy Resources.

---

[2] See https://grants.nih.gov/grants/guide/pa-files/PAR-22-181.html.

The earliest of these mid-scale infrastructure projects started receiving funding in 2019, so it is difficult to assess their success, but the popularity of the program has led to several different requests for proposals by NSF.

In addition, NSF's Divisions of Chemistry and Materials Research fund several chemistry-related centers: CCI, Materials Research Science and Engineering Centers, and two Science and Technology Centers (NSF, 2021). These centers support interdisciplinary research bringing together the expertise of multiple laboratories. For example, CCI's research includes

> major, long-term fundamental chemical research challenges . . . . [that] will produce transformative research, lead to innovation, and attract broad scientific and public interest. CCIs are agile structures that can respond rapidly to emerging opportunities through enhanced collaborations. CCIs integrate research, innovation, education, broadening participation, and informal science communication.[3]

While NSF has been very explicit about its steps to fund mid-scale infrastructure, other federal funding agencies have also been funding collaborative infrastructure and facilities. For example, NIH has invested many millions of dollars in distributed facilities for cryogenic electron microscopy and cryogenic electron tomography. NIH also funds a number of large collaborative research centers with either a chemistry focus or a strong influence from chemical research. One such place is the "Discovery of Chemical Probes and Therapeutic Leads," an NIH Center of Biomedical Research Excellence (COBRE) at the University of Delaware.[4] The COBRE awards are part of a congressionally mandated program of Institutional Development Awards that "build research capacity in states that historically have had low levels of NIH funding."[5] NIH's National Center for Advancing Translational Sciences is building out a platform designed to accelerate the development of therapeutics by transforming "chemistry from an individualized craft to a modern, information-based science."[6] While the focus of this platform, known as ASPIRE (A Specialized Platform for Innovative Research Exploration), is translational biomedical chemistry, the results will include the development of advanced tools in chemical automation and may lead to a greater understanding of reaction mechanisms, reaction kinetics, and biochemical catalysis.

DOE also has many investments in both mid- and large-scale infrastructure projects. Some of DOE's mid-scale opportunities include the Energy Frontiers Research Centers (EFRCs), of which there are currently 46 around the county (DOE, 2019). They focus on cultivating multidisciplinary teams to solve diverse energy challenges and incorporate a large number of chemists and chemical engineers. In addition to the EFRCs, there are also Bioenergy Research Centers, which have a goal similar to the EFRCs' but focus on research to harness and optimize the use of bioenergy (DOE Office of Science, 2020a). DOE has many other examples of mid-scale scientific infrastructure that include the advancement of fundamental chemistry, including the Joint Center for Artificial Photosynthesis[7] where they are looking to develop the science around artificial solar fuel generation.

In addition to these mid-scale facilities, there are also large-scale government-funded research and user facilities. Many of the arguments made during the 2016 NSF workshops advocating for mid-scale infrastructure apply to the building and maintaining of large-scale research infrastructure for chemistry. For example, the growing complexity of chemistry will require advanced instrumentation, much of which is available at these large-scale facilities. Many of the large-scale facilities are designated as federally funded research and development centers (FFRDCs). While a number

---

[3] See https://beta.nsf.gov/funding/opportunities/centers-chemical-innovation-cci.
[4] See https://sites.udel.edu/cobrediscovery/people/.
[5] See https://www.nigms.nih.gov/Research/DRCB/IDeA/Pages/default.aspx.
[6] See https://ncats.nih.gov/aspire/about.
[7] See https://solarfuelshub.org/.

of government agencies run FFRDCs, the DOE's national laboratory system is one of the most prominent for the chemical research community. The national laboratories provide important facilities and research opportunities to advance the chemical sciences, including the housing of x-ray synchrotron radiation light sources, neutron scattering facilities, and nanoscale science research centers. Additionally, there are computational user facilities such as Extreme Science and Engineering Discovery Environment (XSEDE) and the Innovative and Novel Computational Impact on Theory and Experiment (INCITE) that give researchers the opportunity for supercomputing time. These facilities are discussed in detail in Section 4.3.1, but it is important to note that a large percentage of the principal investigators who received computing time are chemical researchers.

The contributions of mid- and large-scale research infrastructure and collaboration to the chemical sciences are all unique and individually important. As the complexity of chemistry continues to increase, advanced instrumentation, facilities, and collaborations will become a more important complementary mechanism of research to the single-investigator research project.

6.1.3 SBIR and STTR Programs: Commercializing Fundamental Chemical Research

The SBIR and STTR programs encourage small U.S. businesses to engage in federal R&D with the "potential for commercialization. Through a competitive awards-based program, SBIR and STTR enable small businesses to explore their technological potential and provide the incentive to profit from its commercialization."[8] Since their inception, these two programs have played a pivotal role in the translation of fundamental research, including in the chemical sciences, into commercialized technology that has propelled the U.S. economy. The SBIR/STTR programs complement venture capital funds in specific strategic areas of R&D, and it is expected that they will continue to play important roles during the energy transition and the expected renovation of the chemical industry. According to the Small Business Administration's (SBA's) website,

the SBIR program is structured in 3 phases:
- Phase I. The objective of Phase I is to establish the technical merit, feasibility, and commercial potential of the proposed R/R&D efforts and to determine the quality of performance of the small business awardee organization prior to providing further Federal support in Phase II. SBIR/STTR Phase I awards are generally $50,000–$250,000 for 6 months (SBIR) or 1 year (STTR).
- Phase II. The objective of Phase II is to continue the R/R&D efforts initiated in Phase I. Funding is based on the results achieved in Phase I and the scientific and technical merit and commercial potential of the project proposed in Phase II. Typically, only Phase I awardees are eligible for a Phase II award. SBIR/STTR Phase II awards are generally $750,000 for 2 years.
- Phase III. The objective of Phase III, where appropriate, is for the small business to pursue commercialization objectives resulting from the Phase I/II R/R&D activities. The SBIR/STTR programs do not fund Phase III. At some Federal agencies, Phase III may involve follow-on non-SBIR/STTR funded R&D or production contracts for products, processes or services intended for use by the U.S. Government.[9]

The SBIR program was established under the Small Business Innovation Development Act of 1982 with the purpose of "strengthen[ing] the role of the small, innovative firms in Federally-funded research and development" (U.S. Congress, 1982). Modeled after the SBIR program, STTR was established by the Small Business Technology Transfer Act of 1992 (U.S. Congress, 1992). According to Tuck and Moeinian (2017), "the goal of the STTR program is to facilitate the transfer

---

[8] See https://www.sbir.gov/about.
[9] See https://solarfuelshub.org/.

of technology developed by a research institution through the entrepreneurship of a small business concern." As noted by the New York State Small Business Development Center, the STTR program requires "the small business to formally collaborate with a research institution in Phase I and Phase II. STTR's most important role is to bridge the gap between performance of basic science and commercialization of resulting innovations. The mission of the [SBIR and] STTR program[s] is to support scientific excellence and technological innovation through the investment of Federal research funds in critical American priorities to build a strong national economy."[10]

Currently, the general criteria for agencies that support SBIR and STTR programs are outlined on the SBA website:

> Each year, Federal agencies with extramural research and development budgets that exceed $100 million are required to allocate 3.2% (since FY2017) of this extramural R&D budget to fund small businesses through the SBIR program. Federal agencies with extramural R&D budgets that exceed $1 billion are required to reserve 0.45% (since FY2016) of this extramural R&D budget for the STTR program. Currently, eleven Federal agencies participate in the SBIR program and five of those agencies also participate in the STTR program [see Figure 6-3]. Each agency administers its own individual program within guidelines established by Congress. These agencies designate R&D topics in their solicitations and accept proposals from small businesses. Awards are made on a competitive basis after proposal evaluation.

Three examples of SBIR/STTR programs are provided to highlight the significant impacts of these federal programs in the commercialization of technologies that otherwise would have taken longer to transfer to the chemical industry or perhaps would never have been commercialized.

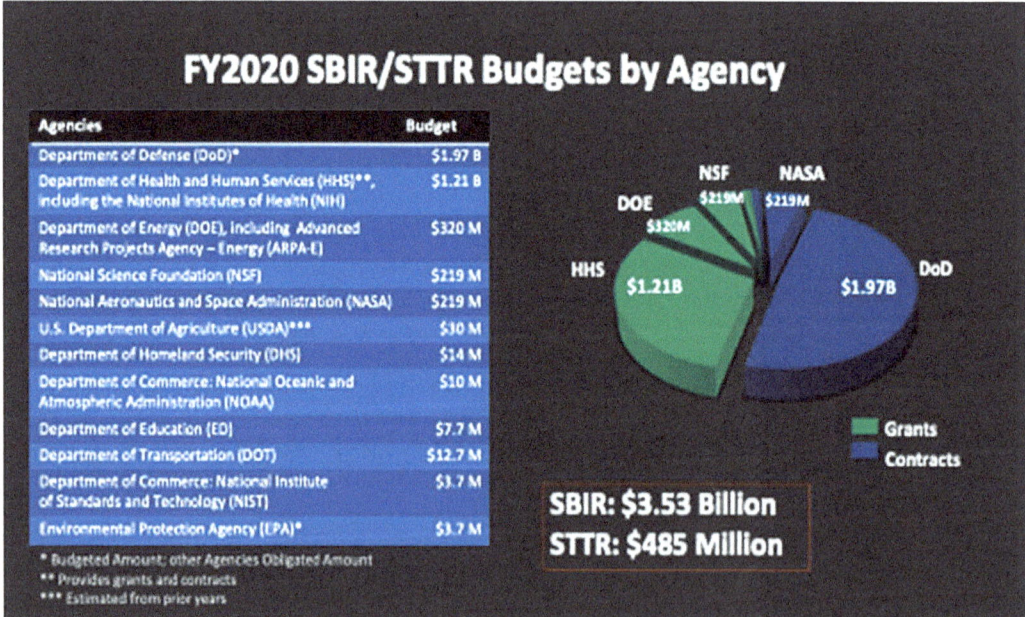

FIGURE 6-3 FY 2020 SBIR and STTR budgets by federal agency. SOURCE: Shieh, 2021, slide #8.

---

[10] See https://sbtdc.org/services/programs/tech/sbirsttr/sttr-facts/.

A Phase I SBIR grant in 2012 from NSF supported Mango Materials, a company that first discovered that methane gas could be used to produce a naturally occurring biopolymer that could compete with conventional oil-based plastics.[11] Mango Materials focused on methane derived from wastewater treatment plants. As the SBA website notes, "They ran the system for 200 days and proved the process could succeed even in the most non-sterile of environments. With a follow-on Phase II and Phase IIB award, the company scaled the technology, and started producing 10 pounds of their polymer each week" (SBIR, 2017b). Expanding on their technology, Mango Materials is currently embarking on a Phase II STTR with NASA in hopes that their material can eventually be manufactured in a microgravity environment.

Another example comes from Exelus,[12] a company that leveraged SBIR grants from NSF and DOE to develop ExSact—an engineered solid-acid catalyst-based alkylation process, which is now being licensed worldwide, generating $11 million in licensing fees (SBIR, 2017a). The ExSact technology produces high-octane alkylate from a variety of feedstocks including fluid catalytically cracked (FCC) olefins, methyl *tertiary*-butyl ether raffinate, FCC off-gas, or olefins derived from natural gas or biomass. The SBA website notes that "the company recently aligned with KBR and has sold two licenses for the technology. Exelus and KBR are currently negotiating with several US refiners to revamp existing alkylation units to replace liquid acids with the ExSact catalyst" (SBIR, 2017a).

Instrumental Polymer Technologies, LLC (iptech) is developing environmentally friendly, low-cost polymers by imitating nature's primary production processes.[13] It is mimicking nature's process using the reversible reactivity of aliphatic polycarbonates, mainly polycarbonate and polyurethane. The company plans to offer its reactive intermediate to other companies, who can extend the intermediate with other molecules to form polymers and plastics to meet their specific needs. Ultimately, these polycarbonate and polyurethane plastics can be digested back into the reactive intermediate by refluxing with alcohol. EPA and NSF SBIR Phase I and II awards supported the development and commercialization of these technologies and polymers. For example, SBIR funding is facilitating iptech's advancement of a sustainable and biodegradable thermoset plastic that can be processed and recycled like a thermoplastic and can be used in the production of biodegradable polymer concrete.

6.2 CORPORATE FUNDING OF CHEMICAL RESEARCH

Organized and large-scale corporate investments in fundamental chemical research began in the United States in the early 20th century and reached their heyday from the 1950s through the 1990s. One of the most revered corporate research centers was DuPont Central Research and Development, which was established in 1957, though its roots date back to 1903 (Tullo, 2016). Dupont announced the closing of these labs in 2015 (Tullo, 2016). Relatedly, for a variety of financial and strategic reasons, chemical companies now rely largely on academic labs to perform the fundamental research that the companies can then feed into their in-house applied R&D programs.

6.2.1 Investments in R&D by Chemical Corporations

Chemical companies spend, on average, 2% to 3% of their annual sales on R&D. Notably, the chemical industry (excluding pharmaceuticals) receives little government money for research that it conducts. In 2020, chemical companies spent $10.1 billion in R&D, of which about 9%

---
[11] See https://www.mangomaterials.com/.
[12] See http://www.exelusinc.com/.
[13] See https://instrumentalpolymer.com/.

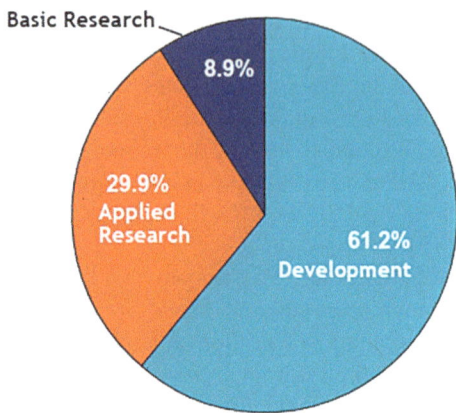

FIGURE 6-4 Distribution of research funding from the chemical industry. SOURCE: ACC, 2021.

($900 million) went to basic research (Figure 6-4) (ACC, 2021). The U.S. pharmaceutical industry invested approximately $72.8 billion into R&D in 2020 (Mikulic, 2021).

Although the chemical industry spends far less in-house on fundamental research, it continues to spend heavily on applied R&D, so it is fair to say that the chemical industry remains an innovation-intensive industry. NSF data on innovation categorized by companies indicate that more than two-thirds of the companies in the chemical industry (including pharmaceuticals) reported introducing product, process, marketing, or organizational innovation. In the economy as a whole, only 43% of companies reported introducing an innovation. The 67% rate of innovation by chemical companies is only slightly lower than that reported for industries such as data processing (69%), although it is lower than computers and electronic products (72%) and software (79%) (Table 6-1). It is important to note that for this survey, the R&D might not have come directly from the company, and there could be funding from the company that is used at an outside institution.

The U.S. chemical industry expenditure of $10.1 billion in 2020, while large, is its lowest level of R&D spending in many years (Figure 6-5). Notably, though, investment in fundamental research in most corporate laboratories is modest compared to an earlier era where many of the largest chemical companies made substantial investments in fundamental research. The decline of fundamental chemical research in corporate laboratories along with the growth of big data, automation, and artificial intelligence is increasing the number of partnerships being formed between universities and companies. These have the potential to drive changes in research and education taking place at academic institutions, as discussed in the next section.

The NSF surveys are further broken down into "Product innovation" and "Process innovation." When looking specifically at product innovations between 2015 and 2017, the leading industries are computers, electrical equipment, software, machinery, pharmaceuticals, and scientific services. All of these innovation-heavy industries, and the products they produce, are dependent on the chemical industry, and some of the research likely includes chemical and materials research. Within the chemical industry, the rate of product R&D is around 83%. Within the category of Chemicals, pharmaceuticals and medicines have rates of product innovation around 90%, while soaps and cleaning products and other chemicals have rates of product innovation of more than 80%. Similarly, nearly 80% of companies in the rubbers and plastics industry report R&D-based product innovation.

These figures are consistent with an R&D-intensive chemical industry that is now maturing. The focus of its innovation nowadays is often improvements in operating efficiency rather than

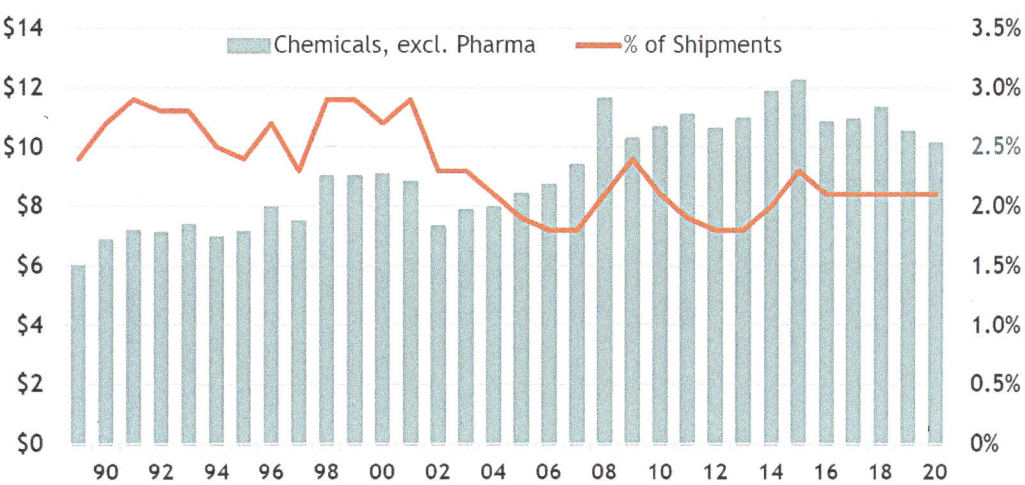

FIGURE 6-5 R&D spending in the U.S. chemical industry. SOURCE: ACC, 2021, p. 43, fig. 6.2.

fundamental or transformative discoveries. This idea is consistent with the relatively slow growth in output of chemical and related industries. It is important to note here that the pharmaceutical sector is a clear exception. As noted above, the pharmaceutical industry invests a significant amount in research and is directly reliant on advances in chemical knowledge, or at the very least, dependent on other parts of the chemical industry.

6.2.2 University–Industry Research Partnerships

Collaborations between universities and industry provide unique opportunities to advance fundamental research, though negotiating the details of a collaboration can come with challenges. For universities, corporate collaborations provide a fresh stream of money to support research; offer a chance to work with people who have the knowledge and resources to scale up research efforts and move those ideas into development and perhaps commercialization; and expand mentorship, internship, and employment opportunities for students and postdocs. All of these benefits are true for departments of chemistry and related sciences.

Universities are hotbeds for start-ups. Between 1995 and 2015, approximately 11,000 start-ups were formed at U.S. universities (autm, n.d.). By forming partnerships with universities, companies have early access to these innovative start-ups. Companies also get the benefit of working with academic research scientists, who may think differently about research problems, providing opportunities for synergy. Companies reduce the risk of their research investment by funding academic research and often get significant leverage from these research dollars. The NSF's Industry–University Cooperative Research Centers program has calculated that through its program, every dollar invested leverages $41 more in research funding. And, as mentioned, companies get to assist in mentoring and working with students and postdocs, some of whom could be their next employees. Students that participate in corporate-sponsored research receive training and practical knowledge that makes them well prepared to transition into industrial jobs.

These benefits are substantial for all involved, but there can be difficulties in establishing and maintaining a university–industry collaboration. The primary reason appears to be that corporate culture fails to map well onto university culture (Frølund et al., 2017). In addition, arrangements have to be made to address intellectual property and nondisclosure agreements. These two topics

TABLE 6-1 Innovating Companies, by Industry: 2015–2017

Industry	NAICS Code	No. of Companies	Any Innovation? (%)
All industries	11, 21–23, 31–33, 42–81	4,603,606	43.2
Manufacturing industries	31–33	220,930	57.9
Petroleum and coal products	324	624	57.6
Chemicals	325	7,910	67.4
Pesticide, fertilizer, and other agricultural chemicals	3253	542	62.4
Pharmaceuticals and medicines	3254	1,412	69.8
Soap, cleaning compound, and toilet preparation	3256	1,615	70.6
Other chemicals	other 325	4,341	66.1
Plastics and rubber products	326	8,412	61.3
Nonmetallic mineral products	327	8,021	51.0
Primary metals	331	2,679	55.9
Fabricated metal products	332	47,340	53.6
Machinery	333	18,792	61.6
Computer and electronic products	334	9,566	71.9
Electrical equipment, appliances, and components	335	4,223	65.1
Transportation equipment	336	8,039	62.4
Furniture and related products	337	12,430	52.2
Miscellaneous	339	22,315	59.8
Nonmanufacturing industries	11, 21–23, 42–81	4,382,677	42.5
Information	51	56,053	60.8
Software publishers	5112	7,438	78.6
Telecommunications	517	6,045	58.7
Data processing, hosting, and related services	518	8,414	69.7
Other information	other 51	24,563	54.6
Finance and insurance	52	192,345	46.1
Real estate and rental and leasing	53	229,923	36.0
Lessors of nonfinancial intangible assets (except copyrighted works)	533	1,701	52.4
Other real estate and rental and leasing	other 53	228,222	35.9
Scientific research and development services	5417	9,293	61.6

SOURCE: NCSES, 2020a, Table 43.

are important points of discussion in collaborations between universities and corporations, even for early-stage research. The potential financial success of some of those collaborations motivates the negotiations, but unfortunately, sometimes lengthy negotiations slow down or prevent university–corporate research partnerships. Other challenges include the propensity for these programs to focus on large research institutions, and to frequently not distribute funding to places with fewer resources. There is a concerted effort in some companies to fix this, but more partnerships with institutions, especially those that have proven that they can help to increase diversity and equity in the chemical workforces, including Historically Black Colleges and Universities and minority-serving institutions such as Hispanic-Serving Institutions, would be a helpful change.

6.3 PHILANTHROPIC FUNDING OF CHEMICAL RESEARCH AND EDUCATION

Over the past 100 years, many foundations have been formed by individuals and corporations, including corporations associated with the chemical industry. Some foundations are focused on supporting university research programs in fundamental science and university professors of science. A small number of foundations support chemistry exclusively, including the Camille and Henry Dreyfus Foundation, the Robert A. Welch Foundation, and the Arnold and Mabel Beckman Foundation. Many other foundations such as the Sloan Foundation, the Research Corporation for Science Advancement, the W.M. Keck Foundation, and the David & Lucile Packard Foundation support scientists more broadly, including chemists. None of these foundations support chemistry at an annual level competitive with the federal government, but as these foundations grow through investment success and other foundations are formed, there is a significant opportunity for them to contribute to the advancement of fundamental chemistry and its application to solve major problems of importance to the nation and to the world.

The Robert A. Welch Foundation, for example, focuses its philanthropy on chemical research and education within the state of Texas. Their support has established 48 endowed professorships in 21 institutions, research grant awards totaling more than $20 million per year, and the recent commitment of $100 million to Rice University to establish the Welch Institute for Advanced Materials (Kuspa, 2021). The foundation also provides undergraduate scholarships and grants to chemistry departments to promote experiential opportunities in the chemical sciences and funds a summer laboratory experience for high school students, pairing them with practicing research chemists.[14] The foundation has chosen to focus much of its support on small- and mid-size Texas schools and has thus been able to support and attract many minority and first-generation college students to study the chemical sciences.

During its initial 10 years, the Arnold and Mabel Beckman Foundation made significant gifts to establish Beckman Institutes at the University of Illinois Urbana-Champaign, California Institute of Technology (Caltech), Stanford University, City of Hope Medical Center, and the University of California, Irvine (and the Beckman Center for the National Academies).[15] The five research centers are each designed to foster collaboration, inspire bold scientific risk taking, and nurture disruptive ideas. The Beckman Institutes are a part of the research infrastructure needed for forefront chemistry and allied areas, the programmatic focus of the Beckman Foundation. The foundation also funds promising young researchers in the chemical sciences by supporting undergraduates, postdoctoral fellows, new faculty, and special research areas, such as cryogenic electron microscopy.

According to the home page of The Giving Pledge,[16] "In August 2010, 40 of America's wealthiest people made a commitment to give the majority of their wealth to address some of

---

[14] See https://welch1.org/.
[15] See https://www.beckman-foundation.org/.
[16] See givingpledge.org.

society's most pressing problems." Today, that number has grown sixfold and includes people from 28 countries. The list includes people who have made major gifts to research universities to support sciences and engineering associated with the challenges that chemists can participate in addressing. Among these is Michael Bloomberg, with substantial support to address climate change, and Gordon and Betty Moore. Gordon Moore graduated from the California Institute of Technology with a Ph.D. in chemistry, and individually and through the Gordon and Betty Moore Foundation, made a total gift of $600 million to Caltech. Founded in 2000, the Gordon and Betty Moore Foundation has already given more than $1.7 billion to important scientific programs.[17] For example, they provided $6 million dollars over 10 years to support research on the photooxidation of water to make fuels via artificial photosynthesis.

The wealth created through technological advances in the past few decades is fueling the development of foundations and other philanthropic actions that present exciting opportunities to advance fundamental chemistry to solve major problems. Problems such as the need for abundant, affordable energy, chemicals, and fuels that do not degrade the environment or contribute to climate change; improvements in agricultural productivity; and the need for access to clean water are all areas where chemistry will play an important role. For example, in 2019, Stewart and Lynda Resnick gave $750 million to Caltech to support environmental sustainability research, which will be used to build a sustainability research institute, create a permanent endowment to support this research, and fund "bold creativity," to use Stewart Resnick's words (Svitil, 2019).

While more than 80% of the funding for research in academia comes from the federal government and corporations (CRS, 2021), that money typically funds research that is either applied or basic but to some degree targeted. Philanthropic support, on the other hand, has the flexibility to support scientists pursuing bold transformative ideas and working in collaborative interdisciplinary environments. Thus, it is a realistic prospect for academic institutions to excite the interest of these philanthropists to provide critical support for infrastructure, innovations, training, and faculty in the pursuit of advancing science and improving society.

6.4 FINANCIAL RESPONSIBILITIES OF ACADEMIC INSTITUTIONS IN SUPPORTING RESEARCH

Academic institutions play a central role in the chemical economy. They perform about half of all basic chemical research, and they are sites for mentoring, educating, and training the next generation of chemical scientists. Support for academic research and education in chemistry comes almost entirely from the government, industry, and philanthropy. But there are additional needs, which have to be covered by the academic institutions.

Chemistry departments at colleges and universities, like all departments that fall under STEM, are charged with teaching and mentoring undergraduates; in some schools, training and mentoring graduate students; and at research institutions, mentoring postdocs. Carrying out these responsibilities is an essential part of enhancing the U.S. chemical economy because (1) these scientists are the next generation of chemists, entrepreneurs, and leaders; and (2) in institutions where research is the priority, undergraduates, grad students, and postdocs make up most of the laboratory workforce. Teaching, training, and mentoring are covered in Chapter 5. This current section looks briefly at some of the financial responsibilities that academic institutions have for maintaining chemistry research programs.

Colleges and universities that conduct research rely primarily on federal dollars to support their research programs. But these institutions must also make substantial financial commitments. Academic institutions are typically responsible for the financial resources needed to launch the

---

[17] See https://www.moore.org/programs/science.

career of a junior professor. These funds are used to acquire instrumentation, prepare laboratory facilities, and provide support for technicians and postdoctoral associates. At universities and at colleges where research is important, but the primary focus is on teaching, start-up funds (minus an assistant professor's salary) range from $20,000 to $100,000 (Argonne Today, 2013). As noted above, some small to mid-size schools have received funding from philanthropic organizations (e.g., the Robert A. Welch foundation) that offset or cover these start-up funds.

Academic institutions are almost completely responsible for facilities development and redevelopment, and significant capital is needed for renovating or establishing facilities for chemistry. Public universities can expect to receive some capital from state governments, but private universities typically rely on philanthropic support for building construction and major renovations. The needs for a department in the chemical sciences to support research vary depending on how much research is prioritized and the aspirations of the institution as described in its strategic plan. All institutions must support teaching laboratories for undergraduates. This requires laboratory space, supporting infrastructure (heat, electric, ventilation, gas, water, and vacuum lines), fume hoods, safety equipment, basic equipment, chemicals, and supplies, as well as support staff.

The costs of running chemical research labs vary considerably depending on the individual needs of the faculty and the expectations of the department. The challenge for most schools is supporting their faculty administratively and financially to maximize opportunities for faculty to obtain sufficient funding from all of the possible sources—in-house, federal, corporate, and philanthropic.

In addition to the start-up costs and capital needs associated with major university chemistry programs, there are ongoing expenses such as research libraries, information technology infrastructure, facilities costs (e.g., electrical energy), safety and environmental health programs, and compliance programs. Universities cover much of the costs of these support services. In the context of scientific research in academia, the expenses for support services and infrastructure are often referred to as "indirect costs" (see Box 6-1). A major problem with the federal system of research support is that the total grant or contract to a university typically does not cover the full costs of the research, because it fails to fully cover the indirect costs.

6.5 CONCLUSIONS

Chemistry funding comes from a diversity of sources in order to fund projects at many different scales. Although it is impossible to assess every funding source for chemistry, this chapter described a number of different funding sources that are critical to advancing the chemical economy and addressed some of their advantages and disadvantages. There are some important facets of funding that the committee wanted to highlight in this chapter.

Conclusion 6-1: Investment in the infrastructure at research universities is not well supported. This diminishes the opportunities for many talented chemical researchers to use the newest tools, technologies, and instrumentation and prevents trainees from having access to the newest technologies being used in the chemical workforce.

Conclusion 6-2: Small Business Innovation Research and Small Business Technology Transfer (SBIR/STTR) programs have proven to be an important mechanism for advancing the chemical enterprise. There are many examples of fundamental chemical research being further pursued as a marketable product or process to contribute to the chemical economy through SBIR/STTR programs, and these programs also foster an emerging area of the chemical workforce where university researchers create and work in these small start-ups that are based on the grants from these programs.

Conclusion 6-3: Partnerships between industry and government as well as those between industry and academia continue to be an important source of funding that provides money for chemical innovation that is necessary to advance fundamental research as well as environmental sustainability in industry.

Conclusion 6-4: In the near term, foundation and individual philanthropic support is likely to grow as a resource for innovations in chemistry. This support provides an important opportunity to use scientific evidence and exploration to address challenges that will benefit all of society, such as climate change and human health.

BOX 6-1
Underrecovery of Research Costs by Universities

The "underrecovery" of expenses associated with academic research is a serious financial challenge that requires universities to spend a significant amount of money to support the facilities and other infrastructure that their research faculty need. This is true for every academic institution supporting the chemical sciences, from large research institutions to small liberal arts colleges and primarily undergraduate institutions.

How much an academic institution pays for the indirect costs is based on a negotiated indirect cost rate. Once negotiated, the indirect cost rate applies to all federal research grants and contracts awarded to a given institution for a period of several years. The indirect cost rate is based on the documented indirect costs and the grant and contract history of the university. The indirect cost rate is applied to certain costs that will be covered as direct costs by the grant or contract. To illustrate, suppose that a university has negotiated a 50% indirect cost rate. And suppose that a scientist in that university's chemistry department applies for a federal grant to continue a research project, requesting $100,000 for salaries, expendable supplies, and certain other expenses that would be charged to the grant. Since these expenses subject to the indirect cost rate total $100,000 and the rate is 50%, then the grant will be awarded at the level of $150,000. The additional $50,000 will be used by the awardee to cover the indirect costs of research.

The problem is that the aggregate shortfall in recovery of indirect costs due to the research infrastructure is typically large, leading to a result that every dollar of research funding costs a university 25 cents. Therefore, a university with $100 million per year in federal research dollars must provide $25 million to cover the research expenses not covered by the grants and contracts. For financially challenged universities, the inevitable result is a weakened infrastructure for research and diminished opportunities for faculty with otherwise competitive ideas.

It has never been true that the federal government has covered all the expenses of federally sponsored research programs at universities. There has always been a partnership between universities and the federal government. The challenge universities face today is that the balance in this partnership has shifted in a way that is increasingly burdensome to the universities, because the negotiated indirect cost rates have been eroded.

Furthermore, this shift has been accompanied by increased compliance costs associated with federal programs without commensurate increases in indirect cost rates.

While the federal government has shifted more costs to universities, foundations—which are growing in their importance in funding academic basic research—typically provide much less than the federal government to support the research infrastructure. While the average indirect cost rate for federal programs at research universities is about 50%, foundations typically provide 10% or less of the foundation award for support of the research infrastructure.

The underrecovery of research expenses is a serious financial challenge that requires universities to spend a significant amount of money to support their research infrastructure. Wealthy academic institutions devote a large fraction of their spendable income from endowment to provide this support. Academic institutions lacking large endowments and large gift streams are at a major disadvantage in competing for research support, because they possess relatively weak research infrastructure and lack the ability to support the recruitment of talented faculty.

7

Conclusions and Recommendations

In the process of addressing the Statement of Task, the committee realized that many of its recommendations and key conclusions drew on evidence from different aspects of the chemical economy, including its economics, politics, research goals, and future workforce needs. To make sure there was adequate information for these conclusions, the committee decided to build a body of evidence throughout the report before the final set of recommendations was presented. This chapter is not meant to serve as an exhaustive list of conclusions and recommendations that cover all of fundamental chemical research and the chemical economy but rather targets some big picture areas, as well as a number of specifics that address the committee's charge, which came through as being particularly important during information gathering, discussions, and deliberations. This chapter begins broadly with recommendations for chemical research and the chemical economy, and then narrows to talk about more specific topics: team science, chemistry and environmental sustainability, challenging the assumptions around chemical research, chemical data, the future chemical workforce, and funding for chemical research. The conclusions summarized here may also be found in the individual chapters. To maintain consistency, the numbers used to designate them are identical to those listed in the different chapters.

7.1 THE IMPORTANCE OF CHEMICAL RESEARCH TO THE CHEMICAL ECONOMY

Throughout the information-gathering process, the committee found numerous examples of how chemistry was central and critical to discoveries that improved the quality of life for all people. In Section 2.3.3, the committee outlined six examples where a foundational base of chemical knowledge, along with several breakthrough chemical discoveries, led to dramatic changes in products or processes, and led to ubiquitous products such as the silicon chips in our electronic devices or the small molecules that are used for birth control, diabetes, or preventing deaths from the current SARS-CoV-2 pandemic. Along with these examples, Chapter 2 notes that together, all sectors reliant on the U.S. chemical economy are responsible for $5.2 trillion, or 25%, of the U.S.

gross domestic product, and the entire chemical enterprise supports 4.1 million jobs in the United States of people who are interacting with the chemical economy. Additionally, there is evidence that chemical patents, and patents that rely on chemical knowledge, are more valuable than comparable patents that are not related to chemistry.

Although expansive and critical to many areas of the economy, the U.S. chemical economy is not the undisputed global leader that it once was. Other countries have seen large sustained investments in their chemical and overall research enterprises, including China and several others, and their rapid advances are starting to threaten U.S. leadership in chemistry in specific areas. Based on all of this information, the committee landed on several important conclusions.

Conclusion 2-3: Chemistry is a foundational and central scientific discipline, and sustained investment in fundamental chemical research provides the chemical knowledge for technology development, generating unexpected discoveries that are the basis for innovation. These innovations directly influence the chemical economy, environment, and quality of life, and also advance knowledge and discovery in many other scientific and technological disciplines, such as the life sciences, information technology, earth sciences, and engineering.

Conclusion 2-4: The chemical economy is critically important for our national economy and our leadership in the international chemical enterprise. This leadership relies heavily on advances in fundamental chemistry that drive the creation of new tools, technologies, processes, and products and enables environmental considerations. However, our nation's leadership in the chemical industry cannot be taken for granted, and this leadership needs continued and sustained nurturing and support.

Conclusion 2-2: It is challenging to directly link chemical research to economic impact because each chemical product or process relies on a broad body of chemical knowledge and discovery that is built over decades or centuries, and chemical knowledge is also deeply integrated into other disciplines, making the specific impacts of chemistry in the broad scientific enterprise difficult to deconvolute. Additionally, analyzing the economic impact of chemical research suffers from a lack of data, including patent value estimations, widely available licensing terms data, and government grant data.

To take action based on this set of broad conclusions, the committee suggests a set of wide-ranging actions, mostly directed at the U.S. government but with advice, as well, for other actors within the chemical economy such as industry practitioners and academic researchers. In relation to international competitiveness and its national security implications, the committee understands the importance of security but notes that U.S. competitiveness in the sciences relies heavily on attracting international talent and ensuring an open exchange of people and ideas. These recommendations emphasize growing and strengthening the U.S. chemical economy and U.S. competitiveness.

Recommendation 1: To foster fundamental chemical research and maintain U.S. competitiveness in the chemical economy, the U.S. chemical enterprise should support funding, workforce, and policy structures that attract international researchers and create a nurturing environment for all research talent.

Sub-Recommendation 1-1: Because it is not possible to predict where the next fundamental breakthroughs will come from, funding agencies that support the chemical sciences, such as the U.S. Department of Energy, National Science Foundation, National Institutes

of Health, U.S. Department of Defense, National Institute of Standards and Technology, and U.S. Department of Agriculture, should fund the largest breadth of fundamental chemical research projects possible. This should include funding for a large range of topics in chemistry, as well as different scales of research projects, ranging from small grants for individual laboratories to large-scale collaborations and facilities.

Sub-Recommendation 1-2: Participants in the chemical economy including chemical industry, pharmaceutical companies, and instrumentation developers should continue to invest in research and development at universities and scientific research institutions in the United States and should increase investments in broad areas of fundamental chemical research, including a focus on environmental sustainability.

Sub-Recommendation 1-3: The U.S. government should continue to produce policies that support international and open exchange of ideas in the chemical sciences and should engage policy and security experts, academic researchers, and industry professionals when considering any limitations on open engagement that are meant to mitigate economic or security risks to the U.S. chemical enterprise.

Sub-Recommendation 1-4: To help guide policy and funding decisions around chemical research, federal agencies who fund and track data related to scientific research should collaborate to collect, and make available, the tools and data needed to understand the impact of fundamental chemical research on the chemical economy. As a part of this initiative, large-scale evidence-building efforts to collect, standardize, use, and interpret these data should be funded.

7.2 ROLE OF CHEMISTRY IN TEAM SCIENCE

Another important outcome from the report is that major chemical discoveries that have large economic and societal impacts do not happen in a vacuum. There are large bodies of scientific research in areas such as the life sciences, physics, engineering, and the social sciences that contribute to all important chemical and technological advances. Additionally, the inverse is true: chemical knowledge contributes to many diverse fields of science and technology. When teams of researchers with diverse expertise gather to solve a central problem in a critical area, chemistry drives basic knowledge and practical application in order to help teams accomplish major advances. This is highlighted in Chapters 2 and 3 of the report but is very strongly emphasized in Chapter 4, where we discuss the emerging areas of measurement, automation, computation, and catalysis, all of which rely heavily on different fields of research in order to successfully advance. To emphasize this, the committee noted the following conclusion and recommendation.

Conclusion 4-1: Chemistry is an enabling scientific discipline that will continue to have the largest impact on society when chemists collaborate with experts from other areas such as engineering, biology, physics, computation, and data science to generate new fundamental knowledge and create translational impact at larger scales.

Recommendation 2: Research groups across all scales—small-to-medium interdisciplinary teams, large-scale collaborations, and facilities—should reflect the centrality of chemistry to science and engineering. Because of the central and enabling nature of chemistry, experts across chemistry and its subdisciplines should be considered when there are large interdisciplinary projects, highly collaborative institutions, national lab research, and other team-based scientific activities.

7.3 CHEMICAL RESEARCH AND SUSTAINABILITY

With all the advances that chemistry has made to improve the quality of life of all individuals, there is the competing fact that chemistry has also caused a lot of harm to the environment and human health. As noted in Chapter 1, it is ironic that the solutions to many of these environmental issues might very well come from chemistry and chemical engineering. In talking to professionals and researchers from industry, academia, and government, it is clear that the next chapter of fundamental chemical research will look to address the climate crisis and look toward solutions for global sustainability. These solutions will need to come from many different areas of engineering and science, but fundamental chemistry research has already played, and will continue to play, a critical role in advancing some technologies and helping with the co-design of other tools and technologies. Chapter 3 highlights areas where chemistry has already had an impact in solving environmental problems and identifies some of the policies that helped to drive these solutions. Additionally the chapter points out a number of areas of sustainability, many aligned with the United Nations Sustainable Development Goals, where fundamental chemical research could have the largest impact in the near future. These ideas are emphasized in many of the conclusions from that chapter, as stated below.

Conclusion 3-1: To implement a circular economy, the future will require a paradigm shift in the way products are designed, manufactured, and used, and how the waste products are collected and reused. These new processes, and the use of clean energy and new feedstocks to enable these processes, will require novel chemistries, tools, and new fundamental research at every stage of design.

Conclusion 3-2: Transitioning the chemical economy into a new paradigm around sustainable manufacturing, in which environmental sustainability is balanced with the need for products that will improve quality of life, enhance security, and increase U.S. competitiveness, will require substantial investment and innovation from industry, government, and their academic partners to create and implement new chemical processes and practices.

To accomplish this paradigm shift in the chemical sciences many steps will need to be taken, and a concerted effort will be needed from government, industry, and academia. While some of the specifics of this transition are explained in Chapter 3, collaboration between different groups involved in design and implementation will be critical for moving forward.

Recommendation 3: The chemical industry and its partners at universities, scientific research institutions, and national labs should create opportunities to collaborate so that the objectives of fundamental research can directly assist in the design process of companies implementing new processes or practices toward environmental stewardship, sustainability, and clean energy.

While the conclusions and recommendations above take a broad look at environmental sustainability, and think through a shift of the entire chemical economy, there are specific conclusions and recommendations that apply to researchers. For chemical research to evolve with, and help advance, the moving landscape of the chemical economy toward sustainability, there are two key conclusions that the committee came to, based on the evidence gathered in Chapter 3.

Conclusion 3-3: As fundamental chemical research continues to evolve, the next generation of research directions will prioritize the future of environmental sustainability and

new energy technologies. Keeping sustainability principles in mind during every stage of research and development will be critical to accomplishing this goal.

Conclusion 3-4: Chemical research will have the greatest impact addressing energy and environmental sustainability if researchers and practitioners develop and use tools to quantify and mitigate environmental and human health impacts of new discoveries and are aware of the societal implications of their work, and if the research is driven by policies that identify specific environmental sustainability outcomes.

To encourage academic researchers to keep environmental sustainability in mind at every stage of research, the committee noted that grant mechanisms usually do not ask researchers to consider the environmental impact of their work unless it is directly related to the grant or contract for which the researcher is applying. Although not ubiquitous in grant writing, "broader impacts statements" have been an important mechanism in encouraging researchers to think through how their labs and research are interacting with the community around them. To similarly encourage all academic chemical researchers to keep environmental sustainability and environmental stewardship at the front of their minds when considering all different types of research endeavors, the committee thought an optional "environmental impacts" statement would be the best way to accomplish this.

There are some inherent downsides to this kind of statement. There is frequently resistance from the scientific community toward the need for additional writing and information to be included in the research proposal writing process. Additionally, it can sometimes be difficult to consider environmental impacts of basic research that does not yet have a described application. Instead of being a burden for researchers, the committee envisions the environmental impacts statement as an optional opportunity to think about how their research aims fit into the broader context of climate change and other environmental issues. Adding an environmental impacts statement would also allow researchers to gain experience applying a systems-level thinking approach to understand their own impact, and how their research might be used to reduce environmental impact or address grand challenges such as climate change. These kinds of statements would strengthen research proposals through broader thinking about the environment.

Recommendation 4: All chemistry-related research grants and proposals should have an option to explain the "environmental impacts" of the proposed research, as an option under the "broader impacts" statement. The "environmental impacts" statement should include a summary of the possible environmental impacts, what is being done to mitigate those impacts, and any outcomes from the research that will directly impact environmental sustainability.

Recommendation 4 is intentionally vague because the committee understands that different funders will have different funding requirements and missions, and might approach such a statement in a wide variety of ways. Similar to the way that the National Science Foundation asks about broader impacts, an environmental impacts statement would benefit from being broad and allowing researchers to include what they think is appropriate. Some possible topics that could be addressed in an environmental impacts statement include, but are not limited to,

- how the methods, chemicals, or technologies used in the research will help to reduce the environmental footprint in comparison to what is currently available;
- what aspects of environmental impact could be measured or tested using the methods under development in the proposed research;

- understanding how the research will be applied to technologies such as carbon capture that could help mitigate climate change; and
- showing a systems-level understanding of how the research might have an impact on the environment and explaining steps that were taken to mitigate any negative impacts.

7.4 CHALLENGING THE UNDERLYING ASSUMPTIONS OF CHEMICAL RESEARCH

Complementary to the paradigm shift toward sustainability described in the previous section, there is a transition happening in the way we collect and use energy. This change is happening rapidly and, as noted in Chapter 3, is affecting the materials, technologies, and processes that are needed to effectively implement the energy landscape. As shown throughout this report, chemical research and the chemical economy have been critical to the implementation of the current energy landscape. Today and in the near future, there are many areas ripe for chemical discovery in a new energy landscape that prioritizes clean alternatives and decarbonization. To accomplish these scientific advances more effectively, there are several factors to take into account, such as the changing needs for metals and minerals that arise with the increase in electric vehicles and complications with acquiring metals based on shifting international politics (see Sections 1.3.2 and 3.4.2.2). Researchers, economists, and industry professionals must constantly assess the assumptions being made about the different aspects of the chemical economy and decide if those assumptions are true or if they are likely to change. By understanding what is most likely for the energy landscape in the future, chemical researchers can make decisions about what the pressing needs will be to help move sustainability forward.

Conclusion 3-5: As the world moves deeper into its current energy transition—including the switch to electric vehicles, the implementation of clean energy alternatives, and the use of new feedstock sources—coupled with an increasing focus on circularity, the committee expects that decarbonization, computation, measurement, and automation will significantly alter the operations and processes of current industries, creating new opportunities and challenges that will benefit from fundamental chemistry and chemical engineering advances.

Recommendation 5: Changes in energy sources complemented by the technology and processes offered by chemical companies will lead to entire industries being created, transformed, and terminated. A group of experts from chemistry and other impacted disciplines, who represent the chemical economy and academic research, should be convened to assess the implications of these industrial shifts and understand their impacts on current chemical research paradigms. Based on the information from these discussions, funding agencies and the chemical industry should put money toward interesting opportunities for chemical research that might emerge based on these trends.

7.5 CHEMICAL DATA AND ANALYSIS

In Chapter 4, we explore the emerging tools and technologies of measurement, automation, computation, and catalysis. These four technologies will not only be critical to addressing the future of sustainability, as noted in Chapter 3, but will also continue to accelerate research in all areas of chemistry. These technologies are also exciting to fundamental chemistry research because, in addition to advancing chemistry, the tools themselves have all benefited from the accumulation of chemical knowledge. Additionally important is that chemical and chemistry-related data are critical to all of these emerging tools and technologies. In assessing these emerging tools and technologies, there was a common thread noted by the committee on the importance of well-curated and accessible data to benefit all aspects of chemistry. While there have been some efforts to collect and share chemical data, as noted in Chapter 4, the practice is not universal, and there is a particularly large dearth of negative data in chemistry. A large supply of both positive and negative data would be particularly helpful for developing models and for understanding different molecular properties and interactions, as well as learning how to more accurately measure chemical systems. For progress to continue, a concerted effort must be made to establish standards for how chemical data are collected, stored, and distributed, leading to the following conclusion and recommendation.

Conclusion 4-3: The ability to collect, document, store, share, and use chemistry-related data is needed to advance the use of new tools such as computation and automation in fundamental chemical research, and increase the accessibility of chemical research to a larger community of practitioners. This information architecture will produce an indispensable tool for the chemical sciences research community to increase the pace and efficiency of innovation by fully harnessing advances made with previous research investments.

Recommendation 6: The National Institute of Standards and Technology (NIST), in consultation with the International Union of Pure and Applied Chemistry, the American Chemical Society, and other global chemistry professional societies, should lead an effort to explore pathways that provide an open-source, accessible, and standardized way for chemical researchers to store, share, and use data from chemical experiments. In establishing these pathways, NIST should seek input from professional societies and stakeholders from different areas of chemical research and data science so that they can best understand the infrastructure needs of different research communities such as inorganic, organic, and analytical chemists. Once standards and data repositories are established, publishers should require researchers to submit all data related to reactions, measurements, or other chemical experiments to these established open-source repositories.

7.6 CHEMICAL WORKFORCE

Throughout this report, there is an emphasis on the individuals who are responsible for chemical advances and those who are driving the chemical economy. Chapter 2 noted how expansive the chemical economy is and how it supports millions of jobs. While the committee understands that not all of these employees will go through chemistry or chemical engineering programs, the persons who are trained as chemists and chemical engineers are connected to and dependent upon all other people in the chemical economy. Formulating conclusions and recommendations for employees of the chemical economy that are not trained in chemistry and chemical engineering is beyond the scope of this report. However, in Chapter 5, there is an emphasis on important areas that will help to build the next generation of the chemical workforce who make their way through chemistry and chemical engineering programs. Importantly, there is a strong need for a diverse workforce that is developed through equitable training practices, such as well-developed mentorship and professional development programs. Additionally, the chapter emphasizes the need for chemistry curricula to be adaptable to the future needs of the chemical enterprise. There are three conclusions from that chapter that cover these principles.

Conclusion 5-1: A skilled science and engineering workforce paired with a diverse, inclusive, and equitable science and engineering research enterprise is central to a thriving, nimble chemical economy equipped to respond to emerging challenges and maintain U.S. competitiveness.

Conclusion 5-2: The current structures and systems governing funding, promotion, retention, and professional development are in conflict and can stymie holistic career advancement for students, faculty, and research staff.

Conclusion 5-5: Creating an equitable and inclusive learning environment that exposes trainees of the future chemical workforce to new and innovative chemical tools, technologies, and instrumentation, as well as interdisciplinary knowledge and critical collaboration skills, will require a serious and sustained investment from funding agencies, universities, industry partnerships, and accreditation programs. This investment is critical because the tools and practices that enable chemical research are constantly evolving, and training programs must be able to adapt to best facilitate the learning of basic-to-advanced chemical principles that will help students succeed.

To properly address these conclusions on a practical level, the committee recommends that steps be taken to fund research in chemical education, continually reassess chemistry curricula, and continue to provide opportunities for professional development. The following recommendations lay out these ideas in more detail.

Recommendation 7-1: Funding agencies that support chemical research should put a substantial investment toward education research to continue enabling the development of innovative ways of teaching students about new and emerging concepts, tools, technologies, and instrumentation in chemistry while creating an inclusive learning environment for all students.

Recommendation 7-2: Universities, colleges, and accreditation programs should continually reassess their curriculum requirements and pedagogical practices to ensure that chemistry students in the chemical sciences are receiving state-of-the-art inclusive training and the most current chemical information and advances.

Recommendation 7-3: Universities and agencies that fund and support education in the chemical sciences should provide professional development at all levels, allowing for opportunities that are specific to the needs of each educational or career stage, such as programs that connect students with internships or resources for career exploration and providing faculty with professional development opportunities aimed at advancing their scholarship and teaching.

Recommendation 7-4: To continue progress in improving the diversity and equity of the chemical workforce, universities and chemical sciences departments should regularly assess their recruitment and retention practices related to trainees, faculty, and research staff. These assessments should be guided by relevant experts in research-informed equitable recruitment and retention practices of higher education institutions and units that also understand the nuances and details of the particular institution or entity. Institutions and units should continually take action and make meaningful investments based on their assessments. This work should be reported in a timely and transparent fashion to the institutional community.

7.7 FUNDING CHEMICAL RESEARCH

The funding landscape for chemical research in the United States is quite broad, and as described in Chapter 6, chemical research is funded by a diverse set of private and public sources. One of the major advantages of this broad network of funding opportunities is that many different types and scales of chemical research are able to seek out and secure funding. One downside is that it can be challenging for new researchers to navigate this complicated landscape. While there are many other challenges with the funding landscape, overall, it is able to provide a wide variety of opportunities, and many aspects do not require immediate change. There are a number of conclusions in Chapter 6 that outline the main findings, but the final recommendations focus on three areas where there is the greatest possibility to be impactful to the U.S. chemical research enterprise and the chemical economy.

First, the committee wanted to highlight the importance of the Small Business Innovation Research and Small Business Technology Transfer (SBIR/STTR) programs in chemistry. This is one of the few government funding mechanisms where there is an opportunity to build fundamental chemical research into a product, process, or technology that will impact the broader chemical economy. In Section 6.1.3 of the report, we highlight some of the success stories of SBIR/STTR programs that funded chemistry-related projects. For these reasons, the following conclusion and recommendation highlight SBIR/STTR programs.

Conclusion 6-2: Small Business Innovation Research and Small Business Technology Transfer (SBIR/STTR) programs have proven to be an important mechanism for advancing the chemical enterprise. There are many examples of fundamental chemical research being further pursued as a marketable product or process to contribute to the chemical economy through SBIR/STTR programs, and these programs also foster an emerging area of the chemical workforce where university researchers create and work in these small start-ups that are based on the grants from these programs.

Recommendation 8: Funding agencies should continue to support and publicize innovations in the chemical sciences through Small Business Innovation Research and Small Business Technology Transfer programs in order to leverage their previous investments in fundamental research and allow researchers the opportunity to bring new products or processes to market.

In the landscape of public and private funding, one area that has become more prominent over the past several years is the rise of philanthropic support. Section 6.3 outlines the many ways that philanthropies have contributed to fundamental chemical research. But relative to federal funding for basic research, philanthropy supports far fewer schools and research projects. The future of scientific funding will continue to include larger percentages of money from philanthropic organizations and independent donations. To ensure that science works to address big societal issues such as climate change and human health, funders will need to invest in the fundamental chemistry that informs and is critical to so many other areas of science. This will be particularly impactful if foundations and organizations continue to have minimal influence on the specifics of funded projects, instead allowing scientists to identify approaches and solutions to global problems. The following conclusion and recommendation start to address these issues.

Conclusion 6-4: In the near term, foundation and individual philanthropic support is likely to grow as a resource for innovations in chemistry. This support provides an important opportunity to use scientific evidence and exploration to address challenges that will benefit all of society, such as climate change and human health.

Recommendation 9: The American Chemical Society, along with other chemistry-related professional societies, universities, and their academic leaders, should explore mechanisms to be more proactive in communicating to philanthropists and foundations about the promise of fundamental chemistry in addressing national and global problems. University and academic leaders should emphasize the importance of funding structures between philanthropic and federal funding mechanisms that ensure balance and complementarity.

To accomplish the chemical research described in this report, there is a critical need for laboratory space, instrumental facilities and support, access to computation, and much more. Chapter 4 covered different emerging areas of research that will be needed for continued advances of chemical science, and all of these areas rely on institutional infrastructure. Infrastructure is also critical for training the next generation of the chemical workforce. In Chapter 5, we showed that there is a need to continually rethink and adapt curricula, based on new tools and technologies that will be critical to the future of research and industry. Having infrastructure in place gives institutions the ability to train students, faculty, and other professionals on these emerging technologies.

In its final section, Chapter 6 explains some of the current issues with how institutional infrastructure is supported. This is particularly important to trainees who will be entering the chemical economy, because chemical research requires a working knowledge of many different facets of technology and research, all of which are supported through infrastructure. These findings are summarized with a conclusion and a supporting recommendation.

Conclusion 6-1: Investment in the infrastructure at research universities is not well supported. This diminishes the opportunities for many talented chemical researchers to use the newest tools, technologies, and instrumentation and prevents trainees from having access to the newest technologies being used in the chemical workforce.

Recommendation 10: The federal government should invest more to support research infrastructure at research institutions to ensure that talented chemical experts and trainees with outstanding ideas can be competitive for research awards.

References

Abinader, L. G. 2021. US government rights in patents on Molnupiravir, based upon funding of R&D at Emory University. *Knowledge Ecology International*. https://www.keionline.org/36648#:~:text=The%20U.S.%20government%20has%20rights%20in%20molnupiravir%20patents,to%20derivatives%20of%20n4-hydroxycytidine%2C%20the%20molnupiravir%20parent%20compound.

ACC (American Chemistry Council). 2020. *2020 Guide to the Business of Chemistry*. https://www.americanchemistry.com/chemistry-in-america/data-industry-statistics/resources/2020-guide-to-the-business-of-chemistry.

ACC. 2021. *2021 Guide to the Business of Chemistry*. https://www.americanchemistry.com/chemistry-in-america/data-industry-statistics/resources/2021-guide-to-the-business-of-chemistry.

ACC. 2022. Shale gas is driving new chemical industry investment in the U.S. https://www.americanchemistry.com/better-policy-regulation/energy/resources/shale-gas-is-driving-new-chemical-industry-investment-in-the-us.

ACS (American Chemical Society). 2021. Diversity Data Report 2021. https://axial.acs.org/acs-diversity-data-2021/?utm_source=pgm&utm_medium=axial&utm_campaign=PUBS_1221_JHS__ALL_BRO_DEIR_White_Paper_1021_&src=PUBS_1221_JHS__ALL_BRO_DEIR_White_Paper_1021_&ref=pgm_axial_PUBS_1221_JHS__ALL_BRO_DEIR_White_Paper_1021_.

ACS, NSF, and NIH (American Chemical Society, National Science Foundation, and National Institutes of Health). 2006. Workshop on Building Strong Academic Chemistry Departments Through Gender Equity. https://www.acs.org/content/dam/acsorg/funding/awards/national/gender-equity-report-cover.pdf.

Adams, F., and M. Adriaens. 2020. The metamorphosis of analytical chemistry. *Analytical and Bioanalytical Chemistry* 412:3525-3537.

Aden, N., M. Stork, and K. Chang. 2020. Barriers, Challenges, and Opportunities for Chemical Companies to Set Science-Based Targets. Science Based Targets. https://sciencebasedtargets.org/events/publication-launch-barriers-challenges-and-opportunities-for-chemical-companies-to-set-science-based-targets.

Adidas. 2019. Adidas unlocks a circular future for sports with FUTURECRAFT.LOOP: A performance running shoe made to be remade. Press Release. https://www.adidas-group.com/en/media/news-archive/press-releases/2019/adidas-unlocks-circular-future-sports-futurecraftloop/.

Adidas. 2021. Launching 'Choose to Give Back', a resale program enabled by ThredUP. Press Release. https://news.adidas.com/made-with-recycled-materials/launching--choose-to-give-back---a-resale-program-enabled-by-thredup/s/ffbe00bd-0507-4262-8457-1c6f2ac9de08.

Agrawaal, H., and J. E. Thompson. 2021. Additive manufacturing (3D printing) for analytical chemistry. *Talanta Open* 3:100036.

Alamri, M. S., A. A. A. Qasem, A. A. Mohamed, S. Hussain, M. A. Ibraheem, G. Shamlan, H. A. Alqah, and A. S. Qasha. 2021. Food packaging's materials: A food safety perspective. *Saudi Journal of Biological Sciences* 28:4490-4499.

Alcasabas, A., P. R. Ellis, I. Malone, G. Williams, and C. Zalitis. 2021. A comparison of different approaches to the conversion of carbon dioxide into useful products. Part I: CO_2 reduction by electrocatalytic, thermocatalytic and biological routes. *Johnson Matthey Technology Review* 65:180-196.

Aldridge, S., J. Parascandola, and J. L. Sturchio. 1999. *Discovery and Development of Penicillin.* Washington, DC: American Chemical Society and London: Royal Society of Chemistry.

Aleahmad, T. 2009. Pasteur's and Edison's quadrants. Open Education Research. https://openeducationresearch.org/2009/01/pasteurs-and-edisons-quadrants/.

Alimi, O. A., and R. Meijboom. 2021. Current and future trends of additive manufacturing for chemistry applications: A review. *Journal of Materials Science* 56:16824-16850.

Allen, D. T., Q. Chen, and J.B. Dunn. 2021. Consistent metrics needed for quantifying methane emissions from upstream oil and gas operations. *Environmental Science & Technology Letters* 8(4):345-349.

Allen, D. T., and D. R. Shonnard. 2001. *Green Engineering: Environmentally Conscious Design of Chemical Processes.* Upper Saddle River: Prentice Hall.

Amar, J. G. 2006. The Monte Carlo method in science and engineering. *Computing in Science & Engineering* 8(2):9-19. http://astro1.panet.utoledo.edu/~jamar/ph/mcmethod.pdf.

Amaro-Soriano, A., F. Hernández-Aldana, and A. Rivera. 2021. Photochemical treatments (UV/H2O2, UV/O3 and UV/H2O2/O3) and inverse osmosis in wastewater: Systematic review. *World Journal of Advanced Research and Reviews* 10(2):229-240.

American Chemical Society National Historic Chemical Landmarks. 1993. Bakelite: The World's First Synthetic Plastic. http://www.acs.org/content/acs/en/education/whatischemistry/landmarks/bakelite.html.

Anastas, P. T., and J. C. Warner. 1998. *Green Chemistry: Theory and Practice.* Oxford University Press.

Appel, A. M., J. E. Bercaw, A. B. Bocarsly, H. Dobbek, D. L. DuBois, M. Dupuis, J. G. Ferry, E. Fujita, R. Hille, P. J. A. Kenis, C. A. Kerfeld, R. H. Morris, C. H. F. Peden, A. R. Portis, S. W. Ragsdale, T. B. Rauchfuss, J. N. H. Reek, L. C. Seefeldt, R. K. Thauer, and G. L. Waldrop. 2013. Frontiers, opportunities, and challenges in biochemical and chemical catalysis of CO_2 fixation. *Chemical Reviews* 113:6621-6658.

Aramco. 2016. Saudi Aramco acquires Novomer's polyol business and associated technologies, enhancing its downstream expansion strategy. https://www.aramco.com/en/news-media/news/2016/acquires-novomers-polyol-business-downstream-expansion.

Ardagh, M. A., M. Shetty, A. Kuznetsov, Q. Zhang, P. Christopher, D. G. Vlachos, O. A. Abdelrahman, and P. J. Dauenhauer. 2020. Catalytic resonance theory: Parallel reaction pathway control. *Chemical Science* 11:3501-3510.

Argonne Today. 2013. Guest Column: Ask the Academic. Argonne Postdoctoral Blog. https://blogs.anl.gov/postdoc/2013/04/30/guest-column-ask-the-academic/.

Arora, A., and A. Gambardella. 2010. Implications for Energy Innovation from the Chemical Industry. Working Paper 15676. National Bureau of Economic Research. https://www.nber.org/system/files/working_papers/w15676/w15676.pdf.

autm. n.d. STATT: Statistics Access for Technology Transfer Database. Autm.net. https://autm.net/surveys-and-tools/databases/statt.

Autor, D. H., D. Dorn, G. H. Hanson, G. Pisano, and P. Shu. 2019. Foreign Competition and Domestic Innovation: Evidence from U.S. Patents. CESifo Working Paper No. 7865. https://www.econstor.eu/bitstream/10419/207256/1/cesifo1_wp7865.pdf.

Ayanoglu, E. 2019. Energy efficiency in data centers. IEEE ComSoc Technical Committees Newsletter, November. https://www.comsoc.org/publications/tcn/2019-nov/energy-efficiency-data-centers#:~:text=The%20United%20States%20Department%20of%20Energy%20currently%20states,1%25%20of%20demand%20for%20electricity%20in%20the%20world.

Azadi, M., S. A. Northey, S. H. Ali, and M. Edraki. 2020. Transparency on greenhouse gas emissions from mining to enable climate change mitigation. *Nature Geoscience* 13:100-104.

Bagnato, G., A. Sanna, E. Paone, and E. Catizzone. 2021. Recent catalytic advances in hydrotreatment processes of pyrolysis bio-oil. *Catalysts* 11:157.

Bardin, B. 2021. Presentation to the committee at the Public Session on Automation, July 7.

Barteau, M., and S. Kota. 2014. *Shale Gas: A Game-Changer for U.S. Manufacturing.* University of Michigan. https://www.nist.gov/system/files/documents/2017/05/09/PDF-Shale-Gas-FINAL-web-version-1.pdf.

Bauer, A. S., M. Tacker, I. Uysal-Unalan, R. M. S. Cruz, T. Varzakas, and V. Krauter. 2021. Recyclability and redesign challenges in multilayer flexible food packaging—A review. *Foods* 10:2702.

BBC News. 2019. UN resolution pledges to plastic reduction by 2030. https://www.bbc.com/news/science-environment-47592111.

Beacham, W. 2021. INSIGHT: EU's Fit for 55 plan to add costs for chemicals, but could spur innovation. ICIS.com, July 14. https://www.icis.com/explore/resources/news/2021/07/14/10663196/insight-eu-s-fit-for-55-plan-to-add-costs-for-chemicals-but-could-spur-innovation/.

REFERENCES

Beaumont, N. J., M. Aanesen, M. C. Austen, T. Börger, J. R. Clark, M. Cole, T. Hooper, P. K. Lindeque, C. Pascoe, and K. J. Wyles. 2019. Global ecological, social and economic impacts of marine plastic. *Marine Pollution Bulletin* 142:189-195.

Beketov, M. A., B. J. Kefford, R. B. Schäfer, and M. Liess. 2013. Pesticides reduce regional biodiversity of stream invertebrates. *Proceedings of the National Academy of Sciences of the United States of America* 110:11039-11043.

Bellona. 2015. Factsheet: Climate Action in the Cement Industry. Bellona.org. https://bellona.org/publication/factsheets-serie-1-climate-action-in-the-cement-industry.

Benavides, P. T., D. C. Cronauer, F. Adom, Z. Wang, and J. B. Dunn. 2017. The influence of catalysts on biofuel life cycle analysis (LCA). *Sustainable Materials and Technologies* 11:53-59.

Bender, T. A., J. A. Dabrowski, and M. R. Gagné. 2018. Homogeneous catalysis for the production of low-volume, high-value chemicals from biomass. *Nature Reviews Chemistry* 2:35-46.

Bento, A. M., and R. Klotz. 2014. Climate policy decisions require policy-based lifecycle analysis. *Environmental Science & Technology* 48:5379-5387.

Bergbreiter, D. E., J. Tian, and C. Hongfa. 2009. Using soluble polymer supports to facilitate homogeneous catalysis. *Chemical Reviews* 109:530-582.

Bergeson & Campbell, P.C. 2021. Sustainable Chemistry Research and Development Act passed as part of National Defense Authorization Act. *National Law Review* 11(20).

Bernal, A. J., M. M. Gomes da Silva, D. B. Musungaie, E. Kovalchuk, A. Gonzalez, V. Delos Reyes, A. Martín-Quirós, Y. Caraco, A. Williams-Diaz, M. L. Brown, J. Du, A. Pedley, C. Assaid, J. Strizki, J. A. Grobler, H. H. Shamsuddin, R. Tipping, H. Wan, A. Paschke, J. R. Butterton, M. G. Johnson, C. De Anda, and MOVe-OUT Study Group. 2022. Molnupiravir for oral treatment of COVID-19 in nonhospitalized patients. *New England Journal of Medicine* 386:509-520.

Bhutada, G. 2021. All the metals we mine each year, in one visualization. Weforum.org. https://www.weforum.org/agenda/2021/10/all-tonnes-metals-ores-mined-in-one-year/.

Bi, H., and X. Han. 2019. Chemical sensors for environmental pollutant determination. Pp. 147-160 in *Chemical, Gas, and Biosensors for Internet of Things and Related Applications*, K. Mitsubayashi, O. Niwa, and Y. Ueno, eds. Amsterdam: Elsevier.

Biddy, M. J., C. Scarlata, and C. Kinchin. 2016. *Chemicals from Biomass: A Market Assessment of Bioproducts with Near-Term Potential.* Technical Report NREL/TP-5100-65509. National Renewable Energy Laboratory, U.S. Department of Energy. https://www.nrel.gov/docs/fy16osti/65509.pdf.

Birnbaum, S., M. S. Cheffo, D. E. Fleming III, R. Passaretti-Wu, M. Schwarz, N. Williams, and M. Fazio. 2021. PFAS: Expected litigation trends. Dechert, LLP, April 7. https://www.dri.org/docs/default-source/paper-uploads/2021/4_pfas--expected-litigation-trends.pdf?sfvrsn=4.

Blas, J. 2022. The city of London is the wild west of metals. Bloomberg. https://www.bloomberg.com/opinion/articles/2022-03-09/nickel-s-mad-surge-exposes-the-shortcomings-of-the-london-metal-exchange.

BLS (U.S. Bureau of Labor Statistics). 2021. Chemists and materials scientists, job outlook. In *Occupational Outlook Handbook*. U.S. Department of Labor. https://www.bls.gov/ooh/life-physical-and-social-science/chemists-and-materials-scientists.htm#tab-6.

BLS. 2022. Sectoral Output for Manufacturing: Pharmaceutical and Medicine Manufacturing (NAICS 3254) in the United States. https://fred.stlouisfed.org/series/IPUEN3254T300000000.

Bo, C., F. Maseras, and N. López. 2018. The role of computational results databases in accelerating the discovery of catalysts. *Nature Catalysis* 1:809-810.

Bockris, J. O. M. 2013. The hydrogen economy: Its history. *International Journal of Hydrogen Energy* 38:2579-2588.

Boerner, L. K. 2019. Industrial ammonia production emits more CO_2 than any other chemical-making reaction. Chemists want to change that. *C&EN* 97(24).

Bogdan, A. R., and A. W. Dombrowski. 2019. Emerging trends in flow chemistry and applications to the pharmaceutical industry. *Journal of Medicinal Chemistry* 62:6422-6468.

Bogue, R. 2018. Remote chemical sensing: A review of techniques and recent developments. *Sensor Review* 38(4).

Bonheure, M., L. A. Vandewalle, G. B. Marin, and K. M. Van Geem. 2021. Dream or reality? Electrification of the chemical process industries. *CEP Magazine*. https://www.aiche.org/resources/publications/cep/2021/march/dream-or-reality-electrification-chemical-process-industries#:~:text=Electrification%20of%20the%20chemical%20process%20industries%20%28CPI%29%20is,drastically%20reducing%20capital%20costs%20and%20waste%20generation%20%286%29.

Borunda, A. 2021. The origins of environmental justice—and why it's finally getting the attention it deserves. *National Geographic*. https://www.nationalgeographic.com/environment/article/environmental-justice-origins-why-finally-getting-the-attention-it-deserves.

Bottoms, R. R. 1930. Separating acid gases, Girdler Corp. *US Patent, 1783901.*

Bourzac, K. 2020. COVID-19 lockdowns had strange effects on air pollution across the globe. *C&EN* 98(37). https://cen.acs.org/environment/atmospheric-chemistry/COVID-19-lockdowns-had-strange-effects-on-air-pollution-across-the-globe/98/i37.

Bowker, M. 2019. Methanol synthesis from CO_2 hydrogenation. *ChemCatChem* 11:4238-4246.

Brauns, J., and T. Turek. 2020. Alkaline water electrolysis powered by renewable energy: A review. *Processes* 8:248.

Breckel, A., A. Canavati, J. DiStefano, T. Green, M. Jeong, A. Kirzer, and A. Maranville. 2021. Building to Net-Zero: A U.S. Policy Blueprint for Gigaton-Scale CO2 Transport and Storage Infrastructure. Energy Futures Initiative and AFL-CIO. https://www.ourenergypolicy.org/resources/building-to-net-zero-a-u-s-policy-blueprint-for-gigaton-scale-co2-transport-and-storage-infrastructure/.

Bridle, H., D. Balharry, B. Gaiser, and H. Johnston. 2015. Exploitation of nanotechnology for the monitoring of waterborne pathogens: State-of-the-art and future research priorities. *Environmental Science & Technology* 49:10762-10777.

Britt, P. F. C., W. Geoffrey, K. I. Winey, J. Byers, E. Chen, B. Coughlin, C. Ellison, J. Garcia, A. Goldman, J. Guzman, J. Hartwig, B. Helms, G. Huber, C. Jenks, J. Martin, M. McCann, S. Miller, H. O'Neill, A. Sadow, S. Scott, L. Sita, D. Vlachos, and R. Waymouth. 2019. Report of the Basic Energy Sciences Roundtable on Chemical Upcycling of Polymers. Technical Report. U.S. Department of Energy Office of Science.

Brock, D. C. 2007. Patterning the world: The rise of chemically amplified photoresists. Science History Institute. https://www.sciencehistory.org/distillations/patterning-the-world-the-rise-of-chemically-amplified-photoresists.

Brookes, G., F. Taheripour, and W. E. Tyner. 2017. The contribution of glyphosate to agriculture and potential impact of restrictions on use at the global level. *GM Crops & Food* 8:216-228.

Brown, T. 2020. Saudi Arabia to export renewable energy using green ammonia. Ammonia Energy Association. https://www.ammoniaenergy.org/articles/saudi-arabia-to-export-renewable-energy-using-green-ammonia/.

Browne, D. L., J. L. Howard, and C. Schotten. 2017. Continuous flow processing as a tool for medicinal chemical synthesis. Pp. 135-185 in *Comprehensive Medicinal Chemistry III*, S. Chackalamannil, D. Rotella, and S. E. Ward, eds. Oxford, UK: Elsevier.

Brudermüller, M. 2020. How to build a more climate-friendly chemical industry. World Economic Forum. https://www.weforum.org/agenda/2020/01/how-to-build-a-more-climate-friendly-chemical-industry/.

Budde, F. 2011. Chemicals' changing competitive landscape. Mckinsey.com. https://www.mckinsey.com/industries/chemicals/our-insights/chemicals-changing-competitive-landscape.

Buglioni, L., F. Raymenants, A. Slattery, S. D. A. Zondag, and T. Noël. 2022. Technological innovations in photochemistry for organic synthesis: Flow chemistry, high-throughput experimentation, scale-up, and photoelectrochemistry. *Chemical Reviews* 122:2752-2906.

Burkart, M. D., N. Hazari, C. L. Tway, and E. L. Zeitler. 2019. Opportunities and challenges for catalysis in carbon dioxide utilization. *ACS Catalysis* 9:7937-7956.

Burke, M. D. 2021. Presentation to the committee at the Public Session on Automation, July 7.

Byrne, F. P., S. Jin, G. Paggiola, T. H. M. Petchey, J. H. Clark, T. J. Farmer, A. J. Hunt, C. R. McElroy, and J. Sherwood. 2016. Tools and techniques for solvent selection: Green solvent selection guides. *Sustainable Chemical Processes* 4:7.

Cafferty, B. J., A. S. Ten, M. J. Fink, S. Morey, D. J. Preston, M. Mrksich, and G. M. Whitesides. 2019. Storage of information using small organic molecules. *ACS Central Science* 5:911-916.

Cambié, D., C. Bottecchia, N. J. Straathof, V. Hessel, and T. Noël. 2016. Applications of continuous-flow photochemistry in organic synthesis, material science, and water treatment. *Chemical Reviews* 116(17):10276-10341.

Carocho, M., P. Morales, and I. C. F. R. Ferreira. 2018. Antioxidants: Reviewing the chemistry, food applications, legislation and role as preservatives. *Trends in Food Science & Technology* 71:107-120.

Caruthers, M. H. 2013. The chemical synthesis of DNA/RNA: Our gift to science. *Journal of Biological Chemistry* 288(2):1420-1427.

CBO (Congressional Budget Office). 2013. Effects of a Carbon Tax on the Economy and the Environment. https://www.cbo.gov/publication/44223.

CBO. 2020. An Update to the Budget Outlook: 2020 to 2030. https://www.cbo.gov/publication/56517.

CEFIC (European Chemical Industry Council). 2020. *2020 Facts & Figures of the European Chemical Industry.* https://www.francechimie.fr/media/52b/the-european-chemical-industry-facts-and-figures-2020.pdf.

CEFIC. 2022a. *2022 Facts and Figures of the European Chemical Industry.* https://cefic.org/a-pillar-of-the-european-economy/facts-and-figures-of-the-european-chemical-industry/.

CEFIC. 2022b. The Carbon Border Adjustment Mechanism (CBAM) Proposal Needs Upgrading for Chemicals. Position Paper. https://cefic.org/app/uploads/2022/01/Cefic-position-on-CBAM-The-Carbon-Border-Adjustment-Mechanism-CBAM-proposal-needs-upgrading-for-chemicals.pdf.

Celik, G., R. M. Kennedy, R. A. Hackler, M. Ferrandon, A. Tennakoon, S. Patnaik, A. M. LaPointe, S. C. Ammal, A. Heyden, F. A. Perras, M. Pruski, S .L. Scott, K. R. Poeppelmeier, A. D. Sadow, and M. Delferro. 2019. Upcycling single-use polyethylene into high-quality liquid products. *ACS Central Science* 5:1795-1803.

Centi, G., and S. Perathoner. 2009. Opportunities and prospects in the chemical recycling of carbon dioxide to fuels. *Catalysis Today* 148:191-205.

Chemical Watch. 2020. Impact of chemicals still a worry for majority of EU citizens. https://chemicalwatch.com/98037/impact-of-chemicals-still-a-worry-for-majority-of-eu-citizens.

REFERENCES

Chen, C. 2018. Designing catalysts for olefin polymerization and copolymerization: Beyond electronic and steric tuning. *Nature Reviews Chemistry* 2:6-14.

Chen, S., S. Brahma, J. Mackay, C. Cao, and B. Aliakbarian. 2020. The role of smart packaging system in food supply chain. *Journal of Food Science* 85(3):517-525.

Chheda, J. N., G. W. Huber, and J. A. Dumesic. 2007. Liquid-phase catalytic processing of biomass-derived oxygenated hydrocarbons to fuels and chemicals. *Angewandte Chemie International Edition* 46:7164-7183.

Clancy, P. 2012. Chemical engineering in the electronics industry: Progress towards the rational design of organic semiconductor heterojunctions. *Current Opinion in Chemical Engineering* 1:117-122.

Clark, J. H., T. J. Farmer, L. Herrero-Davila, and J. Sherwood. 2016. Circular economy design considerations for research and process development in the chemical sciences. *Green Chemistry* 18:3914-3934.

Clarke, C. J., W. C. Tu, O. Levers, A. Brohl, and J. P. Hallett. 2018. Green and sustainable solvents in chemical processes. *Chemical Reviews* 118(2):747-800.

ClinCal. 2019. The Top 200 Drugs of 2019. https://clincalc.com/DrugStats/Top200Drugs.aspx.

CMU (Carnegie Mellon University). 2021. Carnegie Mellon University and Emerald Cloud Lab to build world's first university cloud lab. News, August 30. https://www.cmu.edu/news/stories/archives/2021/august/first-academic-cloud-lab.html.

CO_2 Sciences. 2016. *Global Roadmap for Implementing CO_2 Utilization.* Global CO_2 Initiative. Innovation for Cool Earth Forum. https://assets.ctfassets.net/xg0gv1arhdr3/27vQZEvrxaQiQEAsGyoSQu/44ee0b72ceb9231ec53ed180cb759614/CO2U_ICEF_Roadmap_FINAL_2016_12_07.pdf.

Codexis. 2021. Codexis and Merck amend and extend supply agreement for enzyme used in manufacture of Sitagliptin. Press Release. https://www.codexis.com/investors/news-events/press-releases/detail/303/codexis-and-merck-amend-and-extend-supply-agreement.

Coelho, P. S., E. M. Brustad, A. Kannan, and F. H. Arnold. 2013. Olefin cyclopropanation via carbene transfer catalyzed by engineered cytochrome P450 enzymes. *Science* 339:307-310.

Cole-Hamilton, D. J. 2003. Homogeneous catalysis—New approaches to catalyst separation, recovery, and recycling. *Science* 299:1702-1706.

Coley, C. W., D. A. Thomas, J. A. M. Lummiss, J. N. Jaworski, C. P. Breen, V. Schultz, T. Hart, J. S. Fishman, L. Rogers, H. Gao, R. W. Hicklin, P. P. Plehiers, J. Byington, J. S. Piotti, W. H. Green, A. J. Hart, T. F. Jamison, and K. F. Jensen. 2019. A robotic platform for flow synthesis of organic compounds informed by AI planning. *Science* 365:eaax1566.

Computer History Museum. 2022. 1955: Photolithography Techniques Are Used to Make Silicon Devices. https://www.computerhistory.org/siliconengine/photolithography-techniques-are-used-to-make-silicon-devices/.

Condon, M. 2021. BASF warns on "excessive" emissions levies for chemicals plants under EU plans. Icis.com. https://www.icis.com/explore/resources/news/2021/07/14/10663232/basf-warns-on-excessive-emissions-levies-for-chemicals-plants-under-eu-plans/.

Corma, A., S. Iborra, and A. Velty. 2007. Chemical routes for the transformation of biomass into chemicals. *Chemical Reviews* 107:2411-2502.

Cowley, A. 2021. Pgm market report: February 2021. Johnson Matthey. https://matthey.com/documents/161599/509428/pgm-market-report-february-english-2021.pdf/c8d1bb71-caf8-65e0-ef62-761d5c25ebd6?t=1655877358674.

Creamer, A. E., and B. Gao. 2016. Carbon-based adsorbents for postcombustion CO_2 capture: A critical review. *Environmental Science & Technology* 50.7276-7289.

CRS (Congressional Research Service). 2021. U.S. Research and Development Funding and Performance: Fact Sheet. https://crsreports.congress.gov/product/pdf/R/R44307.

Cui, X., W. Li, P. Ryabchuk, K. Junge, and M. Beller. 2018. Bridging homogeneous and heterogeneous catalysis by heterogeneous single-metal-site catalysts. *Nature Catalysis* 1:385-397.

Cutler, D. M., and L. H. Summers. 2020. The COVID-19 pandemic and the $16 trillion virus. *JAMA* 324:1495-1496.

Daniell, J., M. Köpke, and S. D. Simpson. 2012. Commercial biomass syngas fermentation. *Energies* 5:5372-5417.

Dauter, Z., and A. Wlodawer. 2016. Progress in protein crystallography. *Protein & Peptide Letters* 23:201-210.

Davis, C. 2021. ExxonMobil eyeing $100B carbon capture project in Houston area. Natural Gas Intelligence, April 20. https://www.naturalgasintel.com/exxonmobil-eyeing-100b-carbon-capture-project-in-houston-area-with-burial-at-sea/.

Davis, N. 2009. Dow Chemical to shut down more plants in Louisiana. Independent Commodity Intelligence Services. https://www.icis.com/explore/resources/news/2009/07/01/9229325/dow-chemical-to-shut-down-more-plants-in-louisiana/.

Daza, Y. A., and J. N. Kuhn. 2016. CO_2 conversion by reverse water gas shift catalysis: Comparison of catalysts, mechanisms and their consequences for CO_2 conversion to liquid fuels. *RSC Advances* 6:49675-49691.

de la Guardia, M., and S. Garrigues. 2020. Past, present and future of green analytical chemistry. Pp. 1-18 in *Challenges in Green Analytical Chemistry,* 2nd ed., S. Garrigues and M. de la Guardia, eds. Cambridge, UK: Royal Society of Chemistry.

De Yoreo, J., D. Mandrus, L. Soderholm, T. Forbes, M. Kanatzidis, J. Erlebacher, J. Laskin, U. Wiesner, T. Xu, S. Billinge, S. Tolbert, M. Zaworotko, G Galli, J. Chan, J. Mitchell, L. Horton, A. Kini, B. Gersten, G. Maracas, R. Miranda, M. Pechan, and K. Runkles. 2016. *Basic Research Needs Workshop on Synthesis Science for Energy Relevant Technology.* Technical Report. U.S. Department of Energy, Office of Science.

Deng, X., S. Cao, and A. L. Horn. 2021. Emerging applications of machine learning in food safety. *Annual Review of Food Science and Technology* 12:513-538.

Deuss, P. J., K. Barta, and J. G. de Vries. 2014. Homogeneous catalysis for the conversion of biomass and biomass-derived platform chemicals. *Catalysis Science & Technology* 4:1174-1196.

DiChristopher, T. 2021. Experts explain why green hydrogen costs have fallen and will keep falling. S&P Global Market Intelligence. https://www.spglobal.com/marketintelligence/en/news-insights/latest-news-headlines/experts-explain-why-green-hydrogen-costs-have-fallen-and-will-keep-falling-63037203.

Dickson, D., K. Sullivan, K. Tanger, A. Hussain, and B. Kumpf. 2022. The changing single-use plastics landscape: Is the chemical industry prepared? Deloitte Insights. https://www2.deloitte.com/jo/en/pages/energy-and-resources/articles/changing-single-use-plastics-landscape.html.

Dilberoglu, U. M., B. Gharehpapagh, U. Yaman, and M. Dolen. 2017. The role of additive manufacturing in the era of industry 4.0. *Procedia Manufacturing* 11:545-554.

Dilkes-Hoffman, L. S., S. Pratt, P. A. Lant, and B. Laycock. 2019. The role of biodegradable plastic in solving plastic solid waste accumulation. Pp. 469-505 in *Plastics to Energy,* S. M. Al-Salem, ed. New York: William Andrew.

Ding, Y., D. Harvey, and N. H. L. Wang. 2020. Two-zone ligand-assisted displacement chromatography for producing high-purity praseodymium, neodymium, and dysprosium with high yield and high productivity from crude mixtures derived from waste magnets. *Green Chemistry* 22:3769-3783.

Djerassi, C. 2006. Chemical birth of the pill. *American Journal of Obstetrics and Gynecology* 194(1):290-298. https://www.ajog.org/article/S0002-9378%2805%2900859-8/pdf.

Djurišić, A. B., Y. He, and A. M. C. Ng. 2020. Visible-light photocatalysts: Prospects and challenges. *APL Materials* 8:030903.

DOE (U.S. Department of Energy). 2016. *2016 Billion-Ton Report: Advancing Domestic Resources for a Thriving Bio-economy, Volume 1: Economic Availability of Feedstocks.* ORNL/TM-2016/160. Oak Ridge, TN: Oak Ridge National Laboratory.

DOE. 2019. Energy Frontier Research Centers: Science for Our Nation's Energy Future. https://science.osti.gov/-/media/bes/pdf/brochures/2019/EFRC_Booklet_2019.pdf.

DOE Office of Science. 2017. *Basic Research Needs for Catalysis Science.* Technical Report. https://www.osti.gov/biblio/1545774-basic-research-needs-catalysis-science.

DOE Office of Science. 2020a. *Bioenergy Research Centers: 2020 Program Update.* DOE/SC-0201. https://genomicscience.energy.gov/wp-content/uploads/2021/09/BRC_Booklet_2020LR.pdf.

DOE Office of Science. 2020b. Oil and water almost mix in novel neuromorphic computing components. https://www.energy.gov/science/bes/articles/oil-and-water-almost-mix-novel-neuromorphic-computing-components.

Doll, J., and D. L. Freeman. 1994. Monte Carlo methods in chemistry. *IEEE Computational Science and Engineering* 1:22-32.

Drake, T., P. Ji, and W. Lin. 2018. Site isolation in metal–organic frameworks enables novel transition metal catalysis. *Accounts of Chemical Research* 51(9):2129-2138.

Dufour, J., D. P. Serrano, J. L. Gálvez, J. Moreno, and C. García. 2009. Life cycle assessment of processes for hydrogen production. Environmental feasibility and reduction of greenhouse gases emissions. *International Journal of Hydrogen Energy* 34:1370-1376.

Dumitru, A., B. Kölbl, and M. Wijffelaars. 2021. The carbon border adjustment mechanism explained. Rabobank. https://economics.rabobank.com/publications/2021/july/cbam-carbon-border-adjustment-mechanism-eu-explained/.

Duncan, C. 2020. Supporting women in the chemical sciences. CIC News. Chemical Institute of Canada. https://www.cheminst.ca/magazine/article/supporting-women-in-the-chemical-sciences/.

Dunham, S. J. B., J. F. Ellis, B. Li, and J. V. Sweedler. 2017. Mass spectrometry imaging of complex microbial communities. *Accounts of Chemical Research* 50:96-104.

Dunleavy, K. 2022. Despite early doubts, Merck's COVID antiviral beats sales expectations and drives big growth. Fierce Pharma, April 28. https://www.fiercepharma.com/pharma/merck-has-big-revenue-quarter-thanks-part-surprisingly-strong-sales-covid-pills.

Durand, D. J., and N. Fey. 2021. Building a toolbox for the analysis and prediction of ligand and catalyst effects in organometallic catalysis. *Accounts of Chemical Research* 54:837-848.

Earl, D. J., and M. W. Deem. 2008. Monte Carlo simulations. Pp. 25-36 in *Methods in Molecular Biology,* Vol. 443, *Molecular Modeling of Proteins,* A. Kukol, ed. Totowa, NJ: Humana Press. https://dasher.wustl.edu/chem478/reading/methmolbio-443-25-08.pdf.

EC (European Commission). 2020a. Chemicals strategy: The EU's chemicals strategy for sustainability towards a toxic-free environment. https://ec.europa.eu/environment/strategy/chemicals-strategy_en.

REFERENCES

EC. 2020b. Communication from the Commission to the European Parliament, the Council, the European Economic and Social Committee and the Committee of the Regions: Chemicals Strategy for Sustainability Towards a Toxic-Free Environment. https://ec.europa.eu/environment/pdf/chemicals/2020/10/Strategy.pdf.

EC Directorate-General for Taxation and Customs Union. 2021. Proposal for a regulation of the European Parliament and of the Council establishing a carbon border adjustment mechanism. COM/2021/564 final. https://eur-lex.europa.eu/legal-content/en/TXT/?uri=CELEX:52021PC0564.

ECCC (Environment and Climate Change Canada). 2019. A proposed integrated management approach to plastic products to prevent waste and pollution. Government of Canada. https://www.canada.ca/en/environment-climate-change/services/canadian-environmental-protection-act-registry/plastics-proposed-integrated-management-approach.html.

EDR Group. 2019. *Economic Contribution of the U.S. Lead Battery Industry.* Chicago: Battery Council International. https://essentialenergyeveryday.com/wp-content/uploads/2019/10/Economic-Impact-of-Lead-Batteries-in-the-United-States-Sept-2019.pdf.

EEB (European Environmental Bureau). 2018. The problem of pharmaceutical pollution. https://eeb.org/the-problem-of-pharmaceutical-pollution/.

EIA (U.S. Energy Information Administration). 2019. *International Energy Outlook 2019 with Projections to 2050.* Washington, DC: U.S. Department of Energy. https://www.eia.gov/outlooks/ieo/pdf/ieo2019.pdf.

EIA. 2020. *Annual Energy Outlook 2020 with Projections to 2050.* Washington, DC: U.S. Department of Energy. https://www.eia.gov/outlooks/aeo/pdf/AEO2020%20Full%20Report.pdf.

Eigen, M. 1961. Book Review: *Kinetics and Mechanism—A Study of Homogeneous Chemical Reactions* by A. A. Frost and R. G. Pearson, John Wiley & Sons, Inc. *Angewandte Chemie* 73(21):719. https://onlinelibrary.wiley.com/doi/abs/10.1002/ange.19610732115.

Elliott, R., and L. Santiago. 2019. A decade in which fracking rocked the oil world. *Wall Street Journal,* December 12. https://www.wsj.com/articles/a-decade-in-which-fracking-rocked-the-oil-world-11576630807.

Elnabawy, A. O., J. Schumann, P. Bothra, A. Cao, and J. K. Nørskov. 2020. The challenge of CO hydrogenation to methanol: Fundamental limitations imposed by linear scaling relations. *Topics in Catalysis* 63:635-648.

Elsevier Analytical Services. 2021. *Pathways to Net Zero: The Impact of Clean Energy Research..* https://www.elsevier.com/__data/assets/pdf_file/0006/1214979/net-zero-2021.pdf.

EMA (European Medicines Agency). 2019. ICH Q3D Elemental Impurities. https://www.ema.europa.eu/en/ich-q3d-elemental-impurities.

Engelberth, A. S., S. P. Ventura, F. Vilaplana, P. Venkatesu, J. Y. Zhu, and D. J. Carrier. 2021. ACS sustainable chemistry & engineering welcomes manuscripts on the circular economy of biomass. *ACS Sustainable Chemistry & Engineering* 9(6):2410-2411.

EPA (U.S. Environmental Protection Agency). 2010. Presidential Green Chemistry Challenge: 2010 Greener Reaction Conditions Award: Merck & Co. Inc., Codexis, Inc. https://www.epa.gov/greenchemistry/presidential-green-chemistry-challenge-2010-greener-reaction-conditions-award.

EPA. 2012. Presidential Green Chemistry Challenge: 2012 Academic Award (Coates). https://www.epa.gov/greenchemistry/presidential-green-chemistry-challenge-2012-academic-award-coates.

EPA. 2013. Presidential Green Chemistry Challenge: 2013 Greener Synthetic Pathways Award. https://www.epa.gov/greenchemistry/presidential-green-chemistry-challenge-2013-greener-synthetic-pathways-award.

EPA. 2017. Green Chemistry Challenge: 2017 Designing Greener Chemicals Award. https://www.epa.gov/greenchemistry/green-chemistry-challenge-2017-designing-greener-chemicals-award.

EPA. 2020. Glyphosate. https://www.epa.gov/ingredients-used-pesticide-products/glyphosate.

EPA. 2021. International Efforts on Wasted Food Recovery. https://www.epa.gov/international-cooperation/international-efforts-wasted-food-recovery#:~:text=In%20the%20United%20States%2C%2040%25%20of%20food%20is,of%20human-related%20methane%20emissions%20in%20the%20United%20States.

EPA. 2022. Chlorpyrifos. https://www.epa.gov/ingredients-used-pesticide-products/chlorpyrifos.

Eramo, M. 2014. Investment risk for chemical producers. IHS Markit, May 19. https://ihsmarkit.com/research-analysis/q12-investment-risk-for-chemical-producers.html.

Eramo, M. 2019. State of the US chemical industry: Is the resurgence in growth sustainable? Presentation for Rice Global E&C Annual Forum. https://cpb-us-e1.wpmucdn.com/blogs.rice.edu/dist/3/9169/files/2019/11/Eramo-AF19.pdf.

Erickson, B. E. 2020. Corteva to stop producing chlorpyrifos. *C&EN.* https://cen.acs.org/environment/pesticides/Corteva-stop-producing-chlorpyrifos/98/web/2020/02.

Erickson, B. E. 2021. Bayer to end glyphosate sales to US consumers. *C&EN.* https://cen.acs.org/environment/pesticides/Bayer-end-glyphosate-sales-US/99/web/2021/07.

Esposito, F. 2013. Dow Chemical restarts La. ethylene plant. Plastics News, January 8. https://www.plasticsnews.com/article/20130108/NEWS/301089992/dow-chemical-restarts-la-ethylene-plant.

Esposito, S., and A. Naddeo. 2014. The genesis of the quantum theory of the chemical bond. *Advances in Historical Studies* 3(5).

Eurostat. 2021. Production and international trade in chemicals. https://ec.europa.eu/eurostat/statistics-explained/index.php?title=Production_and_international_trade_in_chemicals#:~:text=The%20value%20of%20EU-27%20exports%20of%20chemicals%20was,in%202009%20to%20EUR%20333%20billion%20in%202019.

Evagorou, M., K. Korfiatis, C. Nicolaou, and C. Constantinou. 2009. An investigation of the potential of interactive simulations for developing system thinking skills in elementary school: A case study with fifth graders and sixth graders. *International Journal of Science Education* 31:655-674.

Everts, S. 2015. A brief history of chemical war. Distillations, May 11. Science History Institute.

ExxonMobil. 2022. Industry support for large-scale carbon capture and storage continues to gain momentum in Houston. News, January 20. https://corporate.exxonmobil.com/News/Newsroom/News-releases/2022/0120_Industry-support-for-large-scale-carbon-capture-and-storage-gains-momentum-in-Houston.

Eygeris, Y., M. Gupta, J. Kim, and G. Sahay. 2022. Chemistry of lipid nanoparticles for RNA delivery. *Accounts of Chemical Research* 55:2-12.

Fabbrizzi, L. 2019. Strange case of Signor Volta and Mister Nicholson: How electrochemistry developed as a consequence of an editorial misconduct. *Angewandte Chemie International Edition* 58:5810-5822.

FAO (Food and Agriculture Organization of the United Nations). 2020. *Agricultural Markets and Sustainable Development: Global Value Chains, Smallholder Farmers and Digital Innovations*. Rome: FAO.

Farago, F., and M. Waslin. 2021. Five of the 2021 Nobel Laureates from the United States are Foreign Born. George Mason University Institute for Immigration Research. https://d101vc9winf8ln.cloudfront.net/documents/41281/original/Nobel_Prize_Summary_2021.pdf?1634227855.

Farrauto, R. J., M. Deeba, and S. Alerasool. 2019. Gasoline automobile catalysis and its historical journey to cleaner air. *Nature Catalysis* 2:603-613.

FDA (U.S. Food and Drug Administration). 2017. Medical applications of 3D printing. https://www.fda.gov/medical-devices/3d-printing-medical-devices/medical-applications-3d-printing.

FDA. 2019. *Quality Considerations for Continuous Manufacturing: Guidance for Industry*. U.S. Department of Health and Human Services. https://www.fda.gov/media/121314/download.

FDA. 2022. Emergency Use Authorization 108. Letter to Merck Sharp & Dohme Corp., March 23. https://www.fda.gov/media/155053/download.

FECM (Office of Fossil Energy and Carbon Management, U.S. Department of Energy). 2011. Hydraulic Fracturing Technology. https://www.energy.gov/fecm/hydraulic-fracturing-technology.

Fehrmann, R., M. Haumann, and A. Riisager. 2014. Introduction. Pp. 1-10 in *Supported Ionic Liquids: Fundamentals and Applications*, R. Fehrmann, A. Riisager, and M. Haumann, eds. John Wiley & Sons. https://onlinelibrary.wiley.com/doi/abs/10.1002/9783527654789.ch1.

Finger, S. R., and S. Gamper-Rabindran. 2013. Testing the effects of self-regulation on industrial accidents. *Journal of Regulatory Economics* 43(2):115-146.

Fiorentino, G., M. Ripa, and S. Ulgiati. 2017. Chemicals from biomass: Technological versus environmental feasibility. A review. *Biofuels, Bioproducts and Biorefining* 11:195-214.

Fleming, L., and D. Basco. 2021. Economic Analysis of the U.S. Chemical Economy: Final Report to the National Academy of Sciences. Vertex Evaluation and Research, LLC. Available in the Public Access File for this study.

Flury, M., and R. Narayan. 2021. Biodegradable plastic as an integral part of the solution to plastic waste pollution of the environment. *Current Opinion in Green and Sustainable Chemistry* 30:100490.

Frankel, T. C. 2016. The cobalt pipeline. *The Washington Post,* September 30.

Frølund, L., F. Murray, and M. Riedel. 2017. Developing successful strategic partnerships with universities. *MIT Sloan Management Review*, December 6. https://sloanreview.mit.edu/issue/2018-winter/.

Frontera, P., A. Macario, M. Ferraro, and P. Antonucci. 2017. Supported catalysts for CO_2 methanation: A review. *Catalysts* 7:59.

Furukawa, H., K. E. Cordova, M. O'Keeffe, and O. M. Yaghi. 2013. The chemistry and applications of metal-organic frameworks. *Science* 341:1230444.

Gagliarducci, S., M. D. Paserman, and E. Patacchina. 2019. Hurricanes, Climate Change Policies and Electoral Accountability. Working Paper 25835. National Bureau of Economic Research. https://www.nber.org/papers/w25835.

Gambetta, J. 2020. IBM's roadmap for scaling quantum technology. IBM blog, September 15. https://research.ibm.com/blog/ibm-quantum-roadmap.

Gamper-Rabindran, S. 2022. *America's Energy Gamble: People, Economy and Planet*. Cambridge: Cambridge University Press. https://www.cambridge.org/core/books/americas-energy-gamble/83DDAEE834233BE82134E274444CB09C.

Gamper-Rabindran, S., and S. R. Finger. 2013. Does industry self-regulation reduce pollution? Responsible care in the chemical industry. *Journal of Regulatory Economics* 43:1-30.

GAO (U.S. Government Accountability Office). 2018. *Chemical Innovation: Technologies to Make Processes and Products More Sustainable*. GAO-18-307. https://www.gao.gov/products/gao-18-307.

Gao, P., L. Zhang, S. Li, Z. Zhou, and Y. Sun. 2020. Novel heterogeneous catalysts for CO_2 hydrogenation to liquid fuels. *ACS Central Science* 6:1657-1670.

REFERENCES

Gao, Y., L. Neal, D. Ding, W. Wu, C. Baroi, A. M. Gaffney, and F. Li. 2019. Recent advances in intensified ethylene production—A review. *ACS Catalysis* 9:8592-8621.

García-Granados, R., J. A. Lerma-Escalera, and J. R. Morones-Ramírez. 2019. Metabolic engineering and synthetic biology: Synergies, future, and challenges. *Frontiers in Bioengineering and Biotechnology* 7:36.

Gázquez, M. J., J. P. Bolivar, R. Garcia-Tenorio, and F. Vaca. 2014. A review of the production cycle of titanium dioxide pigment. *Materials Sciences and Applications* 5(7):441-458.

GCCA (Global Cement and Concrete Association). 2020. *Concrete Future: GCCA 2050 Cement and Concrete Industry Roadmap for Net Zero Concrete.* https://gccassociation.org/concretefuture/.

Gebre, S. H., M. G. Sendeku, and M. Bahri. 2021. Recent trends in the pyrolysis of non-degradable waste plastics. *ChemistryOpen* 10:1202.

Geueke, B., K. Groh, and J. Muncke. 2018. Food packaging in the circular economy: Overview of chemical safety aspects for commonly used materials. *Journal of Cleaner Production* 193:491-505.

Geyer, R., J. R. Jambeck, and K. L. Law. 2017. Production, use, and fate of all plastics ever made. *Science Advances* 3:e1700782.

Gibbs, J. W. 2010. *Elementary Principles in Statistical Mechanics: Developed with Especial Reference to the Rational Foundation of Thermodynamics*, Reissue ed. Cambridge, UK: Cambridge University Press.

Golden, J., R. Handfield, J. Daystar, R. Kronthal-Sacco, and J. Tickner. 2021. Green chemistry: A strong driver of innovation, growth, and business opportunity. Sustainable Chemistry Catalyst, Lowell Center for Sustainable Production, University of Massachusetts Lowell. https://greenchemistryandcommerce.org/documents/uml-rpt-GreenChem-1.22-12.pdf.

González-Torralva, F., J. Gil-Humanes, F. Barro, I. Brants, and R. De Prado. 2012. Target site mutation and reduced translocation are present in a glyphosate-resistant *Lolium multiflorum* Lam. biotype from Spain. *Plant Physiology and Biochemistry* 58:16-22.

Gopeesingh, J., M. A. Ardagh, M. Shetty, S. T. Burke, P. J. Dauenhauer, and O. A. Abdelrahman. 2020. Resonance-promoted formic acid oxidation via dynamic electrocatalytic modulation. *ACS Catalysis* 10:9932-9942.

Gottlieb, R. L., C. E. Vaca, R. Paredes, J. Mera, B. J. Webb, G. Perez, G. Oguchi, P. Ryan, B. U. Nielsen, M. Brown, A. Hidalgo, Y. Sachdeva, S. Mittal, O. Osiyemi, J. Skarbinski, K. Juneja, R. H. Hyland, A. Osinusi, S. Chen, G. Camus, M. Abdelghany, S. Davies, N. Behenna-Renton, F. Duff, F. M. Marty, M. J. Katz, A. A. Ginde, S. M. Brown, J. T. Schiffer, and J. A. Hill. 2022. Early remdesivir to prevent progression to severe Covid-19 in outpatients. *New England Journal of Medicine* 386:305-315.

Govender, S., and H. B. Friedrich. 2017. Monoliths: A review of the basics, preparation methods and their relevance to oxidation. *Catalysts* 7(2):62.

Green, J., J. Hadden, T. Hale, and P. Mahdavi. 2021. Transition, hedge, or resist? Understanding political and economic behavior toward decarbonization in the oil and gas industry. *Review of International Political Economy.* https://doi.org/10.1080/09692290.2021.1946708.

Grim, R. G., Z. Huang, M. T. Guarnieri, J. R. Ferrell, L. Tao, and J. A. Schaidle. 2020. Transforming the carbon economy: Challenges and opportunities in the convergence of low-cost electricity and reductive CO_2 utilization. *Energy & Environmental Science* 13:472-494.

Guil-López, R., N. Mota, J. Llorente, E. Millán, B. Pawelec, J. L. G. Fierro, and R. M. Navarro. 2019. Methanol synthesis from CO_2: A review of the latest developments in heterogeneous catalysis. *Materials* 12:3902.

Hagen, J. 2015. *Industrial Catalysis: A Practical Approach,* 3rd ed. Singapore: Wiley.

Haigh, T., M. Priestley, and C. Rope. 2014. Los Alamos bets on ENIAC: Nuclear Monte Carlo simulations, 1947-1948. *IEEE Annals of the History of Computing* 36:42-63.

Hals, T. 2021. Bayer to rethink Roundup in U.S. residential market after judge nixes $2 bln settlement. Reuters, May 27. https://www.reuters.com/business/healthcare-pharmaceuticals/us-judge-rejects-bayers-2-bln-deal-resolve-future-roundup-lawsuits-2021-05-26/.

Hammett, L. P. 1937. The effect of structure upon the reactions of organic compounds. Benzene derivatives. *Journal of the American Chemical Society* 59:96-103.

Hammond, J., H. Leister-Tebbe, A. Gardner, P. Abreu, W. Bao, W. Wisemandle, M. Baniecki, V. M. Hendrick, B. Damle, A. Simón-Campos, R. Pypstra, and J. M. Rusnak. 2022. Oral nirmatrelvir for high-risk, nonhospitalized adults with Covid-19. *New England Journal of Medicine* 386:1397-1408.

Han, J. W., L. Ruiz-Garcia, J. P. Qian, and X. T. Yang. 2018. Food packaging: A comprehensive review and future trends. *Comprehensive Reviews in Food Science and Food Safety* 17:860-877.

Hand, S., and R. D. Cusick. 2021. Electrochemical disinfection in water and wastewater treatment: Identifying impacts of water quality and operating conditions on performance. *Environmental Science & Technology* 55:3470-3482.

Harper, K. C., E. X. Zhang, Z. Q. Liu, T. Grieme, T. B. Towne, D. J. Mack, J. Griffin, S. Y. Zheng, N. N. Zhang, S. Gangula, J. L. Yuan, R. Miller, P. Z. Huang, J. Gage, M. Diwan, and Y. Y. Ku. 2022. Commercial-scale visible light trifluoromethylation of 2-chlorothiophenol using CF3I gas. *Organic Process Research & Development* 26:404-412.

Hartman, R. L. 2020. Flow chemistry remains an opportunity for chemists and chemical engineers. *Current Opinion in Chemical Engineering* 29:42-50.

Hauch, A., R. Küngas, P. Blennow, A. B. Hansen, J. B. Hansen, B. V. Mathiesen, and M. B. Mogensen. 2020. Recent advances in solid oxide cell technology for electrolysis. *Science* 370:eaba6118.

Hayler, J. D., D. K. Leahy, and E. M. Simmons. 2019. A pharmaceutical industry perspective on sustainable metal catalysis. *Organometallics* 38:36-46.

Heldebrant, D. J., P. K. Koech, V. A. Glezakou, R. Rousseau, D. Malhotra, and D. C. Cantu. 2017. Water-lean solvents for post-combustion CO_2 capture: Fundamentals, uncertainties, opportunities, and outlook. *Chemical Reviews* 117:9594-9624.

Henderson, R. K., A. P. Hill, A. M. Redman, and H. F. Sneddon. 2015. Development of GSK's acid and base selection guides. *Green Chemistry* 17:945-949.

Hennessey, W. O. 1996. Sustainable development: A "win-win" for licensing and for the environment. Franklin Pierce Law Center. https://law.unh.edu/sites/default/files/media/2018/09/sustainable-development-win-win-licensing.pdf.

Himanen, L., A. Geurts, A. S. Foster, and P. Rinke. 2019. Data-driven materials science: Status, challenges, and perspectives. *Advanced Science* 6(21):1900808.

Hochreiter, R. 2021. The PGM markets. Edison, December 13. https://www.edisongroup.com/wp-content/uploads/2021/12/PGM-themes-1221-new-template_CR2_0-2.pdf.

Hogue, C. 2019. Chemical companies spar over PFAS pollution liability in US. *C&EN* 97(36):14.

Hogue, C. 2021. Sustainable chemistry legislation enacted by US Congress. C&EN. https://cen.acs.org/environment/green-chemistry/Sustainable-chemistry-legislation-enacted-US/99/web/2021/01.

Holladay, J. D., J. Hu, D. L. King, and Y. Wang. 2009. An overview of hydrogen production technologies. *Catalysis Today* 139:244-260.

Holladay, J. E., J. J. Bozell, J. F. White, and D. Johnson. 2007. *Top Value-Added Chemicals from Biomass. Volume II—Results of Screening for Potential Candidates from Biorefinery Lignin.* PNNL-16983. U.S. Department of Energy. https://www.pnnl.gov/main/publications/external/technical_reports/PNNL-16983.pdf.

Hong, C., Q. Zhang, Y. Zhang, S. J. Davis, D. Tong, Y. Zheng, Z. Liu, D. Guan, K. He, and H. J. Schellnhuber. 2019. Impacts of climate change on future air quality and human health in China. *Proceedings of the National Academy of Sciences of the United States of America* 116:17193-17200.

Hosseini, S., H. Moghaddas, S. Masoudi Soltani, and S. Kheawhom. 2020. Technological applications of honeycomb monoliths in environmental processes: A review. *Process Safety and Environmental Protection* 133:286-300.

Hounshell, D. A., and J. K. Smith. 1988. *Science and Corporate Strategy: Du Pont R&D, 1902-1980.* Cambridge, UK: Cambridge University Press.

Howard, P., G. Morris, and G. Sunley. 2006. Introduction: Catalysis in the chemical industry. Pp. 1-22 in *Metal-Catalysis in Industrial Organic Processes.* Editors: Gian Paolo Chiusoli, Peter M. Maitlis. London: Royal Society of Chemistry.

Hsiang, S., R. Kopp, A. Jina, J. Rising, M. Delgado, S. Mohan, D. J. Rasmussen, R. Muir-Wood, P. Wilson, M. Oppenheimer, K. Larsen, and T. Houser. 2017. Estimating economic damage from climate change in the United States. *Science* 356:1362-1369.

Hu, A. G. Z., and I. P. L. Png. 2013. Patent rights and economic growth: Evidence from cross-country panels of manufacturing industries. *Oxford Economic Papers* 65:675-698.

Huang, J., A. J. Gates, R. Sinatra, and A. L. Barabási. 2020. Historical comparison of gender inequality in scientific careers across countries and disciplines. *Proceedings of the National Academy of Sciences of the United States of America* 117:4609-4616.

Huber, G. W., J. N. Chheda, C. J. Barrett, and J. A. Dumesic. 2005. Production of liquid alkanes by aqueous-phase processing of biomass-derived carbohydrates. *Science* 308:1446-1450.

Hydrocarbon Processing. 2021. Accelerating electrification with the "Cracker of the Future" consortium. https://www.hydrocarbonprocessing.com/news/2021/09/accelerating-electrification-with-the-cracker-of-the-future-consortium.

ICCA and OE (International Council of Chemical Associations and Oxford Economics). 2019. The Global Chemical Industry: Catalyzing Growth and Addressing Our World's Sustainability Challenges. https://icca-chem.org/wp-content/uploads/2020/10/Catalyzing-Growth-and-Addressing-Our-Worlds-Sustainability-Challenges-Report.pdf.

IEA (International Energy Agency). 2013. *Technology Roadmap: Energy and GHG Reductions in the Chemical Industry via Catalytic Processes.* https://icca-chem.org/wp-content/uploads/2020/05/Technology-Roadmap.pdf.

IEA. 2018. *The Future of Petrochemicals.* https://www.iea.org/reports/the-future-of-petrochemicals.

IEA. 2019. Direct CO2 emissions from selected heavy industry sectors, 2019. https://www.iea.org/data-and-statistics/charts/direct-co2-emissions-from-selected-heavy-industry-sectors-2019.

IEA. 2020. *Energy Technology Perspectives 2020: Part of Energy Technology Perspectives.* https://www.iea.org/reports/energy-technology-perspectives-2020.

IEA. 2021a. *Ammonia Technology Roadmap: Towards More Sustainable Nitrogen Fertiliser Production.* https://iea.blob.core.windows.net/assets/6ee41bb9-8e81-4b64-8701-2acc064ff6e4/AmmoniaTechnologyRoadmap.pdf.

IEA. 2021b. Mineral requirements for clean energy transitions. In *The Role of Critical Minerals in Clean Energy Transitions.* https://www.iea.org/reports/the-role-of-critical-minerals-in-clean-energy-transitions/mineral-requirements-for-clean-energy-transitions.

REFERENCES

IHS Markit. 2022. Quickly compare growth, cost, capex and profitability to know which industries will thrive. https://ihsmarkit.com/products/global-industry-forecasts-analysis.html.
Iles, A., and A. N. Martin. 2013. Expanding bioplastics production: Sustainable business innovation in the chemical industry. *Journal of Cleaner Production* 45:38-49.
In, S. Y., and K. Schumacher. 2021. Carbonwashing: ESG data greenwashing in a post-Paris world. Pp. 39-58 in *Settling Climate Accounts: Navigating the Road to Net Zero*, T. Heller and A. Seiger, eds. Cham, Switzerland: Springer International.
IOM (Institute of Medicine). 2012. *Improving Food Safety Through a One Health Approach: Workshop Summary*. Washington, DC: The National Academies Press.
IPBES (Intergovernmental Science-Policy Platform on Biodiversity and Ecosystem Services). 2019. *Global Assessment Report on Biodiversity and Ecosystem Services*. https://ipbes.net/global-assessment.
IPCC (Intergovernmental Panel on Climate Change). 2021. *Climate Change 2021: The Physical Science Basis. Working Group I Contribution to the Sixth Assessment Report of the Intergovernmental Panel on Climate Change*. Cambridge, UK: Cambridge University Press. https://www.ipcc.ch/report/ar6/wg1/.
IPCC. 2022. *Climate Change 2022: Impacts, Adaptation and Vulnerability. Working Group II Contribution to the Sixth Assessment Report of the Intergovernmental Panel on Climate Change*. Cambridge, UK: Cambridge University Press. https://www.ipcc.ch/report/ar6/wg2/.
Ivleva, N. P. 2021. Chemical analysis of microplastics and nanoplastics: Challenges, advanced methods, and perspectives. *Chemical Reviews* 121:11886-11936.
Jackson, L. S. 2009. Chemical food safety issues in the United States: Past, present, and future. *Journal of Agricultural and Food Chemistry* 57:8161-8170.
Jalbert, M. 2021. GC3 applauds enactment of Sustainable Chemistry R&D Act. Cision PR Newswire, January 4. https://www.prnewswire.com/news-releases/gc3-applauds-enactment-of-sustainable-chemistry-rd-act-301200282.html.
Janssen, M., C. Müller, and D. Vogt. 2011. Recent advances in the recycling of homogeneous catalysts using membrane separation. *Green Chemistry* 13:2247-2257.
JHU (Johns Hopkins University). 2017. OXIDE 2015 faculty demographics survey: Under-represented minority results for AY2015-16. OXIDE 2015 faculty demographics. http://oxide.jhu.edu/src/data/urm/AY15-16/2015-2016_URM_NSF2015_Alpha.pdf.
Jia, H. 2018. Strong spending compounds chemistry prowess. *Nature Index*, December 12. https://www.nature.com/articles/d41586-018-07693-3#:~:text=Strong%20spending%20compounds%20chemistry%20prowess%20The%20discipline%E2%80%99s%20historic,the%20way%20in%20emerging%20areas%2C%20such%20as%20nanomaterials.
Jin, R., G. Li, S. Sharma, Y. Li, and X. Du. 2021. Toward active-site tailoring in heterogeneous catalysis by atomically precise metal nanoclusters with crystallographic structures. *Chemical Reviews* 121(2):567-648.
Jing, Y., L. Dong, Y. Guo, X. Liu, and Y. Wang. 2020. Chemicals from lignin: A review of catalytic conversion involving hydrogen. *ChemSusChem* 13(17):4181-4198.
Kähler, F., M. Carus, O. Porc, and C. vom Berg. 2021. *Turning off the Tap for Fossil Carbon*. Hürth, Germany: nova-Institute GmbH. https://www.unilever.com/files/5a9d4ed5-36ba-4bf1-af56-42367841343a/turning-off-the-tap-for-fossil-carbon-tcm244-561342-en.pdf.
Kali, S., M. Khan, M. S. Ghaffar, S. Rasheed, A. Waseem, M. M. Iqbal, M. Bilal khan Niazi, and M. I. Zafar. 2021. Occurrence, influencing factors, toxicity, regulations, and abatement approaches for disinfection by products in chlorinated drinking water: A comprehensive review. *Environmental Pollution* 281:116950.
Kan, J., R. D. Lewis, K. Chen, and F. H. Arnold. 2016. Directed evolution of cytochrome c for carbon-silicon bond formation: Bringing silicon to life. *Science* 354:1048-1051.
Kätelhön, A., R. Meys, S. Deutz, S., Suh, and A. Bardow. 2019. Climate change mitigation potential of carbon capture and utilization in the chemical industry. *Proceedings of the National Academy of Sciences of the United States of America* 116(23):11187-11194.
Katella, K. 2022. 13 Things to know about Paxlovid, the latest COVID-19 pill. Yale Medicine News, June 6. https://www.yalemedicine.org/news/13-things-to-know-paxlovid-covid-19.
Kazemi, M., and M. Mohammadi. 2020. Magnetically recoverable catalysts: Catalysis in synthesis of polyhydroquinolines. *Applied Organometallic Chemistry* 34:e5400.
Keijer, T., V. Bakker, and J. C. Slootweg. 2019. Circular chemistry to enable a circular economy. *Nature Chemistry* 11:190-195.
Kennedy, R. T. 2017. The 2017 annual review issue (editorial). *Analytical Chemistry* 89(1):1.
Kent, S. 2006. Bruce Merrifield (1921–2006). *Nature* 441:824.
Kim, E. J., R. L. Siegelman, H. Z. H. Jiang, A. C. Forse, J. H. Lee, J. D. Martell, P. J. Milner, J. M. Falkowski, J. B. Neaton, J. A. Reimer, S. C. Weston, and J. R. Long. 2020. Cooperative carbon capture and steam regeneration with tetraamine-appended metal–organic frameworks. *Science* 369:392-396.

Kimball, S. 2022. Paxlovid prescriptions to treat Covid increased tenfold in U.S. since late February, Pfizer says. CNBC, May 3. https://www.cnbc.com/2022/05/03/pfizer-paxlovid-prescriptions-to-treat-covid-increased-tenfold-in-us-since-late-february.html.

Klankermayer, J., and W. Leitner. 2016. Harnessing renewable energy with CO_2 for the chemical value chain: Challenges and opportunities for catalysis. *Philosophical Transactions of the Royal Society A: Mathematical, Physical and Engineering Sciences* 374:20150315.

Klippenstein, S. J., V. S. Pande, and D. G. Truhlar. 2014. Chemical kinetics and mechanisms of complex systems: A perspective on recent theoretical advances. *Journal of the American Chemical Society* 136:528-546.

Klosin, J., P. P. Fontaine, and R. Figueroa. 2015. Development of group IV molecular catalysts for high temperature ethylene-α-olefin copolymerization reactions. *Accounts of Chemical Research* 48:2004-2016.

Kobayashi-Solomon, E. 2021a. LanzaTech: Engineering the future. Forbes, September 30. https://www.forbes.com/sites/erikkobayashisolomon/2021/09/30/lanzatech-engineering-the-future/?sh=67be6b8a3f32.

Kobayashi-Solomon, E. 2021b. LanzaTech's paradigm-shifting plan to create carbon-negative industrial chemicals. Forbes, September 21. https://www.forbes.com/sites/erikkobayashisolomon/2021/09/21/lanzatechs-paradigm-shifting-plan-to-create-carbon-negative-industrial-chemicals/?sh=214d59e73bdf.

Kobayashi-Solomon, E. 2021c. SynBio: The science behind LanzaTech's success. Forbes, September 24. https://www.forbes.com/sites/erikkobayashisolomon/2021/09/24/synbio-the-science-behind-lanzatechs-success/?sh=2897107a391e.

Kogan, L., D. Papanikolaou, A. Seru, and N. Stoffman. 2017. Technological innovation, resource allocation, and growth. *Quarterly Journal of Economics* 132:665-712.

Kopplin, J. 2002. *An Illustrated History of Computers*. Computer Science Lab. http://www.computersciencelab.com/ComputerHistory/HistoryPt4.htm.

Kovacev, N., S. Li, S. Zeraati-Rezaei, H. Hemida, A. Tsolakis, and K. Essa. 2021. Effects of the internal structures of monolith ceramic substrates on thermal and hydraulic properties: Additive manufacturing, numerical modelling and experimental testing. *International Journal of Advanced Manufacturing Technology* 112:1115-1132.

Krishnakumar, T., and R. Visvanathan. 2014. Acrylamide in food products: A review. *Journal of Food Processing & Technology* 5:7. https://www.walshmedicalmedia.com/open-access/acrylamide-in-food-products-a-review-2157-7110.1000344.pdf.

Krupnick, A. 2020. *Green Public Procurement for Natural Gas, Cement, and Steel*. Washington, DC: Resources for the Future.

Küngas, R. 2020. Review—Electrochemical CO_2 reduction for CO production: Comparison of low- and high-temperature electrolysis technologies. *Journal of the Electrochemical Society* 167:044508.

Kunkes, E. L., D. A. Simonetti, R. M. West, J. C. Serrano-Ruiz, C. A. Gärtner, and J. A. Dumesic. 2008. Catalytic conversion of biomass to monofunctional hydrocarbons and targeted liquid-fuel classes. *Science* 322:417-421.

Kuspa, A. 2021. Presentation to the committee at the Public Session on Funding Mechanisms, April 28.

Lacombe, A., I. Quintela, Y.-T. Liao, and V. C. H. Wu. 2021. Food safety lessons learned from the COVID-19 pandemic. *Journal of Food Safety* 41:e12878.

Lagowski, K. K., and J. J. Stewart. 2003. Cognitive apprenticeship theory and graduate chemistry education. *Journal of Chemical Education* 80(12):1362-1366. https://pubs.acs.org/doi/pdf/10.1021/ed080p1362.

Laird, K. 2021. Start-up Renuva plant: Mattress recycling project Renuva now a reality. Sustainable Plastics, September 23.

Langston, N. 2010. Toxic inequities: Chemical exposures and indigenous communities in Canada and the United States. *Natural Resources Journal* 50:393-406.

Lauer, M. S., and D. Roychowdhury. 2021. Inequalities in the distribution of National Institutes of Health research project grant funding. *eLife* 10:e71712.

Leal Filho, W., U. Saari, M. Fedoruk, A. Iital, H. Moora, M. Klöga, and V. Voronova. 2019. An overview of the problems posed by plastic products and the role of extended producer responsibility in Europe. *Journal of Cleaner Production* 214:550-558.

Lee, S. K., Y. W. Cho, J. S. Lee, Y. R. Jung, S. H. Oh, J. Y. Sun, S. Kim, and Y. C. Joo. 2021. Nanofiber channel organic electrochemical transistors for low-power neuromorphic computing and wide-bandwidth sensing platforms. *Advanced Science* 8:2001544.

Lee, S. L., T. F. O'Connor, X. Yang, C. N. Cruz, S. Chatterjee, R. D. Madurawe, C. M. V. Moore, L. X. Yu, and J. Woodcock. 2015. Modernizing pharmaceutical manufacturing: From batch to continuous production. *Journal of Pharmaceutical Innovation* 10:191-199.

Leiserowitz, A., E. Mailbach, S. Rosenthal, J. Kotcher, J. Carman, L. Neyens, J. Marlon, K. Lacroix, and M. Goldberg. 2021. Public support for climate action by the president and Congress is rising. Yale Program on Climate Change Communication, September 28. https://climatecommunication.yale.edu/publications/public-support-for-climate-action-by-the-president-and-congress/.

Levi, P. G., and J. M. Cullen. 2018. Mapping global flows of chemicals: From fossil fuel feedstocks to chemical products. *Environmental Science & Technology* 52:1725-1734.

REFERENCES

Levy, S. 2013. The brief history of the ENIAC computer. *Smithsonian Magazine,* November. https://www.smithsonianmag.com/history/the-brief-history-of-the-eniac-computer-3889120/.

Lewis, A. C. 2021. Optimising air quality co-benefits in a hydrogen economy: A case for hydrogen-specific standards for NOx emissions. *Environmental Science: Atmospheres* 1:201-207.

Leznoff, C. C. 1978. The use of insoluble polymer supports in general organic synthesis. *Accounts of Chemical Research* 11:327-333.

Li, J., S. G. Ballmer, E. P. Gillis, S. Fujii, M. J. Schmidt, A. M. E. Palazzolo, J. W. Lehmann, G. F. Morehouse, and M. D. Burke. 2015. Synthesis of many different types of organic small molecules using one automated process. *Science* 347:1221-1226.

Li, J., Z. Peng, and E. Wang. 2018. Tackling grand challenges of the 21st century with electroanalytical chemistry. *Journal of the American Chemical Society* 140:10629-10638.

Li, L., S. Rong, R. Wang, and S. Yu. 2021. Recent advances in artificial intelligence and machine learning for nonlinear relationship analysis and process control in drinking water treatment: A review. *Chemical Engineering Journal* 405:126673.

Liao, P. V., and J. Dollin. 2012. Half a century of the oral contraceptive pill: Historical review and view to the future. *Canadian Family Physician* 58:e757-e760.

Liew, F. E., R. Nogle, T. Abdalla, B. J. Rasor, C. Canter, R. O. Jensen, L. Wang, J. Strutz, P. Chirania, S. De Tissera, A. P. Mueller, Z. Ruan, A. Gao, L. Tran, N. L. Engle, J. C. Bromley, J. Daniell, R. Conrado, T. J. Tschaplinski, R. J. Giannone, R. L. Hettich, A. S. Karim, S. D. Simpson, S. D. Brown, C. Leang, M. C. Jewett, and M. Köpke. 2022. Carbon-negative production of acetone and isopropanol by gas fermentation at industrial pilot scale. *Nature Biotechnology* 40:335-344.

Lindenberg, E. B., and S. Ross. 1981. Tobin's q ratio and industrial organization. *Journal of Business* 54(1):1-32.

Liu, D., H. Yu, and Y. Chai. 2021. Low-power computing with neuromorphic engineering. *Advanced Intelligent Systems* 3:2000150.

Liu, F., M. Wang, and M. Zheng. 2021. Effects of COVID-19 lockdown on global air quality and health. *Science of the Total Environment* 755:142533.

Livingston, M., J. Fernandez-Cornejo, J. Unger, D. Schimmelpfennig, C. Osteen, T. Park, and D. Lambert. 2015. *The Economics of Glyphosate Resistance Management in Corn and Soybean Production.* Economic Research Report 184. U.S. Department of Agriculture. https://www.ers.usda.gov/webdocs/publications/45354/52761_err184.pdf?v=0.

López de Dicastillo, C., E. Velásquez, A. Rojas, A. Guarda, and M. J. Galotto. 2020. The use of nanoadditives within recycled polymers for food packaging: Properties, recyclability, and safety. *Comprehensive Reviews in Food Science and Food Safety* 19:1760-1776.

MacFarlane, D. R., P. V. Cherepanov, J. Choi, B. H. R. Suryanto, R. Y. Hodgetts, J. M. Bakker, F. M. Ferrero Vallana, and A. N. Simonov. 2020. A roadmap to the ammonia economy. *Joule* 4:1186-1205.

Mackenzie, W. 2020. The energy transition will be built with metals. Forbes, October 29. https://www.forbes.com/sites/woodmackenzie/2020/10/29/the-energy-transition-will-be-built-with-metals/?sh=57f4a90d2b23.

Mah, A. 2021. What is carbon capture, utilization and storage (CCUS)? *Context: Energy Examined,* May 12. https://context.capp.ca/infographics/2021/what-is-carbon-capture-utilization-and-storage-ccus/.

Mangalindan, J. P. 2021. A timeline of computing power. CNN Money. https://money.cnn.com/interactive/technology/computing-power-timeline/.

Martens, E., H. Prommer, R. Sprocati, J. Sun, X. Dai, R. Crane, J. Jamieson, P. O. Tong, M. Rolle, and A. Fourie. 2021. Toward a more sustainable mining future with electrokinetic in situ leaching. *Science Advances* 7(18).

Maughon, B. 2021. Presentation to the committee at the Public Session on Corporate Perspectives, May 17.

Maxim, H. 1903. Inventions that ought to be invented. Women's Home Companion. https://www.flyingcarsandfoodpills.com/inventions-that-ought-to-be-invented.

Mayer, M., and A. J. Baeumner. 2019. A megatrend challenging analytical chemistry: Biosensor and chemosensor concepts ready for the Internet of Things. *Chemical Reviews* 119:7996-8027.

Mazzucato, M., and H. L. Li. 2021. A market shaping approach for the biopharmaceutical industry: Governing innovation towards the public interest. *Journal of Law, Medicine & Ethics* 49:39-49.

McBride, F. 2020. *Redefining Value and Risk in Agriculture Policy and Investment Solutions to Scale the Transition to Regenerative Agriculture.* Berkeley Food Institute and Center for Law, Energy & the Environment, University of California, Berkeley. https://food.berkeley.edu/wp-content/uploads/2020/12/BFI_ValueRisk_in_Ag_120920_Digital.pdf.

McHenry, L .B. 2018. The Monsanto Papers: Poisoning the scientific well. *International Journal of Risk & Safety in Medicine* 29:193-205.

McNally, A., C. K. Prier, and D. W. C. MacMillan. 2011. Discovery of an -amino C–H arylation reaction using the strategy of accelerated serendipity. *Science* 334:1114-1117.

Mehta, P., P. Barboun, D. B. Go, J. C. Hicks, and W. F. Schneider. 2019. Catalysis enabled by plasma activation of strong chemical bonds: A review. *ACS Energy Letters* 4:1115-1133.

Mennen, S. M., C. Alhambra, C. L. Allen, M. Barberis, S. Berritt, T. A. Brandt, A. D. Campbell, J. Castañón, A. H. Cherney, M. Christensen, D. B. Damon, J. Eugenio de Diego, S. García-Cerrada, P. García-Losada, R. Haro, J. Janey, D. C. Leitch, L. Li, F. Liu, P. C. Lobben, D. W. C. MacMillan, J. Magano, E. McInturff, S. Monfette, R. J. Post, D. Schultz, B. J. Sitter, J. M. Stevens, I. I. Strambeanu, J. Twilton, K. Wang, and M. A. Zajac. 2019. The evolution of high-throughput experimentation in pharmaceutical development and perspectives on the future. *Organic Process Research & Development* 23:1213-1242.

Mikulic, M. 2021. U.S. pharmaceutical industry—statistics & facts. Statista, September 24. https://www.statista.com/topics/1719/pharmaceutical-industry/#dossierKeyfigures.

Miller, K. 2021. 10 things everyone can do to support women scientists. *ACS Axial*. https://axial.acs.org/2021/04/21/10-things-everyone-can-do-to-support-women-scientists/.

Misiou, O., and K. Koutsoumanis. 2021. Climate change and its implications for food safety and spoilage. *Trends in Food Science & Technology*. https://doi.org/10.1016/j.tifs.2021.03.031.

Mitchell, C. E., U. Terranova, I. Alshibane, D. J. Morgan, T. E. Davies, Q. He, J. S. J. Hargreaves, M. Sankar, and N. H. de Leeuw. 2019. Liquid phase hydrogenation of CO_2 to formate using palladium and ruthenium nanoparticles supported on molybdenum carbide. *New Journal of Chemistry* 43:13985-13997.

Moeller, K. D. 2000. Synthetic applications of anodic electrochemistry. *Tetrahedron* 56:9527-9554.

Møller, K. T., T. R. Jensen, E. Akiba, and H. Li. 2017. Hydrogen—A sustainable energy carrier. *Progress in Natural Science: Materials International* 27:34-40.

Montoya, J. H., C. Tsai, A. Vojvodic, and J. K. Nørskov. 2015. The challenge of electrochemical ammonia synthesis: A new perspective on the role of nitrogen scaling relations. *ChemSusChem* 8:2180-2186.

Motagamwala, A. H., and J. A. Dumesic. 2021. Microkinetic modeling: A tool for rational catalyst design. *Chemical Reviews* 121:1049-1076.

Motta, E. V. S., K. Raymann, and N. A. Moran. 2018. Glyphosate perturbs the gut microbiota of honey bees. *Proceedings of the National Academy of Sciences of the United States of America* 115:10305-10310.

Motta, M., and J. E. Rice. 2022. Emerging quantum computing algorithms for quantum chemistry. *WIREs Computational Molecular Science* 12:e1580.

Myers, K. 2020. The elasticity of science. *American Economic Journal: Applied Economics* 12:103-134.

NAE and NRC (National Academy of Engineering and National Research Council). 2012. *Community Colleges in the Evolving STEM Education Landscape: Summary of a Summit*. Washington, DC: The National Academies Press.

Nameroff, T. J., R. J. Garant, and M. B. Albert. 2004. Adoption of green chemistry: An analysis based on US patents. *Research Policy* 33:959-974.

Nance, K. D., and J. L. Meier. 2021. Modifications in an emergency: The role of N1-methylpseudouridine in COVID-19 vaccines. *ACS Central Science* 7:748-756.

Narancic, T., S. Verstichel, S. Reddy Chaganti, L. Morales-Gamez, S. T. Kenny, B. De Wilde, R. Babu Padamati, and K. E. O'Connor. 2018. Biodegradable plastic blends create new possibilities for end-of-life management of plastics but they are not a panacea for plastic pollution. *Environmental Science & Technology* 52:10441-10452.

NAS, NAE, and IOM (National Academy of Sciences, National Academy of Engineering, and Institute of Medicine). 2011. *Expanding Underrepresented Minority Participation: America's Science and Technology Talent at the Crossroads*. Washington, DC: The National Academies Press.

NASEM (National Academies of Sciences, Engineering, and Medicine). 2016. *The Changing Landscape of Hydrocarbon Feedstocks for Chemical Production: Implications for Catalysis: Proceedings of a Workshop*. Washington, DC: The National Academies Press.

NASEM. 2017a. *Preparing for Future Products of Biotechnology*. Washington, DC: The National Academies Press.

NASEM. 2017b. *Undergraduate Research Experiences for STEM Students: Successes, Challenges, and Opportunities*. Washington, DC: The National Academies Press.

NASEM. 2018. *Sexual Harassment of Women: Climate, Culture, and Consequences in Academic Sciences, Engineering, and Medicine*. Washington, DC: The National Academies Press.

NASEM. 2019a. *A Research Agenda for Transforming Separation Science*. Washington, DC: The National Academies Press.

NASEM. 2019b. *Frontiers of Materials Research: A Decadal Survey*. Washington, DC: The National Academies Press.

NASEM. 2019c. *Gaseous Carbon Waste Streams Utilization: Status and Research Needs*. Washington, DC: The National Academies Press.

NASEM. 2019d. *Minority Serving Institutions: America's Underutilized Resource for Strengthening the STEM Workforce*. Washington, DC: The National Academies Press.

NASEM. 2019e. *Negative Emissions Technologies and Reliable Sequestration: A Research Agenda*. Washington, DC: The National Academies Press.

NASEM. 2019f. *The Science of Effective Mentorship in STEMM*. Washington, DC: The National Academies Press.

NASEM. 2020a. *Promising Practices for Addressing the Underrepresentation of Women in Science, Engineering, and Medicine: Opening Doors*. Washington, DC: The National Academies Press.

NASEM. 2020b. *Safeguarding the Bioeconomy*. Washington, DC: The National Academies Press.

NASEM. 2021a. *A Research Strategy for Ocean Carbon Dioxide Removal and Sequestration.* Washington, DC: The National Academies Press.
NASEM. 2021b. *Accelerating Decarbonization of the U.S. Energy System.* Washington, DC: The National Academies Press.
NASEM. 2021c. *Diversity, Equity, and Inclusion in Chemistry and Chemical Engineering: Proceedings of a Workshop–in Brief.* Washington, DC: The National Academies Press.
NASEM. 2021d. *Federal Government Human Health PFAS Research Workshop: Proceedings of a Workshop–in Brief.* Washington, DC: The National Academies Press.
NASEM. 2021e. *Innovations in Pharmaceutical Manufacturing on the Horizon: Technical Challenges, Regulatory Issues, and Recommendations.* Washington, DC: The National Academies Press.
NASEM. 2022a. *New Directions for Chemical Engineering.* Washington, DC: The National Academies Press.
NASEM. 2022b. *Promotion, Tenure, and Advancement Through the Lens of 2020: Proceedings of a Workshop–in Brief.* Washington, DC: The National Academies Press.
Nasrollahzadeh, M., M. Sajjadi, S. Iravani, and R. S. Varma. 2021. Green-synthesized nanocatalysts and nanomaterials for water treatment: Current challenges and future perspectives. *Journal of Hazardous Materials* 401:123401.
Nayak, S., N. R. Blumenfeld, T. Laksanasopin, and S. K. Sia. 2017. Point-of-care diagnostics: Recent developments in a connected age. *Analytical Chemistry* 89:102-123.
NCHS (National Center for Health Statistics). 2019. Daily updates of totals by week and state: Provisional death counts for coronavirus disease 2019 (COVID-19). Centers for Disease Control and Prevention. https://www.cdc.gov/nchs/nvss/vsrr/covid19/index.htm.
NCSES (National Center for Science and Engineering Statistics). 2020a. *Annual Business Survey: Data Year 2017.* NSF 21-303. Alexandria, VA: National Science Foundation. https://ncses.nsf.gov/pubs/nsf21303/.
NCSES. 2020b. *Survey of Earned Doctorates.* Alexandria, VA: National Science Foundation. https://ncses.nsf.gov/pubs/nsf22300/report.
Newburger, E. 2021. Fed governor anticipates new guidance on climate change for big banks. CNBC Climate, October 7. https://www.cnbc.com/2021/10/07/fed-governor-anticipates-climate-change-guidance-coming-for-big-banks.html.
NIAID (National Institute of Allergy and Infectious Diseases). 2018. Antiretroviral drug discovery and development. https://www.niaid.nih.gov/diseases-conditions/antiretroviral-drug-development.
Nicholson, S. R., N. A. Rorrer, A. C. Carpenter, and G. T. Beckham. 2021. Manufacturing energy and greenhouse gas emissions associated with plastics consumption. *Joule* 5:673-686.
NIMHD (National Institute on Minority Health and Health Disparities). 2021. Solicited and Investigator-Initiated Research Project Grants. https://www.nimhd.nih.gov/programs/extramural/investigator-initiated-research/#expandAll.
Nobel Prize Outreach. 2022a. The Nobel Prize in Chemistry 1918. https://www.nobelprize.org/prizes/chemistry/1918/summary/.
Nobel Prize Outreach. 2022b. The Nobel Prize in Chemistry 2014. https://www.nobelprize.org/prizes/chemistry/2014/summary/.
Nobel Prize Outreach. 2022c. The Nobel Prize in Chemistry 2017. https://www.nobelprize.org/prizes/chemistry/2017/summary/.
Nobel Prize Outreach. 2022d. The Nobel Prize in Physics 1956. https://www.nobelprize.org/prizes/physics/1956/summary/.
Nørskov, J. K., and T. Bligaard. 2013. The catalyst genome. *Angewandte Chemie International Edition* 52:776-777.
NRC (National Research Council). 2000. *The Future Role of Pesticides in US Agriculture.* Washington, DC: The National Academies Press.
NRC. 2003. *Reducing the Time from Basic Research to Innovation in the Chemical Sciences: A Workshop Report to the Chemical Sciences Roundtable.* Washington, DC: The National Academies Press.
NRC. 2006a. *Sustainability in the Chemical Industry: Grand Challenges and Research Needs.* Washington, DC: The National Academies Press.
NRC. 2006b. *Visualizing Chemistry: The Progress and Promise of Advanced Chemical Imaging.* Washington, DC: The National Academies Press.
NRC. 2010. *Research at the Intersection of the Physical and Life Sciences.* Washington, DC: The National Academies Press.
NRC. 2012. *A Framework for K-12 Science Education: Practices, Crosscutting Concepts, and Core Ideas.* Washington, DC: The National Academies Press.
NSB (National Science Board). 2018. Bridging the Gap: Building a Sustained Approach to Mid-scale Research Infrastructure and Cyberinfrastructure at NSF. https://www.nsf.gov/nsb/publications/2018/NSB-2018-40-Midscale-Research-Infrastructure-Report-to-Congress-Oct2018.pdf.
NSB. 2019. Immigration and the S&E workforce. In *Science and Engineering Indicators 2020: Science and Engineering Labor Force.* NSB-2019-8. Alexandria, VA: National Science Foundation. https://ncses.nsf.gov/pubs/nsb20198/immigration-and-the-s-e-workforce.
NSB. 2021. *Publications Output: U.S. Trends and International Comparisons.* NSB-2021-4. Alexandria, VA: National Science Foundation. https://ncses.nsf.gov/pubs/nsb20214/international-collaboration-and-citations.

NSF (National Science Foundation). 2016a. Mid-Scale Instrument Development for the Chemical Sciences: Workshop Report. https://www.nsf.gov/mps/che/workshops/mid-scale_instrument_development_for_the_chemical_sciences_workshop_september_2016.pdf.

NSF. 2016b. Mid-Scale Instrumentation: Regional Facilities to Address Grand Challenges in Chemistry. Arlington, VA. https://www.nsf.gov/mps/che/workshops/msiegionalcenters_workshopreport_5_1_17.pdf.

NSF. 2021. *FY 2022 Budget Request to Congress*. https://www.nsf.gov/about/budget/fy2022/pdf/fy2022budget.pdf.

NSF. n.d. NSF's Definitions of Research Categories. https://www.radford.edu/content/dam/departments/administrative/sponsored-programs/PDFs/NSFdefinitions.pdf.

NSTC (National Science and Technology Council). 2018. *Charting a Course for Success: America's Strategy for STEM Education. A Report by the Committee on STEM Education*. Washington, DC: Executive Office of the President, Office of Science and Technology Policy. https://eric.ed.gov/?id=ED590474.

OECD (Organisation for Economic Co-operation and Development). 2002. *Frascati Manual 2002: Proposed Standard Practice for Surveys on Research and Experimental Development*. Paris: OECD Publishing. https://www.oecd-ilibrary.org/content/publication/9789264199040-en.

OECD. 2006. Fact sheet: Extended producer responsibility. https://www.oecd.org/env/waste/factsheetextendedproducer-responsibility.htm.

OECD. 2015. *Frascati Manual 2015: Guidelines for Collecting and Reporting Data on Research and Experimental Development*. Paris: OECD Publishing. https://www.oecd-ilibrary.org/content/publication/9789264239012-en.

Olah, G. A. 2005. Beyond oil and gas: The methanol economy. *Angewandte Chemie International Edition* 44:2636-2639.

Olofson, R. A., and L. B. Gortler. 1999. *Russell Marker and the Mexican Steroid Hormone Industry International Historic Chemical Landmark*. Washington, DC: American Chemical Society National Historic Chemical Landmarks Program.

Olsen, K. 2012. The first 110 years of laboratory automation: Technologies, applications, and the creative scientist. *SLAS Technology* 17:469-480.

Onat, N. C., M. Kucukvar, A. Halog, and S. Cloutier. 2017. Systems thinking for life cycle sustainability assessment: A review of recent developments, applications, and future perspectives. *Sustainability* 9(5):706.

OPCW (Organisation for the Prohibition of Chemical Weapons). 2020. Chemical Weapons Convention. https://www.opcw.org/sites/default/files/documents/CWC/CWC_en.pdf.

Orgill, M., S. York, and J. MacKellar. 2019. Introduction to systems thinking for the chemistry education community. *Journal of Chemical Education* 96:2720-2729.

OSTP (Office of Science and Technology Policy). 2021. *Progress Report on the Implementation of the Federal Stem Education Strategic Plan*. Washington, DC: Executive Office of the President. https://www.whitehouse.gov/wp-content/uploads/2022/01/2021-CoSTEM-Progress-Report-OSTP.pdf.

Our World in Data. 2015. Share of the rural population with access to improved drinking water, 2015. https://ourworldindata.org/grapher/rural-population-with-improved-water.

Our World in Data. 2017. Moore's Law: The number of transistors per microprocessor. https://ourworldindata.org/grapher/transistors-per-microprocessor.

Our World in Data. 2020. Food waste is responsible for 6% of global greenhouse gas emissions. https://ourworldindata.org/food-waste-emissions.

Our World in Data. 2022. The computational capacity of the largest superocmputers. https://ourworldindata.org/grapher/supercomputer-power-flops.

Painter, G. R., M. G. Natchus, O. Cohen, W. Holman, and W. P. Painter. 2021. Developing a direct acting, orally available antiviral agent in a pandemic: The evolution of molnupiravir as a potential treatment for COVID-19. *Current Opinion in Virology* 50:17-22.

Palmer, J. T., S. M. Gallo, T. R. Furlani, M. D. Jones, R. L. DeLeon, N. Simakov, J. P. White, A. K. Patra, J. Sperhac, T. Yearke, R. Rathsam, C. D. Cornelius, M. Innus, J. C. Browne, W. L. Barth, and R. T. Evans. 2015. Open XDMoD: A tool for the comprehensive management of high-performance computing resources. *Computing in Science & Engineering* 17(4):52-62.

Panzone, C., R. Philippe, A. Chappaz, P. Fongarland, and A. Bengaouer. 2020. Power-to-liquid catalytic CO_2 valorization into fuels and chemicals: Focus on the Fischer-Tropsch route. *Journal of CO_2 Utilization* 38:314-347.

Paolucci, C., I. Khurana, A. A. Parekh, S. Li, A. J. Shih, H. Li, J. R. D. Iorio, J. D. Albarracin-Caballero, A. Yezerets, J. T. Miller, W. N. Delgass, F. H. Ribeiro, W. F. Schneider, and R. Gounder. 2017. Dynamic multinuclear sites formed by mobilized copper ions in NOx selective catalytic reduction. *Science* 357:898-903.

Peplow, M. 2020. Cryo-electron microscopy reaches resolution milestone. *C&EN* 98(37). https://cen.acs.org/analytical-chemistry/microscopy/Cryo-electron-microscopy-reaches-resolution/98/i37.

Perkel, J. M. 2014. Miniaturizing mass spectrometry. *Science*, February 21. https://www.science.org/content/article/miniaturizing-mass-spectrometry.

Petranikova, M., P. L. Naharro, N. Vieceli, G. Lombardo, and B. Ebin. 2022. Recovery of critical metals from EV batteries via thermal treatment and leaching with sulphuric acid at ambient temperature. *Waste Management* 140:164-172.

Petrenko, V. V., P. Martinerie, P. Novelli, D. M. Etheridge, I. Levin, Z. Wang, T. Blunier, J. Chappellaz, J. Kaiser, P. Lang, L. P. Steele, S. Hammer, J. Mak, R. L. Langenfelds, J. Schwander, J. P. Severinghaus, E. Witrant, G. Petron, M. O. Battle, G. Forster, W. T. Sturges, J. F. Lamarque, K. Steffen, and J. W .C. White. 2013. A 60 yr record of atmospheric carbon monoxide reconstructed from Greenland firn air. *Atmospheric Chemistry and Physics* 13:7567-7585.

Piątek, J., S. Afyon, T. M. Budnyak, S. Budnyk, M. H. Sipponen, and A. Slabon. 2021. Sustainable Li-ion batteries: Chemistry and recycling. *Advanced Energy Materials* 11:2003456.

Plante, O. J., E. R. Palmacci, and P. H. Seeberger. 2001. Automated solid-phase synthesis of oligosaccharides. *Science* 291:1523-1527.

Plaza, M. G., S. Martínez, and F. Rubiera. 2020. CO_2 capture, use, and storage in the cement industry: State of the art and expectations. *Energies* 13:5692.

Plett, T., and V. Bernales. 2016. Five cents about nickel catalysts. *Frontiers in Energy Research Newsletter.* https://www.pnnl.gov/science/highlights/highlight.asp?id=4282.

Polterauer, D., D. M. Roberge, P. Hanselmann, P. Elsner, C. A. Hone, and C. O. Kappe. 2021. Process intensification of ozonolysis reactions using dedicated microstructured reactors. *Reaction Chemistry & Engineering* 6:2253-2258.

Porter, M. E. 1991. Towards a dynamic theory of strategy. *Strategic Management Journal* 12:95-117.

Powell, K., R. Terry, and S. Chen. 2020. How LGBT+ scientists would like to be included and welcomed in STEM workplaces. *Nature* 586(7831):813-817.

Preuster, P., C. Papp, and P. Wasserscheid. 2017. Liquid organic hydrogen carriers (LOHCs): Toward a hydrogen-free hydrogen economy. *Accounts of Chemical Research* 50:74-85.

Puthongkham, P., S. Wirojsaengthong, and A. Suea-Ngam. 2021. Machine learning and chemometrics for electrochemical sensors: Moving forward to the future of analytical chemistry. *Analyst* 146:6351-6364.

Qian, J. X., T. W. Chen, L. R. Enakonda, D. B. Liu, J. M. Basset, and L. Zhou. 2020. Methane decomposition to pure hydrogen and carbon nano materials: State-of-the-art and future perspectives. *International Journal of Hydrogen Energy* 45:15721-15743.

Ramberg, P. J. 2000. The death of vitalism and the birth of organic chemistry: Wohler's urea synthesis and the disciplinary identity of organic chemistry. *Ambix* 47:170-195.

Ramonas, A., and J. Rund. 2021. Climate change risks surge in companies' annual reports to SEC. Bloomberg Law, March 25.

Ravikumar, D., D. Zhang, G. Keoleian, S. Miller, V. Sick, and V. Li. 2021. Carbon dioxide utilization in concrete curing or mixing might not produce a net climate benefit. *Nature Communications* 12:855.

Reardon, K. 2019. Double Bind: Women of Color in STEM. Duke University Center for International & Global Studies. https://igs.duke.edu/news/double-bind-women-color-stem-0.

Reisman, S. E., R. Sarpong, M. S. Sigman, and T. P. Yoon. 2020. Organic chemistry: A call to action for diversity and inclusion. *Journal of Organic Chemistry* 85:10287-10292.

Revankar, S. T. 2019. Chemical energy storage. Pp. 177-227 in *Storage and Hybridization of Nuclear Energy*. London: Academic Press.

Richardson, S. D., and T. A. Ternes. 2014. Water analysis: Emerging contaminants and current issues. *Analytical Chemistry* 86:2813-2848.

Richter, L., A. Cordner, and P. Brown. 2021. Producing ignorance through regulatory structure: The case of per- and polyfluoroalkyl substances (PFAS). *Sociological Perspectives* 64:631-656.

Riisager, A., R. Fehrmann, S. Flicker, R. van Hal, M. Haumann, and P. Wasserscheid. 2005. Very stable and highly regioselective supported ionic-liquid-phase (SILP) catalysis: Continuous-flow fixed-bed hydroformylation of propene. *Angewandte Chemie International Edition* 44:815-819.

Rogoff, M. J. 2014. Collection approaches. Pp. 19-42 in *Solid Waste Recycling and Processing,* 2nd ed., M. J. Rogoff, ed. Oxford, UK: William Andrew Publishing.

Roman-White, S. A., J. A. Littlefield, K. G. Fleury, D. T. Allen, P. Balcombe, K. E. Konschnik, J. Ewing, G. B. Ross, and F. George. 2021. LNG supply chains: A supplier-specific life-cycle assessment for improved emission accounting. *ACS Sustainable Chemistry & Engineering* 9(32):10857-10867.

Romm, C. 2015. Before there were home pregnancy tests. *The Atlantic,* June 17. https://www.theatlantic.com/health/archive/2015/06/history-home-pregnancy-test/396077/.

Rosevear, J. 2022. Nickel's price surge could threaten automakers' ambitious electric-vehicle plans. CNBC, March 8.

Rosselot, K. S., D. T. Allen, and A. Y. Ku. 2021. Comparing greenhouse gas impacts from domestic coal and imported natural gas electricity generation in China. *ACS Sustainable Chemistry & Engineering* 9(26):8759-8769.

RSC (Royal Society of Chemistry). 2018. *Breaking the Barriers: Women's Retention and Progression in the Chemical Sciences.* https://www.rsc.org/globalassets/02-about-us/our-strategy/inclusion-diversity/womens-progression/media-pack/v18_vo_inclusion-and-diversity-_womans-progression_report-web-.pdf.

Rubio, S., T. R. P. Ramos, M. M. R. Leitão, and A. P. Barbosa-Povoa. 2019. Effectiveness of extended producer responsibility policies implementation: The case of Portuguese and Spanish packaging waste systems. *Journal of Cleaner Production* 210:217-230.

Saito, Y. 1997. The Monte Carlo simulation of microstructural evolution in metals. *Materials Science and Engineering: A* 223:114-124.

Salmeron, M., and B. Eren. 2021. High-pressure scanning tunneling microscopy. *Chemical Reviews* 121:962-1006.

Sanford, M. S. 2020. Equity and inclusion in the chemical sciences requires actions not just words. *Journal of the American Chemical Society* 142:11317-11318.

SBIR (Small Business Innovation Research). 2017a. SBIR-STTR Success: Exelus, Inc. https://www.sbir.gov/node/1308743.

SBIR. 2017b. SBIR-STTR Success: Mango Materials, Inc. https://www.sbir.gov/node/877615.

Schaaf, T., J. Grünig, M. R. Schuster, T. Rothenfluh, and A. Orth. 2014. Methanation of CO_2—Storage of renewable energy in a gas distribution system. *Energy, Sustainability and Society* 4:2.

Schiffer, Z. J., and K. Manthiram. 2017. Electrification and decarbonization of the chemical industry. *Joule* 1:10-14.

Schlögl, R. 2003. Catalytic synthesis of ammonia—A "never-ending story"? *Angewandte Chemie International Edition* 42:2004-2008.

Schmitz, C. 2015. Real-time analysis is critical to quality. *Chemical Processing,* January 27. https://www.chemicalprocessing.com/articles/2015/real-time-analysis-critical-to-quality/.

Schnoor, J.-K., M. Fuchs, A. Böcking, M. Wessling, and M. A. Liauw. 2019. Homogeneous catalyst recycling and separation of a multicomponent mixture using organic solvent nanofiltration. *Chemical Engineering & Technology* 42:2187-2194.

Schuman, C. D., T. E. Potok, R. M. Patton, J. D. Birdwell, M. E. Dean, G. S. Rose, and J. S. Plank. 2017. A survey of neuromorphic computing and neural networks in hardware. *arXiv*. https://arxiv.org/abs/1705.06963.

Schütte, G., M. Eckerstorfer, V. Rastelli, W. Reichenbecher, S. Restrepo-Vassalli, M. Ruohonen-Lehto, A. G. W. Saucy, and M. Mertens. 2017. Herbicide resistance and biodiversity: Agronomic and environmental aspects of genetically modified herbicide-resistant plants. *Environmental Sciences Europe* 29:5.

Schutyser, W., T. Renders, S. Van den Bosch, S. F. Koelewijn, G. T. Beckham, and B. F. Sels. 2018. Chemicals from lignin: An interplay of lignocellulose fractionation, depolymerisation, and upgrading. *Chemical Society Reviews* 47:852-908.

Scott, G. I., D. E. Porter, R. S. Norman, C. H. Scott, M. I. Uyaguari-Diaz, K. A. Maruya, S. B. Weisberg, M. H. Fulton, E. F. Wirth, J. Moore, P. L. Pennington, D. Schlenk, G. P. Cobb, and N. D. Denslow. 2016. Antibiotics as CECs: An overview of the hazards posed by antibiotics and antibiotic resistance. *Frontiers in Marine Science* 3:24.

Seay, J., W. T. Chen., and M. E. Ternes. 2020. Waste plastic: Challenges and opportunities for the chemical industry. AIChE. https://www.aiche.org/resources/publications/cep/2020/november/waste-plastic-challenges-and-opportunities-chemical-industry#:~:text=%20Waste%20Plastic%3A%20Challenges%20and%20Opportunities%20for%20the,a%20circular%20plastics%20economy.%20Circular%20plastics...%20More%20.

SEC (Securities and Exchange Commission). 2022. Fact Sheet: Enhancement and Standardization of Climate-Related Disclosures. https://www.sec.gov/files/33-11042-fact-sheet.pdf.

Selekman, J. A., J. Qiu, K. Tran, J. Stevens, V. Rosso, E. Simmons, Y. Xiao, and J. Janey. 2017. High-throughput automation in chemical process development. *Annual Review of Chemical and Biomolecular Engineering* 8:525-547.

Selvam, T., A. Machoke, and W. Schwieger. 2012. Supported ionic liquids on non-porous and porous inorganic materials—A topical review. *Applied Catalysis A: General* 445-446:92-101.

Sendlinger, S. C., D. J. DeCoste, T. H. Dunning, D. A. Dummitt, E. Jakobsson, D. R. Mattson, and E. N. Wiziecki. 2008. Transforming chemistry education through computational science. *Computing in Science & Engineering* 10(5):34-39.

Sensiba, J. 2021. Lithium-ion recycling company is going public. CleanTechnica, February 27. https://cleantechnica.com/2021/02/27/lithium-ion-recycling-company-is-going-public/.

Sepulveda, N. A., J. D. Jenkins, A. Edington, D. S. Mallapragada, and R. K. Lester. 2021. The design space for long-duration energy storage in decarbonized power systems. *Nature Energy* 6:506-516.

Shalf, J. 2020. The future of computing beyond Moore's Law. *Philosophical Transactions of the Royal Society A: Mathematical, Physical and Engineering Sciences* 378:20190061.

Shanks, B. H., and P. L. Keeling. 2017. Biopriviledged molecules: Creating value from biomass. *Green Chemistry* 19:3177-3185.

Sharp, B. E., and S. A. Miller. 2016. Potential for integrating diffusion of innovation principles into life cycle assessment of emerging technologies. *Environmental Science & Technology* 50:2771-2781.

Shi, Y., Z. Lyu, M. Zhao, R. Chen, Q. N. Nguyen, and Y. Xia. 2021. Noble-metal nanocrystals with controlled shapes for catalytic and electrocatalytic applications. *Chemical Reviews* 121:649-735.

Shieh, J. 2021. Presentation to the committee at the Public Session on Analyzing Financial Investments in the Chemical Sciences, April 2.

Shinde, S. R., and S. Apte. 2021. A systematic review on advancements in drinking water disinfection technologies: A sustainable development perspective. *Journal of Environmental Treatment Techniques* 9(2):349-360.

Shiva Kumar, S., and V. Himabindu. 2019. Hydrogen production by PEM water electrolysis—A review. *Materials Science for Energy Technologies* 2:442-454.

SIA and OE (Semiconductor Industry Association and Oxford Economics). 2021. *Chipping in: The U.S. Semiconductor Industry Workforce and How Federal Incentives Will Increase Domestic Jobs*. SIA/Oxford Economics Report. https://www.semiconductors.org/chipping-in-sia-jobs-report/.

Siegel, J. B., A. Zanghellini, H. M. Lovick, G. Kiss, A. R. Lambert, J. L. St. Clair, J. L. Gallaher, D. Hilvert, M. H. Gelb, B. L. Stoddard, K. N. Houk, F. E. Michael, and D. Baker. 2010. Computational design of an enzyme catalyst for a stereoselective bimolecular Diels-Alder reaction. *Science* 329:309-313.

Siegelman, R. L., E. J. Kim, and J. R. Long. 2021. Porous materials for carbon dioxide separations. *Nature Materials* 20:1060-1072.

Smalley, M. 2021. Indorama Ventures to build Indonesian PET recycling facility. *Recycling Today*, July 6. https://www.recyclingtoday.com/article/indorama-ventures-builds-indonesian-pet-recycling-facility-karawang/.

Smith, A. B. 2021. 2020 U.S. billion-dollar weather and climate disasters in historical context. Beyond the Data blog, January 8. https://www.climate.gov/disasters2020.

Smith, C., A. K. Hill, and L. Torrente-Murciano. 2020. Current and future role of Haber–Bosch ammonia in a carbon-free energy landscape. *Energy & Environmental Science* 13:331-344.

Smithsonian Magazine. 2019. Development of the lithium-ion battery earns Nobel Prize in chemistry. Smart News, October 10. https://www.smithsonianmag.com/smart-news/development-lithium-ion-battery-earns-nobel-prize-chemistry-180973310/.

Smolinka, T. 2009. Fuels—Hydrogen production: Water electrolysis. Pp. 394-413 in *Encyclopedia of Electrochemical Power Sources*, J. Garche, C. K. Dyer, P. Moseley, Z. Ogumi, D. Rand, and B. Scrosati, eds. Amsterdam: Elsevier.

Snyder, H. D., and T. G. Kucukkal. 2021. Computational chemistry activities with Avogadro and ORCA. *Journal of Chemical Education* 98:1335-1341.

Soil Association. 2016. The Impact of Glyphosate on Soil Health, the Evidence to Date. https://www.soilassociation.org/media/7202/glyphosate-and-soil-health-full-report.pdf.

Stevens, K. R., K. S. Masters, P. I. Imoukhuede, K. A. Haynes, L. A. Setton, E. Cosgriff-Hernandez, M. A. Lediju Bell, P. Rangamani, S. E. Sakiyama-Elbert, S. D. Finley, R. K. Willits, A. N. Koppes, N. C. Chesler, K. L. Christman, J. B. Allen, J. Y. Wong, H. El-Samad, T. A. Desai, and O. Eniola-Adefeso. 2021. Fund Black scientists. *Cell* 184:561-565.

Stinn, C., and A. Allanore. 2022. Selective sulfidation of metal compounds. *Nature* 602:78-83.

Stokes, D. E. 1997. *Pasteur's Quadrant: Basic Science and Technological Innovation*. Washington, DC: Brookings Institution Press.

Stone, D. 2021. Birth of the Petrochemical Industry. American Chemical Society National Historic Chemical Landmarks. https://www.acs.org/content/acs/en/education/whatischemistry/landmarks/petrochemical-industry-birthplace.html.

Straathof, N. J. W., B. J. P. Tegelbeckers, V. Hessel, X. Wang, and T. Noël. 2014. A mild and fast photocatalytic trifluoromethylation of thiols in batch and continuous-flow. *Chemical Science* 5:4768-4773.

Stürzel, M., S. Mihan, and R. Mülhaupt. 2016. From multisite polymerization catalysis to sustainable materials and all-polyolefin composites. *Chemical Reviews* 116:1398-1433.

Subramaniam, B., D. Allen, K. Kuok Hii, J. Colberg, and T. Pradeep. 2021a. Lab to market: Where the rubber meets the road for sustainable chemical technologies. *ACS Sustainable Chemistry & Engineering* 9:2987-2989.

Subramaniam, B., P. Licence, A. Moores, and D. T. Allen. 2021b. Shaping effective practices for incorporating sustainability assessment in manuscripts submitted to *ACS Sustainable Chemistry & Engineering*: An initiative by the editors. *ACS Sustainable Chemistry & Engineering* 9(11):3977-3978.

Sun, C., J. Ge, J. He, R. Gan, and Y. Fang. 2021. Processing, quality, safety, and acceptance of meat analogue products. *Engineering* 7:674-678.

Sun, Z., B. Fridrich, A. de Santi, S. Elangovan, and K. Barta. 2018. Bright side of lignin depolymerization: Toward new platform chemicals. *Chemical Reviews* 118:614-678.

Suntinger, H. 2020. Recycling breakthrough for clothing made with polyamide. Innovation Origins, November 26. https://innovationorigins.com/en/recycling-breakthrough-for-clothing-made-with-polyamide/#:~:text=Recycling%20breakthrough%20for%20clothing%20made%20with%20polyamide%20Sustainability,can%20then%20be%20spun%20into%20new%20textile%20fibers.

Svitil, K. 2019. Stewart and Lynda Resnick Pledge $750 Million to Caltech to Support Environmental Sustainability Research. Caltech, September 26. https://www.caltech.edu/about/news/stewart-and-lynda-resnick-pledge-750-million-caltech-support-environmental-sustainability-research.

Takkellapati, S., T. Li, and M. A. Gonzalez. 2018. An overview of biorefinery-derived platform *Policy* 20:1615-1630.

Tang, J., F. Yuan, X. Shen, Z. Wang, M. Rao, Y. He, Y. Sun, X. Li, W. Zhang, Y. Li, B. Gao, H. Qian, G. Bi, S. Song, J. J. Yang, and H. Wu. 2019. Bridging biological and artificial neural networks with emerging neuromorphic devices: Fundamentals, progress, and challenges. *Advanced Materials* 31:1902761.

Thayer, A. M. 2013. Nobel Prizes recognized notable developments in catalysis. *C&EN*. https://cen.acs.org/articles/91/i36/Nobel-Prizes-Recognized-Notable-Developments.html.

Thayer, G. R., J. F. Roach, and L. Dauelsberg. 2006. *Estimated Energy Savings and Financial Impacts of Nanomaterials by Design on Selected Applications in the Chemical Industry.* Technical Report. U.S. Department of Energy. https://www.osti.gov/servlets/purl/1218765.

Thomas, G., and G. Parks. 2006. Potential Roles of Ammonia in a Hydrogen Economy. White Paper. U.S. Department of Energy. https://www.energy.gov/sites/prod/files/2015/01/f19/fcto_nh3_h2_storage_white_paper_2006.pdf.

Thompson, D. L., J. M. Hartley, S. M. Lambert, M. Shiref, G. D. J. Harper, E. Kendrick, P. Anderson, K. S. Ryder, L. Gaines, and A. P. Abbott. 2020. The importance of design in lithium ion battery recycling—A critical review. *Green Chemistry* 22:7585-7603.

Tian, Y., and G. Zhu. 2020. Porous aromatic frameworks (PAFs). *Chemical Reviews* 120:8934-8986.

Timoshenko, J., and B. Roldan Cuenya. 2021. In situ/operando electrocatalyst characterization by x-ray absorption spectroscopy. *Chemical Reviews* 121:882-961.

Tomkins, P., and T. E. Müller. 2019. Evaluating the carbon inventory, carbon fluxes and carbon cycles for a long-term sustainable world. *Green Chemistry* 21:3994-4013.

Trobe, M., and M. D. Burke. 2018. The molecular industrial revolution: Automated synthesis of small molecules. *Angewandte Chemie International Edition* 57(16):4192-4214.

Tuck, B., and M. Moeinian. 2017. *Economic Contribution of Federal Funding for Small Business Innovation Research (SBIR) and Small Business Technology Transfer (STTR) Programs.* University of Minnesota Extension. https://conservancy.umn.edu/bitstream/handle/11299/197828/2017-economic-contribution-sbir-sttr.pdf?sequence=1&isAllowed=y.

Tudball, M. 2022. Eastman invests $1bn in French hard-to-recycle PET waste chemical recycling facility. ICIS, January 17. https://www.icis.com/explore/resources/news/2022/01/17/10725224/eastman-invest-1bn-in-french-hard-to-recycle-pet-waste-chemical-recycling-facility/.

Tullo, A. H. 2016. Why DuPont shrunk its central research unit. *C&EN* 94. https://cen.acs.org/articles/94/i4/DuPont-Shrunk-Central-Research-Unit.html.

Tullo, A. H. 2021. Eastman will build a $250 million plastics recycling plant. *C&EN*. https://cen.acs.org/environment/recycling/Eastman-build-250-million-plastics/99/web/2021/02.

Turner, N. J. 2003. Directed evolution of enzymes for applied biocatalysis. *Trends in Biotechnology* 21:474-478.

Tyssowski, K. 2018. Pee is for pregnant: The history and science of urine-based pregnancy tests. Harvard University Science in the News, August 31. https://sitn.hms.harvard.edu/flash/2018/pee-pregnant-history-science-urine-based-pregnancy-tests/.

U.S. Congress, House. 2019. The devil they knew: E contamination and the need for corporate accountability. Hearing Before the Subcommittee on Environment of the Committee on Oversight and Reform, 116th Cong., 1st Sess., No. 53. Washington, DC: U.S. Government Printing Office.

U.S. Congress, Senate. 1982. Small Business Innovation Development Act of 1982. S.881. 97th Cong.

U.S. Congress, Senate. 1992. Small Business Technology Transfer Act of 1992. S.3385. 102nd Cong.

Ubando, A. T., C. B. Felix, and W. H. Chen. 2020. Biorefineries in circular bioeconomy: A comprehensive review. *Bioresource Technology* 299:122585.

UCS (Union of Concerned Scientists). 2017. How Dow Chemical influenced the EPA to ignore the scientific evidence on chlorpyrifos. UCS blog, October 11. https://www.ucsusa.org/resources/how-dow-chemical-influenced-epa-ignore-scientific-evidence-chlorpyrifos.

UCS. 2019. DuPont, 3M concealed evidence of PFAS risks. UCS blog, March 22. https://www.ucsusa.org/resources/dupont-3m-concealed-evidence-pfas-risks.

UN (United Nations). 2015. *Transforming our World: The 2030 Agenda for Sustainable Development.* https://sustainabledevelopment.un.org/post2015/transformingourworld/publication.

UNEP (United Nations Environment Programme). 2019. *Global Chemicals Outlook II: From Legacies to Innovative Solutions.* https://www.unep.org/resources/report/global-chemicals-outlook-ii-legacies-innovative-solutions.

Urbina-Blanco, C. A., S. Z. Jilani, I. R. Speight, M. J. Bojdys, T. Friščić, J. F. Stoddart, T. L. Nelson, J. Mack, R. A. S. Robinson, E. A. Waddell, J. L. Lutkenhaus, M. Godfrey, M. I. Abboud, S. O. Aderinto, D. Aderohunmu, L. Bibič, J. Borges, V. M. Dong, L. Ferrins, F. M. Fung, T. John, F. P. L. Lim, S. L. Masters, D. Mambwe, P. Thordarson, M. M. Titirici, G. D. Tormet-González, M. M. Unterlass, A. Wadle, V. W. W. Yam, and Y. W. Yang. 2020. A diverse view of science to catalyse change. *Journal of the American Chemical Society* 142:14393-14396.

USAID (U.S. Agency for International Development). 2021. Family Planning and Reproductive Health. https://www.usaid.gov/global-health/health-areas/family-planning.

USPTO (U.S. Patent and Trade Office). 2002. 2002 Laureates—National Medal of Technology and Innovation. https://www.uspto.gov/learning-and-resources/ip-programs-and-awards/national-medal-technology-and-innovation/recipients/2002.

USPTO. 2003. 2003 Laureates—National Medal of Technology and Innovation. https://www.uspto.gov/learning-and-resources/ip-programs-and-awards/national-medal-technology-and-innovation/recipients/2003.

van der Ent, A., A. Parbhakar-Fox, and P. D. Erskine. 2021. Treasure from trash: Mining critical metals from waste and unconventional sources. *Science of the Total Environment* 758:143673.

Van Geem, K. M., and B. M. Weckhuysen. 2021. Toward an e-chemistree: Materials for electrification of the chemical industry. *MRS Bulletin* 46:1187-1196.

van Leeuwen, G., and P. Mohnen. 2017. Revisiting the Porter hypothesis: An empirical analysis of Green innovation for the Netherlands. *Economics of Innovation and New Technology* 26:63-77.

van Tol, M. 2021. Presentation to the committee at the Public Session on Energy, June 28.

van Tol-Koutstaal, A. 2021. Patent valuation. https://docs.google.com/presentation/d/1eWER57wQMywi_b-YP7jMwkd JSrRG9DzP/edit#slide=id.p1.

Ventola, C. L. 2014. Medical applications for 3D printing: Current and projected uses. *P&T* 39:704-711.

Vericella, J. J., S. E. Baker, J. K. Stolaroff, E .B. Duoss, J. O. Hardin, J. Lewicki, E. Glogowski, W. C. Floyd, C. A. Valdez, W. L. Smith, J. H. Satcher, W. L. Bourcier, C. M. Spadaccini, J. A. Lewis, and R. D. Aines. 2015. Encapsulated liquid sorbents for carbon dioxide capture. *Nature Communications* 6:6124.

Volans and United Nations Global Compact. n.d. Breakthrough Business Models: Closed-Loop. Project Breakthrough. https://breakthrough.unglobalcompact.org/breakthrough-business-models/closed-loop/.

Vural Gürsel, I., T. Noël, Q. Wang, and V. Hessel. 2015. Separation/recycling methods for homogeneous transition metal catalysts in continuous flow. *Green Chemistry* 17:2012-2026.

Wacławek, S., V. V. T. Padil, and M. Černík. 2018. Major advances and challenges in heterogeneous catalysis for environmental applications: A review. *Ecological Chemistry and Engineering* 25:9-34.

Wagner, W. E., and R. Steinzor. 2006. *Rescuing Science from Politics: Regulation and the Distortion of Scientific Research.* Cambridge, UK: Cambridge University Press.

Wainerdi, R. E., ed. 1970. *Analytical Chemistry in Space.* Pergamon Press.

Walter, E. D., L. Qi, A. Chamas, H. S. Mehta, J. A. Sears, S. L. Scott, and D. W. Hoyt. 2018. Operando MAS NMR reaction studies at high temperatures and pressures. *Journal of Physical Chemistry C* 122:8209-8215.

Wan, Z., Y. Tao, J. Shao, Y. Zhang, and H. You. 2021. Ammonia as an effective hydrogen carrier and a clean fuel for solid oxide fuel cells. *Energy Conversion and Management* 228:113729.

Wang, L., Z. Yuan, H. E. Karahan, Y. Wang, X. Sui, F. Liu, and Y. Chen. 2019. Nanocarbon materials in water disinfection: State-of-the-art and future directions. *Nanoscale* 11:9819-9839.

Wang, S., S. Yan, X. Ma, and J. Gong. 2011. Recent advances in capture of carbon dioxide using alkali-metal-based oxides. *Energy & Environmental Science* 4:3805-3819.

Wang, Z., K. G. Burra, T. Lei, and A. K. Gupta. 2021. Co-pyrolysis of waste plastic and solid biomass for synergistic production of biofuels and chemicals—A review. *Progress in Energy and Combustion Science* 84:100899.

Wang, Z., and A. Krupnick. 2015. A retrospective review of shale gas development in the United States: What led to the boom? *Economics of Energy & Environmental Policy* 4:5-18.

Ward, J. 2014. Historic semiconductors research and collecting kit. Transistor Museum. http://semiconductormuseum.com/MuseumStore/TransistorMuseum_Brief_History_of_Early_Semiconductors.pdf.

Wawryk, N. J. P., C. B. Craven, L. K. J. Blackstock, and X. F. Li. 2021. New methods for identification of disinfection byproducts of toxicological relevance: Progress and future directions. *Journal of Environmental Sciences* 99:151-159.

Weick, M., M. Hanford, and S. Tame. 2021. How can net zero in chemicals be profitable? EY, May 24. https://www.ey.com/en_us/chemicals/how-can-net-zero-in-chemicals-be-profitable.

Westerweel, B., R. Basten, J. den Boer, and G. J. van Houtum. 2021. Printing spare parts at remote locations: Fulfilling the promise of additive manufacturing. *Production and Operations Management* 30:1615-1632.

White House. 1985. Directive 189: National Policy on the Transfer of Scientific, Technical and Engineering Information. National Security Decision Directive. https://irp.fas.org/offdocs/nsdd/nsdd-189.htm.

WHO (World Health Organization). 2016. World Health Statistics Data Visualizations Dashboard Air Pollution. https://apps.who.int/gho/data/node.sdg.3-9-viz-1?lang=en.

Widener, A. 2020. Who has the most success preparing Black students for careers in science? Historically Black Colleges and Universities. *C&EN* 98. https://cen.acs.org/education/success-preparing-Black-students-careers/98/i34.

Wilkinson, J. L., A. B. A. Boxall, D. W. Kolpin, K. M. Y. Leung, R. W. S. Lai, C. Galbán-Malagón, A. D. Adell, J. Mondon, M. Metian, R. A. Marchant, A. Bouzas-Monroy, A. Cuni-Sanchez, A. Coors, P. Carriquiriborde, M. Rojo, C. Gordon, M. Cara, M. Moermond, T. Luarte, V. Petrosyan, Y. Perikhanyan, C. S. Mahon, C. J. McGurk, T. Hofmann, T. Kormoker, V. Iniguez, J. Guzman-Otazo, J. L. Tavares, F. G. D. Figueiredo, M. T. P. Razzolini, V. Dougnon, G. Gbaguidi, O. Traoré, J. M. Blais, L. E. Kimpe, M. Wong, D. Wong, R. Ntchantcho, J. Pizarro, G. G. Ying, C. E. Chen, M. Páez, J. Martínez-Lara, J. P. Otamonga, J. Poté, S. A. Ifo, P. Wilson, S. Echeverría-Sáenz, N. Udikovic-Kolic, M. Milakovic, D. Fatta-Kassinos, L. Ioannou-Ttofa, V. Belušová, J. Vymazal, M. Cárdenas-Bustamante, B. A. Kassa, J. Garric, A. Chaumot, P. Gibba, I. Kunchulia, S. Seidensticker, G. Lyberatos, H. P. Halldórsson, M. Melling, T. Shashidhar, M. Lamba, A. Nastiti, A. Supriatin, N. Pourang, A. Abedini, O. Abdullah, S. S. Gharbia, F. Pilla, B. Chefetz, T. Topaz, K. M. Yao, B. Aubakirova, R. Beisenova, L. Olaka, J. K. Mulu, P. Chatanga, V. Ntuli, N. T. Blama, S. Sherif, A. Z. Aris, L. J. Looi, M. Niang, S. T. Traore, R. Oldenkamp, O. Ogunbanwo, M. Ashfaq, M. Iqbal, Z. Abdeen, A. O'Dea, J. M. Morales-Saldaña, M. Custodio, H. de la Cruz, I. Navarrete, F. Carvalho, A. B. Gogra, B. M. Koroma, V. Cerkvenik-Flajs, M. Gombač, M. Thwala, K. Choi, H. Kang, J. L. Celestino Ladu, A. Rico, P. Amerasinghe, A. Sobek, G. Horlitz, A. K. Zenker, A. C. King, J.-J. Jiang, R. Kariuki, M. Tumbo, U. Tezel, T. T. Onay, J. B. Lejju, Y. Vystavna, Y. Vergeles, H. Heinzen, A. Pérez-Parada, D. B. Sims, M. Figy, D. Good, and C. Teta. 2022. Pharmaceutical pollution of the world's rivers. *Proceedings of the National Academy of Sciences of the United States of America* 119:e2113947119.

Willems, P. A. 2009. The biofuels landscape through the lens of industrial chemistry. *Science* 325:707-708.

Wilson, L. J., and D. Liotta. 1990. A general method for controlling glycosylation stereochemistry in the synthesis of 2'-deoxyribose nucleosides. *Tetrahedron Letters* 31:1815-1818.

WIPO (World Intellectual Property Organization). 2021. *World Intellectual Property Indicators 2021*. https://www.wipo.int/publications/en/details.jsp?id=4571.

Witze, A. 2022. Why the Tongan eruption will go down in the history of volcanology. *Nature*, February 9. https://www.nature.com/articles/d41586-022-00394-y.

Wong, L., and T. van Drill. 2020. Decarbonisation Options for Large Volume Organic Chemicals Production, Shell Moerdijk. PBL Netherlands Environmental Assessment Agency and TNO. https://www.pbl.nl/sites/default/files/downloads/pbl-2020-decarbonisation-options-for-large-volume-organic-chemicals-production-shell-moerdijk_3483.pdf.

Wong, R., and H. Tse, eds. 2009. *Lateral Flow Immunoassay*. New York: Humana Press.

Woolston, C. 2019. A message for mentors from dissatisfied graduate students. *Nature,* November 20. https://www.nature.com/articles/d41586-019-03535-y.

Wulf, C., M. Beller, T. Boenisch, O. Deutschmann, S. Hanf, N. Kockmann, R. Kraehnert, M. Oezaslan, S. Palkovits, S. Schimmler, S. A. Schunk, K. Wagemann, and D. Linke. 2021. A unified research data infrastructure for catalysis research—Challenges and concepts. *ChemCatChem* 13:3223-3236.

Xia, Y., C. T. Campbell, B. Roldan Cuenya, and M. Mavrikakis. 2021. Introduction: Advanced materials and methods for catalysis and electrocatalysis by transition metals. *Chemical Reviews* 121(2):563-566.

Xie, S., W. Zhang, X. Lan, and H. Lin. 2020. CO_2 reduction to methanol in the liquid phase: A review. *ChemSusChem* 13:6141-6159.

Xie, Z., G. R. Akien, B. R. Sarkar, B. Subramaniam, and R. V. Chaudhari. 2015. Continuous hydroformylation with phosphine-functionalized polydimethylsiloxane rhodium complexes as nanofilterable homogeneous catalysts. *Industrial & Engineering Chemistry Research* 54:10656-10660.

Yang, J., B. Hou, J. Wang, B. Tian, J. Bi, N. Wang, X. Li, and X. Huang. 2019. Nanomaterials for the removal of heavy metals from wastewater. *Nanomaterials* 9:424.

Yaroshenko, I., D. Kirsanov, M. Marjanovic, P. A. Lieberzeit, O. Korostynska, A. Mason, I. Frau, and A. Legin. 2020. Real-time water quality monitoring with chemical sensors. *Sensors* 20:3432.

Yu, A., G. Ma, J. Ren, P. Peng, and F.-F. Li. 2020. Sustainable carbons and fuels: Recent advances of CO_2 conversion in molten salts. *ChemSusChem* 13:6229-6245.

Yu, C.-J., S. von Kugelgen, D. W. Laorenza, and D. E. Freedman. 2021. A molecular approach to quantum sensing. *ACS Central Science* 7:712-723.

Yu, X., T. Zhong, Y. Zhang, X. Zhao, Y. Xiao, L. Wang, X. Liu, and X. Zhang. 2022. Design, preparation, and application of magnetic nanoparticles for food safety analysis: A review of recent advances. *Journal of Agricultural and Food Chemistry* 70:46-62.

Zainzinger, V. 2020. Is green investing influencing the value of chemical companies? *C&EN*. 98(44). https://cen.acs.org/business/finance/green-investing-influencing-value-chemical/98/i44.

Zecchina, A., and S. Califano. 2017. *The Development of Catalysis: A History of Key Processes and Personas in Catalytic Science and Technology*. Hoboken: John Wiley & Sons.

REFERENCES

Zhang, Z., Y. Lou, C. Guo, Q. Jia, Y. Song, J.-Y. Tian, S. Zhang, M. Wang, L. He, and M. Du. 2021. Metal–organic frameworks (MOFs) based chemosensors/biosensors for analysis of food contaminants. *Trends in Food Science & Technology* 118:569-588.

Zheng, J., and S. Suh. 2019. Strategies to reduce the global carbon footprint of plastics. *Nature Climate Change* 9:374-378.

Zhu, H., T. A. Jackson, and B. Subramaniam. 2021. Highly selective isobutane hydroxylation by ozone in a pressure-tuned biphasic gas–liquid process. *ACS Sustainable Chemistry & Engineering* 9:5506-5512.

Zimmerman, J. B., P. T. Anastas, H. C. Erythropel, and W. Leitner. 2020. Designing for a green chemistry future. *Science* 367(6476):397-400.

Zuth, C., A. L. Vogel, S. Ockenfeld, R. Huesmann, and T. Hoffmann. 2018. Ultrahigh-resolution mass spectrometry in real time: Atmospheric pressure chemical ionization orbitrap mass spectrometry of atmospheric organic aerosol. *Analytical Chemistry* 90(15):8816-8823.

Appendix A

Committee Member Biographical Sketches

Mark S. Wrighton (*Chair*) is currently serving as president of George Washington University. He is concurrently on sabbatical from Washington University in St. Louis where he is the James and Mary Wertsch Distinguished University Professor and Chancellor Emeritus. Dr. Wrighton served as the 14th Chancellor of the University from July 1, 1995, through May 31, 2019. He served as a presidential appointee to the National Science Board (2000–2006), which is the science policy advisor to the President and Congress and is the primary advisory board of the National Science Foundation. He is a past chair of the Business–Higher Education Forum and the Association of American Universities. Dr. Wrighton has received many awards for his research and scholarly writing, including the distinguished MacArthur Prize. He is the author of more than 300 articles in professional and scholarly journals, is the holder of 16 patents, and co-author of a book, *Organometallic Photochemistry*. His research interests are in the areas of transition metal catalysis, photochemistry, surface chemistry, molecular electronics, and photoprocesses at electrodes. He is a fellow of the American Academy of Arts and Sciences and of the American Association for the Advancement of Science and is a member of the American Philosophical Society. Active in public and professional affairs, he has served on numerous government panels and has been a consultant to industry. He is an active member of numerous professional organizations and serves as a director on the boards of national companies and St. Louis organizations. From 1990 until 1995, he served as provost and chief academic officer at the Massachusetts Institute of Technology (MIT). A member of the MIT faculty from 1972 until 1995, Dr. Wrighton became a full professor of chemistry in 1977. He was named Frederick G. Keyes Professor of Chemistry in 1981 and became head of the Chemistry Department in 1987. In 1989, he was named the first holder of the Ciba-Geigy Professorship. Wrighton received his Ph.D. in chemistry from the California Institute of Technology in 1972.

Cathy L. Tway (*Vice Chair*) is the Technology and Applications Director for Catalyst Technologies at Johnson Matthey. In this role, she is responsible for a global team of scientists and engineers specializing in catalysis, process technologies, and engineering design. Additionally, Dr. Tway provides technical input, oversight, and direction as well as ensures that customer-driven research and development (R&D) and engineering are delivered efficiently. Prior to joining Johnson Matthey, Dr. Tway held positions at Dow, Celanese, Solutia, and Akzo Nobel, holding both R&D leadership and individual contributor roles. Dr. Tway has more than 25 years of industrial experience that covers the entire catalyst project life cycle including front-end opportunity identification and creation of new technologies, process scale-up, commercialization, and plant support. Over her career, she has commercialized two new inorganic materials and four catalyst technologies, with two of these processes still in use today. She has served on numerous review panels, boards, and committees including the committee for the National Academies of Sciences, Engineering, and Medicine consensus study report *Gaseous Carbon Waste Streams Utilization*. She earned her Ph.D. in physical inorganic chemistry from the University of Nebraska–Lincoln.

Ashish Arora is the Rex D. Adams Professor of Business Administration at the Fuqua School of Business at Duke University. Previously, he was a faculty member at Carnegie Mellon University, where he held the H. John Heinz Professorship until 2009. His research focuses on the economics of technology and technical change. Dr. Arora's research has included the study of technology-intensive industries such as software, biotechnology, and chemicals; the economics of information security; and the role of patents and licensing in promoting technology start-ups. He has studied the rise of the software industry and the pharmaceutical industry in emerging economies. His current research focuses on the management of intellectual property and licensing in corporations, and innovation-based entrepreneurship. He served as co-editor of *Research Policy* from 2008 to 2014 and currently serves as the department editor for *Management Science* (Innovation and Entrepreneurship) as well as on the editorial board of *Strategic Management Journal*. In the past, he has served on advisory panels to the Secretary of Commerce, the National Academy of Sciences, and the Association for Computing Machinery. Dr. Arora received his Ph.D. in economics from Stanford University in 1992.

Raychelle Burks is an associate professor of chemistry at American University. Her research focuses on developing targeted, tech-tunable sensing systems for use in the lab and in the field to detect compounds of forensic interest. She has contributed to projects that aim to bring sensing systems to the market. Dr. Burks was awarded the 2020 American Chemical Society Grady-Stack Award for excellence in public engagement. Dr. Burks earned her Ph.D. in chemistry from the University of Nebraska–Lincoln in 2011.

Joseph M. DeSimone, NAE, NAM, NAS, is a professor at Stanford University with appointments in the School of Medicine, School of Engineering, and Graduate School of Business (by courtesy). Dr. DeSimone is also the board chair of Carbon, Inc., a 3-D printing company he co-founded in 2013 and of which he served as CEO until 2019. Prior to co-founding Carbon, DeSimone was a professor at the University of North Carolina at Chapel Hill (UNC) and North Carolina State University (NC State) for more than 25 years. Today, he maintains affiliations at UNC as the Chancellor's Eminent Professor of Chemistry Emeritus and at NC State as the William R. Kenan, Jr. Distinguished Professor of Chemical Engineering Emeritus. Dr. DeSimone has made scientific breakthroughs in areas including green chemistry, new polymer materials, medical devices, nanotechnology, and 3-D printing, also co-founding several companies based on his research in addition to Carbon. In recognition of his achievements, he has received major accolades including the U.S. Presidential Green Chemistry Challenge Award, the Lemelson-MIT Prize, the American Chemical Society Award for Creative Invention, the National Academy of Sciences Award for Convergent Science, the EY Entrepreneur of the Year award, and the U.S. National Medal of Technology and Innovation. He has served at the National Academies of Sciences, Engineering, and Medicine on the Committee on Convergence in Biomedical Research, the Committee on Advancing Institutional Transformation for Minority Women in Academia, and the Committee on Effectiveness of National Biosurveillance Systems: BioWatch and the Public Health System, as well as serving as co-chair and member of the Materials Engineering Section Peer Committee, and member of the Board on Chemical Sciences and Technology. Dr. DeSimone received his B.S. in chemistry from Ursinus College and earned his Ph.D. at Virginia Polytechnic Institute and State University in 1990.

Shanti Gamper-Rabindran is an associate professor at the University of Pittsburgh. Dr. Gamper-Rabindran applies her interdisciplinary training by working in the intersection of economic development, public health, and environmental and energy issues. She is the contributing editor of the *Shale Dilemma: A Global Perspective on Fracking and Shale Development* and author of *US Energy Policy: Impacts on the Economy, People and Planet*. Her research examines the effectiveness of policy instruments to improve health and safety and to reduce pollution (e.g., corporate social responsibility programs, information disclosure and regulations) and the impact of investments into public goods (e.g., hazardous waste cleanup on housing values and provision of piped water on infant mortality). She received the Faculty Award for Sustainability from the University of Pittsburgh in 2020. Dr. Gamper-Rabindran earned her Ph.D. in economics from Massachusetts Institute of Technology.

Jeannette M. Garcia is the senior manager for the Quantum Applications, Algorithms and Theory team at IBM Research. Her team's research focuses on computational science applications and theory for quantum computing. Previously, Dr. Garcia's research interests have focused on the rational design of new polymers and materials, targeting recyclable materials with unique mechanical and thermal properties. When she first joined IBM Research in 2012, she worked on high-performance and recyclable materials. She became a research staff member in 2013 and from 2017 to 2018, served as the technical advisor to Dr. Sophie Vandebroek, Chief Operating Officer of IBM Research. From 2018 to 2019, she was a manager and global lead for Quantum Applications in Quantum Chemistry and Science. Dr. Garcia earned her Ph.D. in chemistry from Boston College, where she focused on catalyst design and development.

Javier Guzman is currently the Global Research Guidance and Valuation manager in Strategy and Planning for ExxonMobil Research and Engineering. Dr. Guzman leads a group that provides active guidance to technology program planning, prioritization, and decisions. Dr. Guzman worked in academia for 4 years before starting his industrial career in 2009. During his 11-year industrial career, he has served in a number of technical leadership and management assignments. Dr. Guzman is a leader in research and development, specifically in the area of catalysis and separations for sustainable chemicals and fuels. He obtained his Ph.D. in chemical engineering from the University of California, Davis in 2003.

Martha Head is currently serving as the executive director of Computational and Data Sciences at Amgen. Dr. Head recently served as the director of the Joint Institute for Biological Sciences, a collaborative research effort between Oak Ridge National Laboratory (ORNL) and the University of Tennessee (UT) system. As director, she focused on applying the world-leading capabilities of ORNL and UT to biomedical research and health outcomes of relevance to Tennessee and the Appalachian region. Before joining ORNL, Dr. Head spent 20 years in R&D at GlaxoSmithKline Pharmaceuticals (GSK). For many of those years, Dr. Head led GSK's U.S. Computational Chemistry team, whose accountability was to proactively and creatively apply all relevant computational tools to progressing drug discovery efforts from target selection through to selection of a candidate for clinical trials. While at GSK, Dr. Head was a co-creator of the Accelerating Therapeutics for Opportunities in Medicine (ATOM), a public–private partnership under the auspices of the Cancer Moonshot, and at ORNL continues to be a contributor to ATOM and a member of the ATOM Joint Research Council. Dr. Head also leads the Molecular Design and Analysis to Inform Therapeutics

Related to COVID-19 project sponsored by the U.S. Department of Energy National Virtual Biotechnology Laboratory. Dr. Head received her Ph.D. in physical chemistry from Duke University in 1995.

Russell Moy was general counsel for the Southeastern Universities Research Association (SURA) until January 2022. In this role, Dr. Moy was responsible for managing SURA's legal, regulatory, compliance, and technology transfer matters, and supporting the Thomas Jefferson National Accelerator Facility's legal staff. Dr. Moy also participates in SURA's new technology collaborations and research initiatives. He served previously as attorney-advisor to the National Science Board at the National Science Foundation; as a senior staff officer in the National Academies of Sciences, Engineering, and Medicine's Board on Science, Technology, and Economic Policy; as executive director of the President's Committee of Advisors on Science and Technology for the White House Office of Science and Technology Policy; as a technology policy analyst at the Department of Commerce; and in the U.S. Congress House and Senate offices. Additionally, he previously was a group leader for Energy Storage Programs at Ford Motor Company's Scientific Research Laboratory. He is admitted to practice in Michigan, the District of Columbia, and the U.S. Supreme Court. He is a Licensed Professional Engineer in Michigan and an elected fellow of the American Association for the Advancement of Science. Dr. Moy earned his Ph.D. in chemical engineering from the University of Michigan, his J.D. from Wayne State University School of Law, and his L.L.M from Georgetown University Law Center.

Kristala L. J. Prather is the Arthur D. Little Professor and Executive Officer of the Department of Chemical Engineering at Massachusetts Institute of Technology (MIT). Prior to joining MIT, Dr. Prather worked 4 years in BioProcess Research and Development at the Merck Research Labs. Her research interests are centered on the design and assembly of recombinant microorganisms for the production of small molecules, with additional efforts in novel bioprocess design approaches. She particularly focuses on the elucidation of design principles for the production of unnatural organic compounds with engineered control of metabolic flux within the framework of the burgeoning field of synthetic biology. Dr. Prather is the recipient of an Office of Naval Research Young Investigator Award, a Technology Review "TR35" Young Innovator Award, a National Science Foundation CAREER Award, the Biochemical Engineering Journal Young Investigator Award, and the Charles Thom Award of the Society for Industrial Microbiology and Biotechnology. Additional honors include selection as the Van Ness Lecturer at Rensselaer Polytechnic Institute, and a named fellow at the Radcliffe Institute for Advanced Study, the American Association for the Advancement of Science, and the American Institute for Medical and Biological Engineering. She earned her Ph.D. from the University of California, Berkeley in 1999.

Jason Sello is a professor in the Department of Pharmaceutical Chemistry at the University of California, San Francisco (UCSF). Prior to his appointment at UCSF, Dr. Sello was a professor in the Department of Chemistry at Brown University. Before his first faculty appointment, he investigated RNA processing in *Streptomyces* bacteria using genetic tools as a visiting scientist at the John Innes Centre in Norwich, England, and also studied enzymes catalyzing antibiotic biosynthesis as a postdoctoral research fellow at Harvard Medical School. In his independent career, Dr. Sello has been creatively using experimental methods from chemistry, biophysics, biochemistry, and genetics to study biological phenomena and to develop new therapeutics for infections, cancer, and neurological disorders. He has also worked on technologies for the conversion of plant biomass into commodity chemicals. He has been the recipient of several awards, including career awards from the Burroughs Welcome Fund and the National Science Foundation. In 2013, Dr. Sello was recognized with a year-long appointment at Massachusetts Institute of Technology as the Martin

Luther King, Jr. Visiting Professor of Biology. Currently, he serves on the Antimicrobial Resistance and Drug Discovery Study Section at the National Institute of Allergy and Infectious Diseases, and is a founding co-editor of *Synthetic and Systems Biology*. He earned a Ph.D. in biophysics from Harvard University in 2002.

Bala Subramaniam is the Dan F. Servey Distinguished Professor of Chemical Engineering at the University of Kansas (KU). Dr. Subramaniam is the founding director of the Center for Environmentally Beneficial Catalysis, a unique university–industry consortium that is developing and providing licensing opportunities for novel resource-efficient technologies related to fuels and chemicals. Dr. Subramaniam's primary research interests are in catalysis and reactor engineering with emphasis on developing sustainable processes for making fuels and chemicals from both traditional and renewable feedstocks. In particular, his research has exploited tunable solvents such as supercritical fluids and gas-expanded liquids to develop greener chemical technologies. Dr. Subramaniam is the executive editor of *ACS Sustainable Chemistry and Engineering* and chaired the 2018 Gordon Research Conference on Green Chemistry. His honors include American Society for Engineering Education's Dow Outstanding Young Faculty Award, Indian Institute of Chemical Engineers' Chemcon Lectureship Award, and KU's Higuchi Research Achievement Award. Dr. Subramaniam is a fellow of the American Institute of Chemical Engineers, the American Chemical Society Industrial & Engineering Chemistry Division, and the National Academy of Inventors. He earned his Ph.D. in chemical engineering from the University of Notre Dame in 1984.

Jean W. Tom, NAE, is currently executive director in Chemical Process Development, a group focused on developing enabling and commercial processes to manufacture small-molecule drug substances at Bristol Myers Squibb (BMS). Previously, she worked on the process development of such molecules at Merck. Her research interests are in driving new approaches to more efficiently develop compounds, drawing upon opportunities using modeling, high-throughput experimentation, and enabling technologies. She is active in external organizations driving innovations in the precompetitive collaboration for pharmaceutical development (Enabling Technologies Consortium) and supporting the chemical engineering discipline as well as efforts at BMS to attract women students to STEM fields. Dr. Tom was a member of the National Academies of Sciences, Engineering, and Medicine Board on Chemical Science and Technology from 2010 to 2013, and was elected to the National Academy of Engineering in 2019. She received her Ph.D. in chemical engineering from Princeton University in 1993.

Appendix B

Request for Proposal for Economic Analysis

Background

The National Academy of Sciences ("NAS"), which includes under its Congressional Charter, the National Academy of Engineering and National Academy of Medicine (collectively "the Academies") and its Board on Chemical Sciences and Technology (BCST) is conducting a study funded by the Department of Energy, the National Science Foundation, the National Institute of Standards and Technology, and the American Chemical Society, to examine the current state of the United States chemical economy. The study will recommend targeted investments in fundamental chemical research to stimulate economic growth. The full scope of the study is outlined in the "Study Statement of Task," below. The areas outlined in the Statement of Task will be addressed by members of an ad hoc committee of the Academies ("the Committee") with expertise in diverse areas such as the chemical sciences, biochemistry, energy and sustainability, pharmaceuticals, patent and regulatory laws, and economics.

As a part of the study process, the Committee is tasked with gathering information from external experts and reviewing previously published work to better inform the findings, conclusions, and recommendations published in the final report. While any member of the public has the opportunity to comment on the study and the statement of task, the Committee will seek out specific expertise to help inform the study process.

As a part of this process of information gathering, the Committee is seeking an independent contractor to conduct an economic analysis on the U.S. chemical economy. The analysis and its deliverables will address the "Description of Analysis," outlined below, with the intent of helping the consensus study committee address elements of their own statement of task. The final analysis produced by the independent contractor will either be included as an appendix in the published consensus study report or be made available to the public online. Any analysis work submitted to the Committee for consideration will be included in a public access file list, and members of the public can request the full analysis via the National Academic Public Access Records Office at any time.

Description of Analysis

In accordance with the scope of the statement of task for the consensus study, the economic analysis will examine how investments in fundamental chemical research and development affect the size and scope of the chemical economy.[1] This analysis should:

1. Assess the economic value of the chemical economy.[1]
2. Assess in what sectors or industries chemical research is driving employment opportunities and value creation.
3. Include an assessment of how chemical research is contributing to sustainable economic progress.

The deliverable may also include an analysis of the environmental impact of the U.S. chemical sector on the economy, an analysis of different factors that cause fluctuations in the U.S. chemical economy, or other analyses that directly address the consensus study's statement of task.

---

[1] The chemical economy includes all sectors and industries that rely on chemical knowledge for their advancement and growth. Examples include, but are not limited to, the petroleum industry, the energy sector, materials production, pharmaceuticals, and agrochemicals.

Appendix C

List of Open Session Speakers

Committee Meeting 1: February 4 and 5, 2021
- **Joshua L. Rosenbloom**, University of Iowa
- **Kevin Swift**, American Chemistry Council
- **Thomas Connelly**, American Chemical Society
- **Mary Maxon**, Lawrence Berkeley National Laboratory
- **Jeff Furman**, Boston University

Public Session on Analyzing Financial Investments in the Chemical Sciences: April 2, 2021
- **Vinod Khosla**, Khosla Ventures
- **Jennifer Shieh**, U.S. Small Business Administration, Office of Investment and Innovation, Office of Innovation and Technology
- **Johanna Wolfson**, Prime Impact Fund
- **John Jankowski**, National Science Foundation/National Center for Science and Engineering Statistics

Public Session on Funding Mechanisms: April 28, 2021
- **Michael Yonas**, Pittsburgh Foundation
- **Adam Kuspa**, Welch Foundation

Committee Meeting 4: May 5, 2021
- **Alexa Dembek**, DuPont
- **David McConville**, ExxonMobil Chemical Company
- **Neil Hawkins**, Erb Family Foundation

Public Session on Corporate Perspectives: May 17, 2021
- **Bob Maughon**, SABIC
- **Darlene Solomon**, Agilent

Public Session on Chemistry Education: June 14, 2021
- **Debra Rolison**, Naval Research Laboratory
- **Ginger Shultz**, University of Michigan

Public Session on Energy: June 28, 2021
- **Jennifer Wilcox**, Fossil Energy and Carbon Management, Department of Energy
- **Andrea Ramirez**, Delft University of Technology
- **T. Alan Hatton**, Massachusetts Institute of Technology
- **Maurits van Tol**, Johnson Matthey
- **Bill Tumas**, National Renewable Energy Laboratory
- **Nathan Lewis**, California Institute of Technology

Public Session on Materials: July 1, 2021
- **Elsa Reichmanis**, Lehigh University
- **Karthish Manthiram**, Massachusetts Institute of Technology
- **Andre Benard**, Michigan State University
- **Richard Venditti**, North Carolina State University

Public Session on Automation: July 7, 2021
- **Pierre Baldi**, University of California, Irvine
- **Kevin Naidoo**, University of Cape Town
- **Connor Coley**, Massachusetts Institute of Technology
- **Song Lin**, Cornell University
- **Corey Stephenson**, University of Michigan
- **Doug Fuerst**, GlaxoSmithKline
- **Martin Burke**, University of Illinois
- **Ying Wang**, AbbVie
- **Billy Bardin**, Dow

Public Session on Economics: July 8, 2021
- **Adeliene van Tol-Koutstaal**, Freeline, European Patent Attorney, Dutch Patent Attorney

Public Session on Pharmaceuticals and Agrochemistry: July 29, 2021
- **Derek Lowe**, Novartis Institutes for BioMedical Research
- **Vid Hegde**, Corteva Agriscience

Public Panel on Education: August 3, 2021
- **Polly Arnold**, University of California, Berkeley

Committee Meeting 7: August 6, 2021
- **Bindu Nair**, U.S. Department of Defense
- **Anne Fischer**, Defense Advanced Research Projects Agency
- **David Christian (Chris) Hassell**, U.S. Department of Health and Human Services

Appendix D

Individual Expert Interviews

Call: May 26, 2021
- *Expert interviewed*:
  - **Paul Larsen**, Corteva Agriscience
- *Committee member(s) involved:*
  - **Jean Tom**, Bristol Myers Squibb

Call: May 28, 2021
- *Expert interviewed:*
  - **Steve Evans**, BioMADE
- *Committee member(s) involved:*
  - **Kristala L. J. Prather**, Massachusetts Institute of Technology

Call: June 4, 2021
- *Expert interviewed:*
  - **Greg Vite**, Bristol Myers Squibb
- *Committee member(s) involved:*
  - **Jean Tom**, Bristol Myers Squibb

Call: June 4, 2021
- *Expert interviewed:*
  - **Negar Garizi**, Corteva Agriscience
- *Committee member(s) involved:*
  - **Jean Tom**, Bristol Myers Squibb

Call: June 4, 2021
- *Expert interviewed:*
  - **Hans Renata**, Scripps Research Institute
- *Committee member(s) involved:*
  - **Jean Tom**, Bristol Myers Squibb

Call: June 7, 2021
- *Expert interviewed:*
  - **Alison Narayan**, University of Michigan
- *Committee member(s) involved:*
  - **Jean Tom**, Bristol Myers Squibb

Call: June 7, 2021
- *Expert interviewed:*
  - **Martin Eastgate**, Bristol Myers Squibb
- *Committee member(s) involved:*
  - **Jean Tom**, Bristol Myers Squibb

Call: June 14, 2021
- *Expert interviewed:*
  - **Greg Whiteker**, Corteva
- *Committee member(s) involved:*
  - **Jean Tom**, Bristol Myers Squibb

Call: June 16, 2021
- *Expert interviewed:*
  - **Shannon Stahl**, University of Wisconsin
- *Committee member(s) involved:*
  - **Jean Tom**, Bristol Myers Squibb

Call: June 22, 2021
- *Expert interviewed:*
  - **Tehshik Yoon**, University of Wisconsin
- *Committee member(s) involved:*
  - **Jean Tom**, Bristol Myers Squibb

Call: July 9, 2021
- *Expert interviewed:*
  - **Keary Engle**, Scripps Research Institute
- *Committee member(s) involved:*
  - **Jean Tom**, Bristol Myers Squibb

Call: July 12, 2021
- *Expert interviewed:*
  - **Sarah Reisman**, California Institute of Technology
- *Committee member(s) involved:*
  - **Jean Tom**, Bristol Myers Squibb

APPENDIX D

Call: September 17, 2021
- *Expert interviewed:*
  - **Lynn Walker**, Carnegie Mellon University
- *Committee member(s) involved:*
  - **Jean Tom**, Bristol Myers Squibb
  - **Jason Sello**, University of California, San Francisco

Call: October 1, 2021
- *Expert interviewed:*
  - **Edward Dunlea**, Carnegie Mellon University
- *Committee member(s) involved:*
  - **Jean Tom**, Bristol Myers Squibb
  - **Jason Sello**, University of California, San Francisco

Appendix E

Call for Input from the Chemistry Community

The committee behind the ongoing National Academies of Sciences, Engineering, and Medicine study, *Enhancing the U.S. Chemical Economy through Investments in Fundamental Research in the Chemical Sciences*, invites you to submit input related to their charge.

The committee is particularly interested in hearing about emerging areas of fundamental chemistry research that might have a lasting impact on the chemical enterprise. This research can have implications for pharmaceuticals, agrochemistry, materials, energy, education, or any other sector where chemistry has an impact. The committee appreciates any input you are able to provide, but requests answers to the following guiding questions:

1. How have investments in long-term fundamental chemical research contributed to U.S. priority areas, such as national security, environmental sustainability, thriving manufacturing industries, energy-technology development, or economic growth?
2. What are new and emerging areas of fundamental chemical research that have the potential to make a lasting impact on pharmaceuticals, agrochemicals, materials, energy, education, or any other sectors where chemistry plays a key role?
3. What are different targeted investment strategies in the chemical sciences that would stimulate U.S. economic growth?
4. How can future research investments enhance the chemical economy while also advancing environmentally sustainable practices?
5. How can future research investments enhance the chemical economy while integrating a diverse chemical economy workforce?

You are welcome to type your input into the comment box provided, or attach a file containing any written input you would like to provide. You are also encouraged and welcome to distribute this call for input to colleagues and peers interested in the chemical economy. Respondents should feel free to provide as much or little information as they would like, as well as any accompanying published materials that they think would be helpful.

In accordance with Section 15 of the Federal Advisory Committee Act, any information provided will be placed in and available through the project's public access record. Respondents may, however, choose to submit their input anonymously.